PROGRAMMABLE DEVICES AND SYSTEMS 2001
(PDS 2001)

A Proceedings volume from the 5th IFAC Workshop,
Gliwice, Poland, 22 – 23 November 2001

Edited by

W. CIĄŻYŃSKI, E. HRYNKIEWICZ

and

P. KŁOSOWSKI
Institute of Electronics,
Silesian University of Technology,
Gliwice, Poland

Published for the

INTERNATIONAL FEDERATION OF AUTOMATIC CONTROL

by

PERGAMON
An Imprint of Elsevier Science

ELSEVIER SCIENCE Ltd
The Boulevard, Langford Lane
Kidlington, Oxford OX5 1GB,UK

Elsevier Science Internet Homepage
http://www.elsevier.com

Consult the Elsevier Homepage for full catalogue information on all books, journals and electronic products and services.

IFAC Publications Internet Homepage
http://www.elsevier.com/locate/ifac

Consult the IFAC Publications Homepage for full details on the preparation of IFAC meeting papers, published/forthcoming IFAC books, and information about the IFAC Journals and affiliated journals.

First edition 2002

Library of Congress Cataloging in Publication Data

A catalogue record for this book is available from the Library of Congress

British Library Cataloguing in Publication Data

A catalogue record for this book is available from the British Library

ISBN 0-08-044081 9
ISSN 1474-6670

To Contact the Publisher

Elsevier Science welcomes enquiries concerning publishing proposals: books, journal special issues, conference proceedings, etc. All formats and media can be considered. Should you have a publishing proposal you wish to discuss, please contact, without obligation, the publisher responsible for Elsevier's industrial and control engineering publishing programme:

Dr Martin Ruck
Publishing Editor
Elsevier Science Ltd
The Boulevard, Langford Lane
Kidlington, Oxford
OX5 1GB, UK

Phone: +44 1865 843230
Fax: +44 1865 843920
E.mail: m.ruck@elsevier.co.uk

General enquiries, including placing orders, should be directed to Elsevier's Regional Sales Offices – please access the Elsevier homepage for full contact details (homepage details at the top of this page).

5th IFAC WORKSHOP ON PROGRAMMABLE DEVICES AND SYSTEMS 2001

Sponsored by
International Federation of Automatic Control (IFAC)
Technical Committee on Components and Instruments

Co-sponsored by
Technical Committee on Safety of Computer Control Systems

Organized by
Institute of Electronics, Silesian University of Technology
Gliwice, Poland

In co-operation with
Department of Measurement and Control, FEI,
VSB – Technical University of Ostrava
Ostrava, Czech Republic
Computer Society Chapter, IEEE Poland Section

International Programme Committee (IPC)
Hrynkiewicz, E. (PL) (Chairman)
Boverie, S. (FR) (Co-chairman)
Vlček, K. (CZ) (Co-chairman)

Adamski, M. (PL)
Besançon, A. (FR)
Binder, Z. (FR)
Černohorsky, J. (CZ)
Chapenko, V. (LV)
Chojcan, J. (PL)
Hławiczka, A. (PL)
Horáček, P. (CZ)
Kirianaki, N.V. (UA)
Kopacek, P. (AT)
Łuba, T. (PL)

Nawarecki, E. (PL)
Ollero, A. (ES)
Pogoda, Z. (PL)
Pokorný, M. (CZ)
Raes, P. (BE)
Roach, J. (UK)
Rojek, R. (PL)
Saucier, G. (FR)
Srovnal, V. (CZ)
Taylor, G.E. (UK)
Zalewski, J. (USA)

National Organizing Committee (NOC)
Ciążyński, W. (Chairman)

Drelichowska, M.
Kania, D.
Kidoń, Z.
Kłosowski, P.
Niedziela, T.

PREFACE

The history of scientific meetings on Programmable Devices and Systems (PDS) started in the mid-nineties in Gliwice, Poland with the PDS'95 event organised by the Institute of Electronics, Silesian University of Technology (SUT).

It was noted that many papers on the issues of programmable devices and systems were presented at numerous conferences and workshops devoted to electronics and circuit theory, nevertheless, there were no workshops solely devoted to those particular topics. That fact and the organisers' belief that some specific common problems appeared in the area of PDS sufficiently justified the decision to organise the PDS meeting. The main objective of the PDS'95 event was to provide a forum to present the latest research results and experiences in the field of design and application of PDS. In particular the aim of the meeting was to define the future trends in this field via the interaction of industry, technical research centres and academia representatives. Some 30 participants, coming mainly from the Polish and Czech universities and research centres supported that idea and on the basis of their opinions it was decided to organize further meetings on PDS with similar scope. The result of that first meeting was the co-operation agreement between the SUT Gliwice and VSB Technical University in Ostrava, Czech Republic. According to the agreement the successive PDS events would be organised at 1-2 year intervals in Gliwice and Ostrava alternately.

Starting from the 4th event in the series i.e. PDS2000, organised by the VSB, Department of Measurement and Control and held in Ostrava, Czech Republic - the PDS meetings attained the status of IFAC Workshop and this greatly contributed to the publicity of the event.

The PDS2001 IFAC Workshop, organised again by the Institute of Electronics, SUT, Gliwice, Poland was the 5th event in the series. The IFAC Technical Committees: TC on Components and Instruments and TC on Safety of Computer Control Systems were, as for the previous meeting in Ostrava, correspondingly the Workshop sponsor and co-sponsor. These TC Chairs were the IPC members.

From the 70 preliminarily registered contributions, the National Organising Committee (NOC) finally received 66 papers from 11 countries. The Programme Committee members reviewed all submitted papers and accepted 58 of them for presentation. All registration, paper submission and reviewing operations were carried out electronically through the net. The organisers prepared and distributed the Preprints volume.

This Proceedings volume contains 54 duly presented papers and many of them when compared to the Preprints volume version have been corrected and enriched with the discussion results. The papers are grouped according to the Workshop plenary sessions topics as follows:

- **Communication**
- **Digital Signal Processing**
- **Industrial Programmable Logic Controllers**
- **Field Programmable Logic**

The 15 papers presented in the poster session have been included in the above topic groups to express the fact that, in the opinion of the IPC, they are of similar value. Four of the scheduled papers were not presented because their authors were absent.

We hope discussions during the meeting were fruitful and all participants of the Workshop enjoyed their stay in Gliwice.

Edward HRYNKIEWICZ, IPC (Chair)
Władysław CIĄŻYŃSKI, NOC (Chair)
Institute of Electronics, STU Gliwice, Poland

CONTENTS

COMMUNICATION

DIGITAL SIGNAL PROCESSING

INDUSTRIAL PROGRAMMABLE LOGIC CONTROLLERS

FIELD PROGRAMMABLE LOGIC

IFAC

Publications
www.elsevier.com/locate/ifac

SENSOR/ACTUATOR WEB ORIENTED INTERFACE

Petr Cach* Petr Fiedler* Frantisek Zezulka*

** Dept. of Control and Instrumentation, Faculty of
Electrical Engineering and Computer Science, Brno
University of Technology, Boetchova 2, CZ-61266 Brno,
Czech Republic*

Abstract: This paper describes data acquisition device that was designed to collect data from smart sensors of temperature, pressure, flow, etc. The device contains an Ethernet interface, a web server and a ftp server so all acquired data can be reached over the internet in the form of both actual values and logged data. The device is based on powerful microprocessor and cooperating integrated Ethernet interface. At present time the prototype of device operates as a gateway between smart sensors and an internet. Web server in the device displays web pages generated by the microprocessor and using standard internet browser it is possible to change configuration parameters of connected sensors, monitor logged values, monitor actual values etc. Due to the capabilities of smart sensors user can not only watch present values. Using standard internet browser one can also set up dumping value, unit, tag, date, etc. *Copyright © 2001 IFAC*

Keywords: Ethernet, fieldbus, datalogger

1. INTRODUCTION

Evolution of Internet technologies leads to massive penetration of Internet to almost all parts of human activities. At present time Internet technologies are commonly used to get various types of information for business, social, educational and entertainment purposes. There is general effort to use already developed Internet technologies (such as Ethernet, TCP/IP, HTTP, SMTP) for other purposes then they were originally intended for. The rising popularity of Ethernet and Internet is reflected by rising effort of manufacturers of automation devices to include Ethernet TCP/IP interface into their devices. The amount of products that can be connected to Ethernet and Internet is dramatically increasing and the Internet technologies are used for visualization, configuration and control of the devices.

2. INTERNET

The Internet is built up on vast amount of protocols. To clarify the principles of Internet, it is useful to compare the Internet protocols suite with the OSI model defined by International Standards Organization.

OSI Layers	Internet protocols	
7. Application layer	Application protocols	HTTP, FTP SMTP, Telnet, DCOM, ...
6. Presentation layer		
5. Session layer		
4. Transport layer	TCP/UDP protocol	
3. Network layer	IP protocol	
2. Link layer	Ethernet, Token bus, ...	
1. Physical layer	Twisted pair, fibre, ...	

Fig. 1. OSI model and Internet

It is obvious using 1 that all application layer protocols utilize for transportation of its data either TCP or UDP protocol. In either case data are encapsulated into datagrams of IP protocol. These datagrams are sent from one computer to another using standard network technologies like Ethernet, ATM, etc.

3. COMMUNICATION TECHNOLOGIES IN AUTOMATION

At present time, there are various industrial communication systems used to interconnect a variety of distributed devices, field instrumentation and control systems. To assure real-time operations, the industrial communication bus has to guarantee defined media access time for all connected devices and high reliability of data transmission. The present state reflects past development of many various incompatible proprietary and national standards, each one designed to meet specific requirements. So these industrial communication technologies differ in the complexity of functions, communication speed, flexibility, medium access method, bus cycle, etc.

At present time, there is general effort to use well-proven communication technologies in the field of automation. This effort brings advantage of long experience, broad component range, low price and easier system integration. However, it is necessary to insist on real-time and reliable operations as this requirements cannot be omitted. At first, it means that the media access time and communication delays have to be predictable and well defined.

Today, the most promising networking technology for use in automation is Ethernet. Ethernet is the most widespread technology used in LANs (Local Area Networks) and many manufacturers of automation devices equip their products by Ethernet interface nowadays. In general Ethernet is considered as nondeterministic due the CSMA/CD media access method. The reasons for the non-determinism are possible collisions that occur when two or more devices try to send a frame at the same time. But with careful network design (active switches, full duplex mode, FastEthernet) and with reasonable network traffic the collisions might be avoided and the bus fulfils real-time requirements.

Most manufacturers and organizations concerning industrial communications agree on the Ethernet as suitable transfer medium. However, they don't agree on common transfer protocol for data transmission. The solution they selected differs case by case. The difference between different protocols is what levels of the standard protocol of the layered

structure of internet are used (see 1). There are three groups of protocols:

(1) protocols using only Ethernet
(2) protocols using TCP/IP
(3) protocols using standard protocols above TCP/IP

Protocols belonging to the first group use Ethernet only for the advantage of the high speed. The example of such protocol is PowerLink protocol developed by B+R. This protocol uses special kind of master-slave protocol, which guaranties the response time up to 400 μs. To reach this speed, it is required to have network separated from any other traffic.

There are several protocols belonging to the second group. They originate mostly in already developed industrial communications protocols. Examples of such protocols are Modbus/TCP, Foundation Fieldbus, EthernetIP, many are being developed. The TCP/IP protocol is used for transport of datagrams of standard fieldbus protocol over Ethernet. There are several groups working on completely new protocol, which will allow to use all the advantages offered by Ethernet TCP/IP and will omit all the drawbacks inherited from the previous generations of industrial fieldbuses.

As for the third group of protocols, there are several well established communication protocols used in the information technology. Because there is lot of reliable tools and applications ready for use, the industrial applications can take the advantage of easy integration in the information systems. The simplest example of such application is to use simple web or ftp server to transport data. A more advanced example of such protocol is ProfiNET defined by Profibus organization. This protocol employs DCOM communication technology developed by Microsoft. At the same time it could be an example of not very suitable choice, because the DCOM technology is complicated and it is proprietary. The industrial communication standard should be built on existing, open and reliable technology. OPC Foundation is working on an OPC XML standard, which will utilize SOAP (Simple Object Access protocol) transfer protocol and XML language.

4. SENSVISION - INTERNET COMPATIBLE DATA ACQUISITION SYSTEM

Sensvision is an example of a system that takes advantage of widespread communication technologies like Ethernet, TCP/IP, etc. BD Sensors Company in cooperation with authors started the development of this system in the year 2000. The aim was to develop a platform that would enable to collect data from a technological process and allow access to all this information from Intranet or Internet. It was required that no special software should be needed for access to this data. Only standard software such as a web browser or ftp-client was allowed.

In principle, there are two ways how to achieve this goal. One can embed all the necessary hardware in a sensor and individually connect each sensor to Internet. The other way is to create a universal device with number of universal inputs and connect some standard sensors to them. Both ways have their advantages and disadvantages. In case of single intelligent sensor, it's possible to fully utilize all features of the sensor (full configuration and parameter setting over Internet). The drawback is higher price of such solution. Also there are not available microprocessors that would have complete Ethernet interface integrated as an on-chip peripheral. In consequence it leads to larger size of PCB unsuitable for mounting Into an sensor.

At present time the more perspective solution seems to be an universal acquisition device - a data concentrator. Its main advantage is much lower cost per sensor. This solution has also other advantages like, flexibility, possibility to pre-process collected data directly in the data concentrator, etc.

The Ethernet interface with TCP/IP libraries put fairly high requirements on the used processor's performance. After a careful survey of available tools and products it was decided to use Net+ARM microprocessor manufactured by NET+Silicon company. It is 32-bit microprocessor with well-known ARM core. The main advantage of such processor is integrated 10/100 Ethernet controller module, that serves as the Ethernet Medium Access Control layer. This module, together with powerful ARM core, gives us enough performance for processing of the massive data flow from Ethernet interface and for processing of TCP/IP and application layer protocols.

At the present time, the system can be briefly described as an web server providing dynamically generated web pages. The contents of the pages can reflect actual or past measurements with other additional functions. The functions of the software could be separated in three groups:

- HTTP server - the HTTP server provides dynamically generated html pages. The content of the web pages is affected by command directly in the source html code. These command have form of so-called "server side includes". The web server interprets these commands and replaces them by the result. This way it is possible to display actual measurements or to set parameters of a sensor. There aren't any special limitations of the type of web page. It can contain forms, Java scripts, etc.

- FTP server - the main role of the ftp server is to allow to upload the content of the web server. The web content is stored in a non-volatile memory and can be updated using any standard ftp client. The secondary role of the ftp server is to allow download the stored data in the form of standard text file.

- Database file server - as the system can be easily equipped with sufficient memory, the next logical step is to store measured data in a database, which allows further processing and archiving. The user can define databases according his/her needs. It includes definition of the time sampling period, the length of database, determination of recorded variables (in case of data concentrator), starting conditions (for example after crossing a limit) and so on. The stored data is then accessible either as a web page or as a file on the ftp server.

As addition to the data concentrator, there is a graphical user interface under development. This user interface provides easy way to database definition, as well as the direct graphical representation of the stored data in the form of graphs. Because such functionality can't be achieved by means of standard html language, a more sophisticated software solution has to be chosen. In order to remain platform independent, it was decided to use the well-known JAVA language to create the user-friendly interface. The JAVA technology fits very well for this purpose, as it produces small and portable code, which may be stored in the memory of microprocessor and downloaded to the user's computer on demand.

For the next development of the Sensvision system, there should be considered several aspects. It is necessary to extend it capability in three ways:

(1) Interfacing to intelligent sensors - at present time, the system offers to connect only common sensors with the current output 0/4 - 20mA. It is also possible to connect intelligent sensors equipped by RS-232 interface using HART protocol. In next enhancement there will be possible to connect various analog or digital input, either directly or using some kind of serial communication (RS-232, RS-485, CAN).

(2) User interface - there has to be done much more on the user interface. In the present time, the interface is able to display the stored data in the form of graph. There should be completed the database definition.

(3) Communication ability

There are several trends in the industrial communication, which can't be omitted. There are several attempt to define a specialized industrial communication protocols based on Ethernet TCP/IP. The simplest but most popular is Modbus/TCP. There are many tools available for this protocol. For example various DDE and OPC servers are available from various sources. With utilization of Modbus/TCP in a Sensvision, there is wide field of possible applications, such as interconnection with existing SCADA systems, simple data processing by special Visual Basic programs and many more.

The increasing popularity of the SOUP and XML protocols is also very promising. SOAP is based on the standard http server, the protocols are open and popular (e.g. the Microsoft .NET technology includes SOAP protocol). Once the standard will be defined (OPC XML), it would be simple to implement such functionality in the Sensvision system.

5. CONCLUSION

During development of the Sensvision system it has been found that Ethernet with TCP/IP protocol suite is a good choice as an automation networking technology that allows easy connection to Internet. The system shows several advantages of Internet technologies in the industrial segment. There is no need for special tools, just standard web browser of ftp client. The TCP/IP protocol allows to use several communication protocol to exchange data, independently on each other. The standard Internet application layer protocols like HTTP and FTP are utilized for non-real-time tasks, while the lack of mighty, standardized and open application layer protocol compatible with Internet leads to development of many proprietary automation protocols, so at present only the old Modbus/TCP is used for real-time processing.

www.elsevier.com/locate/ifac

A CAN BASED NETWORK MODEL FOR LABORATORY DEMONSTRATION OF AUTOMOTIVE MULTIPLEXED WIRING SYSTEMS

Jerzy Fiołka

Institute of Electronics, Silesian University of Technology, Gliwice, Poland

Abstract: The Controller Area Network (CAN) is a serial communications protocol which efficiently supports distributed real-time control with very high level of security. It was initially developed for the use in motor vehicles by Robert Bosch GmbH, Germany, in the late 1980s. Apart from motor vehicles, other applications for CAN are industrial automation, medical equipment, building automation, household appliances. The CAN bus model presented in this paper is a most-versatile aid consisting of a ready-to-use hardware and software module. The software is designed for users with different experience. A beginner is able to activate the demonstration software for Windows, which allows to monitor and set contents of the SJA1000 (CAN controller) registers. A more experienced user can load his software (assembler, "C" for '51 family) and create some CAN applications. *Copyright © 2001 IFAC*

Keywords: Automotive control, Computer-aided engineering, Communication networks, Educational aids, Microcomputer based systems, Networks.

1. INTRODUCTION

Many vehicles already have a large number of electronic control systems. Examples of such systems include engine management systems, active suspension, ABS, gear control, lighting control, air conditioning, airbags and central locking. The complexity of the functions implemented in these systems necessitates an exchange of data between them. With conventional systems, data is exchanged by means of dedicated signal lines, but this is becoming increasingly difficult and expensive as control functions become ever more complex. The solution to this problem was the connection of the control systems via a serial bus system (see Fig. 1). A comparison of the requirements for vehicle and industrial bus systems shows more similarities: low connection cost per station, high transmission reliability, real-time capabilities, ease of use. For this reason CAN bus is also well suited to industrial applications.

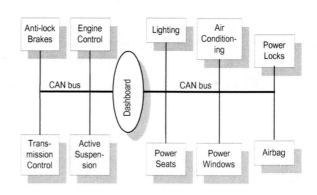

Fig.1. Architectures of the vehicle electronic systems. (Siemens, 1998).

2. BASIC CONCEPT

The Controller Area Network (CAN) is a multi-master bus with an open, linear structure with one logic bus line and equal nodes. The number of nodes is not limited by the protocol. In the CAN protocol, the bus nodes do not have a specific address. Instead, the address information is contained in the identifiers of the transmitted messages, indicating the message content and the priority of the message. The number of nodes may be changed dynamically without disturbing the communication of the other nodes. Furthermore, the CAN protocol provides sophisticated error and signalling detection mechanisms. Multicasting and Broadcasting is supported by CAN. More information on CAN protocol can be found in Bosch, (1991). The CAN allows the flexible configuration of networks with different types of microprocessors and microcontrollers. Fig.2 shows the principal configurations of CAN nodes (module 1) and different configurations with microcontroller plus the stand-alone controller (module 2), microcontroller with on-chip CAN (module 3) and SLIO node contains bus controller with I/O facilities (module 4). The connection to the CAN bus lines is usually built with a CAN transceiver (**Physical interface**) optimized for the applications. The transceiver controls the logic level signals from the CAN controller into the physical levels on the bus and vice versa. **Bus Controller** implements the complete CAN protocol defined in the CAN specification. All CAN functions are controlled by a **Module Controller** which performs the functionality of the **Application Interface** (for example actuators, sensors, pushbuttons).

3. PHYSICAL REALISATION

The CAN based network model presented in this paper consists of one host node (operating in conjunction with PC) and a few (max. 16) SLIO nodes. The choice of the configurations of CAN nodes (module 2 and module 4, see Fig. 2) has been based on general specifications of requirements:
- system based on the well-known '51 kernel,
- optimisation of design from the point of view of educational features,
- simplicity,
- low cost.

3.1 *Host node*

The block diagram of the host node (Multi-Chip Solution, see Fig.2) is shown in Fig. 3. U1 (8051), U2 ('573), U3 (external program memory) are a minimum system with an 8051 family controller. The external circuits of the microcontroller U1 are as usual, a latch U2 to separate address and data, and an external program memory U3 (in this case RAM

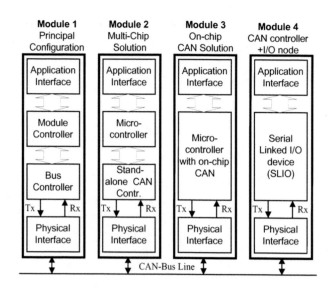

Fig.2. CAN module set-up.

memory) to store the object code. The SJA1000 (U4) is a stand-alone controller for the CAN used within automotive and general industrial environments. It is designed to be hardware and software compatible to the PCA82C200 CAN controller from Philips Semiconductors (Philips Semiconductors, 1997). Additionally, a new mode of operation is implemented which supports the CAN 2.0B protocol specification with several new features. The SJA1000 is mapped in the external data memory area. Because in the host node no external data memory or other mapped peripherals are used, no extra address decoding components are necessary. The PCA82C250 (U5) is the interface between the SJA1000 and the physical bus. The device provides differential transmit capability to the bus and differential receive capability to the CAN controller (Philips Semiconductors, 1997). The AT89C51 (U6) is a 8-bit microcontroller with 4K bytes of Flash Programmable and Erasable Read Only Memory (PEROM). The control program contained in the on-chip Flash allows the user to download application software into the system. The program file is downloaded from the PC's parallel port through the U6 into the external program memory (U3). The program file is expected to contain binary data; hex files are not accepted. The first byte in the file is written into the first location in the memory. Successive bytes are written into successive locations until the data in the file has been exhausted. When downloading is complete the U6 resets the system and next application is allowed to begin execution of the new program. The host node also includes the MAX232 (U7) line driver/receiver which produces RS-232 levels at the serial interface, a reset logic circuit resets the AT89C51 after power-on (hardware reset of the SJA1000 is driven by U6) and a simple application interface. The application interface consists of four pushbuttons and four discrete LEDs connected to the P1 port of the microcontroller.

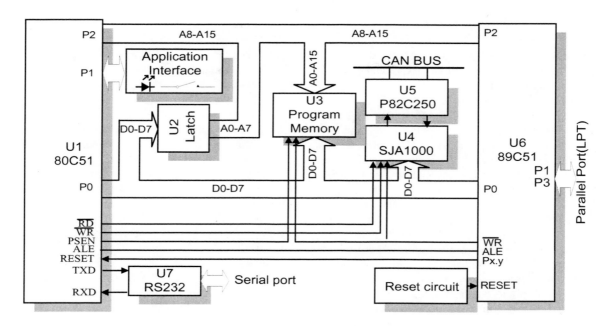

Fig.3. The block diagram of the host node.

The discrete LEDs are used to indicate the logic level on the P1.0 - P1.3 I/O pins. Pushbuttons connected to P1.4 - P1.7 lines of P1 port enable to set corresponding pins to logic "0" (normally a logic "1" is set by internal pull-ups). The host node is powered from wall mount, 9V-12V, 200mA supply. A 5V regulator is on the board.

3.2 SLIO Node

A Serial Linked I/O device is a port device directly interfacing to the CAN-bus. It can be used as an extension of digital and analog port function for a remote microcontroller. The SLIO node consists of the P82C150, the P82C250 (described above), a reset logic circuit and an application interface. The P82C150 is a single-chip 16-bit I/O device including a CAN protocol controller with automatic bit rate detection and calibration. It features 16 I/O pins (see Fig. 4a) which are individually configurable (with respect to I/O port function) via CAN-bus: port direction, port mode and event capture facilities for inputs. For more details see the data sheets of the P82C150 in Philips Semiconductors, (1997). A network containing one or more SLIO nodes must have at least one node operating autonomously (host node). In order to be able to operate the P82C150 needs to be calibrated by an operating host node sending specific calibration messages (the P82C150 calibrates its internal bit clock to the bit timing of the bus). After a successful calibration of the SLIO node, the host node has to send configuration messages, which prepare the ports for specific needs of the connected application at the node. Fig.4b gives an overview of the P82C150 pins configuration used in CAN model. To distinguish between the different SLIOs (in one network can be maximum sixteen

P82C150 circuits) each device has its own identifier, which is set by applying voltage level (+5V or 0V) at the pins of the I/O ports P0-P3 during the reset phase. After the end of the reset these pins may be used as inputs or outputs.

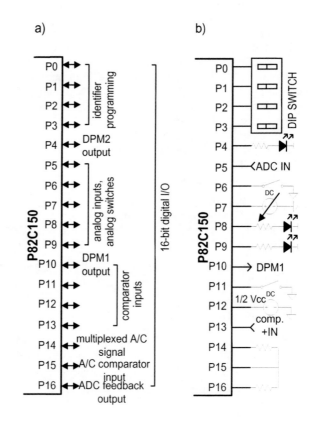

Fig.4. Pin configuration of the P82C150
    a)   functional diagram,
    b)   configuration used in the SLIO node.

4. SOFTWARE

The software provided with the CAN based network model supports „learning about CAN" and assist in prototype networks. It provides:

● demonstration software

The demonstration software (CAN-MON) is an easy-to-use software designed for Microsoft Windows 95/98 environment. The CAN-MON provides an intuitive, object-oriented graphical user interface (GUI) that simplifies control strategy and display setup. A menu-driven software allows the contents of the SJA1000 registers to be altered and thereby the user is able to configure the CAN controller and prepare the CAN-protocol frame to be transmitted over the bus. The model is also ready to be used as a bus monitor, receiving messages from the bus for display on a screen. Figure 5 shows a screen shot of the CAN-MON environment. This window displays in binary and decimal format the content of the SJA1000 transmit and receive buffer. The content of transmit buffer can be changed by clicking on the selected bit. In order for the new selections to take effect, the user has to click on the SEND button after a new frame has been sent into CAN bus. The content of the receive buffer is changed by the messages received from the bus.

● ready-to-use software for standard tasks

Currently, thanks to more powerful microcontrollers and advanced compiler technology, it is feasible to program control applications using a High-Level Language (HLL), like "C". The C language is a general-purpose programming language which combines code efficiency, structured programming, comfortable data structures and a wide variety of operations. Furthermore, C is based on a "minimalist" philosophy in which the language performs only those functions explicitly requested by the programmer. This approach is well-suited for a beginner. Thanks to this the user can create some CAN applications in an easy way.

The software, written in the C language supports:
- setting up the SJA1000 with respect to hardware and software links to the microcontroller,

- setting up the CAN controller for the communication with respect to the selection of mode, acceptance filtering, bit timing etc.,
- interrupt and polling controlled transmission of a message,
- interrupt and polling controlled reception of a message,
- P82C150 start-up bit time calibration procedure,
- setting up the P82C150 with respect to I/O port function.

In addition, each program includes the flowchart which shows the basic processing and detailed description of the operations. Application programs written in the C language should be converted into the '51 compatible code before downloaded into the external program memory of the system. An adaptation of the source code can be done by using the ANSI-C version of the development package (e.g. Franklin Software C compiler, IAR). The C program is compiled, debugged and linked on the IBM PC (and compatibles) to provide the required absolute object code in the form of a .HEX file. Next, the error-free (syntax and semantic) code is converted into the .BIN format and downloaded into the system external program memory. After a successful downloading the host node is automatically reset and the application program is executed in real-time.

● an example of CAN application

The overall operation of CAN application is as follows. Normally, the host node periodically transmits Calibration Message in order to calibrate the on-chip RC-osciliator of the P82C150. After pressing the P1.7 pushbutton the host node sends to the SLIO a frame containing the data for Data Output Register. It allows remote driving the P7 LED of the SLIO. When the P1.6/P1.5 pushbutton is pressed the host node increments/decrements the value for DPM1 Register (this register contains data for quasi-analog output signal, which is generated by Distributed Pulse Modulation) and transmits Data Frame with a new value for the DPM1 output. It allows the user to vary the brightness of the P4 LED connected to the output port (by changing the average voltage level on port). The P82C150 includes a 10-bit analog-to-digital converter. After pressing the P1 pushbutton, the host node transmits a frame, addressing the ADC Register. After that, the SLIO node starts an analog-to-digital conversion cycle ended with the transmission of a message containing the result. The message received by the host node is transmitted via the serial port to a PC and displayed on the screen.

Fig.4. A screen shot of the CAN-MON.

5. CONCLUSIONS

The CAN based network model for laboratory operation is a versatile tool, being a ready-to-use hardware and software module, supports "learning about CAN" and assists in prototype networks. The demonstration software provides basic information necessary for understanding of the CAN protocol as well as for the configurations of CAN nodes and the realization of CAN communications. Furthermore, the tool also offers a special software package, written in the C language, which provides a function for transmission and reception of a message, communication with serial port, keyboard, etc. Due to this, the user is able to set up his own application within a short time. To ensure comfortable usage it is very important to provide the user with a simple and fast means of downloading application software into the system. In the presented device it is done by using only one additional chip. Thus, no external, usually expensive devices (like RAM emulator) are necessary. Furthermore, thanks to loading an application program into the system using the parallel port, downloading is very fast. It reduces the testing efficiently.

REFERENCES

Bosh (1991). CAN Specification Version 2.0. Robert Bosch GmbH, Stuttgart.

Philips Semiconductors (1997). *Semiconductors for in-car electronics. Data handbook IC18*. Philips Electronics N.V.

Siemens (1998). Canpress Version 2.0. Siemens Microelectronics, Inc.

IFAC
Publications
www.elsevier.com/locate/ifac

DYNAMICALLY RECONFIGURABLE WIRELESS LOCAL NET IN ON-FIELD APPLICATION

Bogdan Olech, Mariusz Kapruziak

*Faculty of Computer Science & Information Technology, Technical University of Szczecin,
49, Zolnierska, 71-210 Szczecin*

Abstract: There are many quite fundamental differences between proposed solutions for the wireless network technology; the standardization is not yet completed. It is hard to find a pattern to follow when building a dedicated network with reasonable cost. This paper presents an approach of building customised wireless system evaluated from a simple one to the complex system with dynamical reconfiguration ability. *Copyright © 2001 IFAC*

Keywords: communication networks, simulators, systems engineering

1. INTRODUCTION

The term „wireless networking" refers to the technology that enables two or more devices to communicate using proper network protocol, but without network cabling. This technology supports variety of popular wireless applications from cellular Personal Communications System (PCS), and Group System for Mobile communication (GSM) networks to typical networking systems. At first, wireless networks were used only in specific applications where the need of supporting mobility or long-range communication made it impossible to use wires. During last few years however new research and technological achievements have made wireless networks much more powerful, widening functionality maintained. Wired communication is gradually displaced from more and more applications. Nowadays in many cases the wireless technology is fast, reliable and cheap enough to be used in spite of wired one.

Wireless systems can be implemented following different configuration patterns. In general, it is possible to specify three main wireless network configurations: ad-hoc wireless network, single and multiple access point network. For the ad-hoc configuration network nodes are temporarily arranged; there are no extra tasks for a privileged node. The two other configurations distinguish one or more nodes (access points) to control the network resources. Implementation based on units that control the system requires an infrastructure to be built. In some cases however building an infrastructure is too expensive or even impossible. These limitations can be met when building communication system for battlefield use, security or disaster recovery team, but also for more and more civilian needs of recent years. In the result, the most convenient way of building wireless network is to base it on the ad-hoc architecture.

To organize network without any central infrastructure it is a challenging task. In this case there is no defined point of control of the network packet traffic. It is required therefore from any node in the network to support fast end efficient communication, in the same time, minimally disturbing other communication links. There are many ways of resolving this task depending on the application the system is supposed to be used for. Currently, two main standards of building the Physical (PHY, hardware and hard-wired radio transmission handling) and Medium Access Control (MAC, firmware access to the physical layer) layers are defined: IEEE 802.11 and HiperLAN1.

In the base version the IEEE 802.11 (IEEE, 1997; Linda, 2000) makes transmission speed up to 2 Mbps available, or up to 11 Mbps when extended (IEEE 802.11b). Node devices can be established either to peer-to-peer network or there are fixed access points supported with mobile nodes to communicate. This standard is designed to operate in so-called industrial, scientific and medical (ISM) unlicensed 2.4Ghz frequency band. On the physical layer, the original IEEE 802.11 uses either FHSS (Frequency-Hopping Spread Spectrum) or DSSS (Direct-Sequence Spread Spectrum) technologies. For security, authentication and encryption including 128-bit WEP (Wired Equivalent Privacy) cryptography are supported there.

HiperLAN.1 offers up to 23.5 Mbps throughput (Anastasi, *et al.*, 2000). So high data rate is the result of efficient power amplifier and 5-GHz frequency band used. On the physical layer GMSK (Gaussian Minimum Shift Keying) is used there. HiperLAN.1 supports also Quality of Service (QoS, some real time constraints) in some implementations. For security, the technology supports encryption, including WEP algorithm.

Standards described above define only two lowest layers of ISO OSI (Open System Interconnection Reference Model) model, leaving higher layers to be implemented by system developer. For ad-hoc networks the routing layer is probably as important as two lower ones. Unfortunately, routing protocols are generally still in progress. Protocols defining physical and MAC layers were created much earlier than routing protocols because of some technological obstacles. Wireless communication was not so popular, expenses and bandwidth limitations where not possible to overcome so successful before. In order to raise the throughput of network a central unit was generally used. Expenses paid when building central unit was then not so high comparing with the cost of the whole system. All efforts where concentrated on the physical and MAC layer to make them more efficient. During last years the technological advancement and cost reductions of wireless systems change this point of view. Currently the base station of wireless network is the most expensive part of the system and, because of much more bandwidth allowed today, often useless. Without a central node network nodes are forced to realize all tasks of this special one. That is the reason why so much attention is paid recently to routing layer of wireless networks.

Routing, as the method of finding communication paths between nodes is well known for wired networks. There are many protocols and algorithms capable of completing this task. It seems therefore to be possible to adopt one of existing wired protocol to wireless routing. The advantage of this approach is that the existing network software could be directly applied to wireless networks. Achieved results where not promising however, showing that this approach is not efficient for most of wireless applications. The main reason is that topology of wireless network is usually much more prone to change than of wired one. Also communication between nodes, required to maintain topological information, generates too much overhead and increases power consumption of the network. Wired algorithms usually are not prepared to cope with unidirectional communication between nodes and so many possible communication paths to consider. Routing in ad-hoc networks is particularly challenging because of the assumption that there is no control infrastructure of the network. In addition, any node at any time can move or disappear.

Wireless communication is being widely used for many years, but only a few last years would be characterized as the wireless local network era. The technological progress made methodology, tools and semiconductors available and good enough to build wireless local networks with reasonable costs. But problems to be overcome still exist; there are many quite fundamental differences between proposed solutions. So, it makes almost impossible to build a homogenous physical and logical standard in the nearest future. In many published materials there are disclosed different examples of practical and only theoretical approaches. Movable and stationary nodes of the network are analysed; even network functions changeable on the fly are adopted in some cases. The consequence is that there is no pattern to follow. As it can be observed, like in other industry brunches, customisation is usually the only way out to achieve practical results. This paper presents an approach of building customized wireless system evaluated from a simple one, already working and verified, to the complex system with dynamical reconfiguration ability. There is the motivation also presented with same practical aspects already recognized.

2. TARGET IMPLEMENTATION

Wireless systems not supported by the standard high volume production often can be found where on-field application is undertaken. The authors of this paper are working on training and security systems for many years (for example: Olech and Poplawski, 2000). For the training systems, those discussed in this paper, the artificial battlefield is created. Real vehicles perform manoeuvres and shoot to artificial aims on a restricted area. This behaviour is made as much as possible similar to the real combat. The locations of vehicles and the aims actuation are under control from a headquarters central computer. This control ability is as good and reliable as the communication system is. However, in case of failure there are precaution procedures and command communication systems so, there is the guarantee that nothing unexpected can happen.

In case of security systems usually there is no guaranteed communication redundancy, over the redundancy strictly required, so the reliable work is the must. In many cases high volume of data is to be transferred on-line with the multiple threats flow. The multiple threats flow is exploited intensively here not only to optimise path between nodes, but also to avoid sabotage, failure or unexpected dynamical net topology changes.

The on-field application causes some extra demands. Devices must stand very low and very high temperatures, high humidity, water and under water pressure, vibration and acceleration, electric and magnetic interferences - also from thundering. To build a complete system functions of different kind have to be integrated. Radio subsystem, analog and digital parts and other components have to work in one enclosure or even in one chip.

These already mentioned and many other difficult requirements must be conformed and many trade-offs resolved to build reliable training or security system. Authorized functional demands requested by a customer are the base of the system specification. The co-design method is then used to postulate the system architecture, also hardware/software partitioning and the way of simulation and verification.

3. SYSTEMS ALREADY IMPLEMENTED

There are two most characteristic systems for the discussion presented here, redesigned and improved with participation of the authors, standardized systems, used for years as continuation of WSB-03 and WSB-04. Both work on VHF band. The first is the typical simplex with handy transmitter. The second is the semi-duplex system with a central transceiver. The improvements where concentrating basically on better transmission ranges, on communication quality and the overall reliability. In both cases system architecture should have stayed the same because of the backwards compliance.

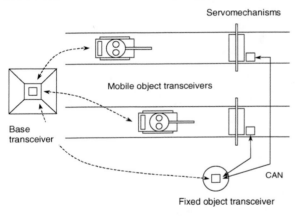

Fig. 1. Example of WSB-04 field application.

Both systems are provided with redesigned power supplies working up to 60V and tolerating 200V of continuous voltage. All RF and digital electronics of WSB-03 has been redesigned. Much more general improvements for the WSB-04 system where expected. Direct Minimum Shift Keying (MSK) modulation scheme has been replaced with two-level linear Gaussian MSK. Pre-modulation Gaussian filtering reduces the occupied spectrum for wireless data transmission. Channel spacing consistency is then quite better. The radio frequency bandwidth is literally comparable to the transmission speed. The range and reliability is getting better too, since more power is converted into the effective spectrum, also bit error rate (BER) significantly improves. Another very attractive feature of GMSK is the possibility to use of class C power amplifiers.

The WSB-04 is a hierarchical wireless/wired two level communication system. The wired sub-system has been completely redesigned also. The Control Area Network (CAN) standard was chosen as the most recommended in the industry. Very reliable and functionally enhanced wired communication has been implemented this way. Figure 1 depicts simplified example of WSB-04 field application. There is the base transceiver located on a tower, two mobile object transceivers located on vehicles, one fixed object transceiver wired with two servomechanisms actuating the artificial aims.

4. NEW DIRECTIONS UNDERTAKEN

Implemented systems mentioned above have some disadvantages already stated. Working on VHF band the transmission bandwidth is far not enough. It is important especially for WSB-04 where the real time response is recommended. When dozens of objects are controlled the object transceiver is placed directly on an object. The problem starts when hundreds of objects are on the field and the radio transmission appears to be the only bottleneck. Now, it is recompensed with the hierarchy of wireless and wired communication. Any object transceiver is connected with an individual object with wire, up to 16 objects can be linked this way. But it makes the system not movable in case when many objects have to be controlled.

The obvious way out is to faster the radio communication. To do that the radio frequency has to be changed to a higher one, from VHF to UHF bandwidth. Taking into account the available technology on the market, cost and potential requirement in the future, the 2.4 GHz was selected.

Not only the higher bandwidth characterizes the new undertaking. For many transceivers randomly and dynamically located on a field the wireless network idea should be introduced. Some implementation specific restrictions are then added: net nodes are movable or mobile built with a considerable cost, there is no base station and low power consumption is preferred over transmission speed. Also, the new wireless system should benefit from features matching the security system demands.

The IEEE 802.11 standardization is the most convenient to this approach. The spread spectrum technique applied there, being so pervasive in military applications, is ready to use. Two main properties of spread spectrum are worth to mention here: low power spectral density and high immunity to jamming and interference. Thanks to low power spectral density the information signal is a noise-like for eavesdropping or other radios. Even if recognized is difficult to cover by any other radio signal.

The key problem of customisation is the granularity. How small building blocks should be, preserving flexibility and minimizing non-recurring engineering (NRE) extra costs. Following the IEEE 802.11 standardization also this dilemma seams to be resolved – the physical and MAC layers are stated. The economically reasonable choice is to adopt a ready to use complete radio hardware with MAC layer implemented. There are many competitive offers on the marked. Taking into account the availability and technological progression ratio for the last few years the products from Intersil Co. look most attractive. Their Programmable Radio for ISM band (PRISM) features the programmable data rate capability with high rates of 11, 5.5 and 4Mbps and IEEE 802.11 fallback rates of 2 and 1Mbps (Intersil, 1998).

More theoretical part of the work presented in this paper refers to the routing in wireless networks. This problem not yet standardized is indispensable to solve in the target application. As the rule, the routing level procedures do not influence the lower and higher level components of the architecture. Being soft-wiring can be developed somewhat apart from the others.

5. ROUTING IN WIRELESS NETWORKS

Recent years has brought a real breakthrough on the field of routing in ad-hoc networks. In general, current routing algorithms for wireless networks can be classified into three main categories (IEEE, 1997):
- Global pre-computed routing.
- On-demand routing.
- Flooding.

In the global pre-computed routing it is assumed that routes to all destinations are computed before and maintained then in the background via a periodic background process. The global pre-routing can be divided furthermore to flat (the same tasks for all nodes) routing and hierarchical routing. The hierarchical routing resolves some scalability problems of proactive (preserving data destination information) routing scheme. In on-demand routing, on the other hand, the route to a specific destination is computed only when needed. The on-demand routing relay on the reactive scheme and does scale well to large populations of nodes. However, it introduces initial search latency and makes it impossible to know in advance the quality of path (delay, bandwidth, error level, etc.). The flooding relay on a packet broadcast. Packets are broadcast to all possible destinations, with the expectation that at least one copy of the packet will reach the intended destination.

Most of the routing proposals have been evaluated by the IETF MANET Working Group. Algorithms named DSDV, AODV, TORA, LSR or WRP origins

there (Iwata, 1999). Routing algorithms are still under development and there is no international standard published. Depending on the application different routing proposals appears to be privileged.

Two lower layers (PHY and MAC of IEEE 802.11) defined are the good base for extended system development. For the routing layer new approach has been considered. In this work a hybrid protocol between on-demand and proactive routing is proposed. In general, the model of the protocol can be characterized by three main mechanisms:
- Route discovery.
- Route maintenance.
- Network information maintenance.

Route discovery mechanism is initialised when a node needs to transmit data to another node and when position of destination node is unknown to the source node (fig. 2a). Source node (S) in that case transmit RREQ (Route REQuest) packet that informs other nodes in the network that information about position of destination node (D) is required. RREQ packet is than flooding through the network.

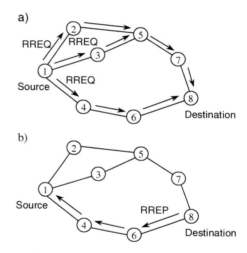

Fig. 2. Control packets flow scheme.

When a node receives RREQ packed with current information about path to node D, as the answer RREP (Route REPly) packet is transmitted to the network (fig. 2b). RREP packet is sent along the path learned from RREQ transmission from the S node. Every node that receives this packet stores the path information to D node. Therefore, when RREP packet reaches node S bi-directional path between node S and D is defined. Only when this path is known data transmission can take place.

Route maintenance mechanism concentrates on maintaining current information about paths used by the system. If any events occur that could break path used, this mechanism is responsible for noticing that event and for repairing that route. State of links in the network is determined through periodically RHELLO packet transmission. Each node keeps the list of actively used links by that node. If the node will not receive three consecutive RHELLO packets from any of nodes, link breakage information is transmitted along the active path. Source node receiving this information initialise route request mechanism if still wants to communicate with that destination node.

Network information maintenance mechanism is required to keep current information about link state between neighbouring nodes in the network. This mechanism was separated from the previous one in order to expand system performance by including information about direction of links, channel parameters and topological information of the neighbouring area. Future directions could include also movement and error statistics of a particular link.

The new routing protocol described does not require changing higher and lower layers of the network, what was assumed as the rule.

6. RESULTS ACHIEVED

To simulate and verify some key concepts of the project specialized simulation environment has been built. The need for this particular tool results from the original requirements not possible to be fulfilled by any other, public domain software. The possibility to model a wireless ad-hoc network for on-field applications was the purpose. Modelling features include data acquisition, control and communication tasks. Simultaneously, it was required to compute all propagation and interference properties of a wireless channel.

In order to simulate the model the area of 500 square meters was generated and inside that area network nodes were placed. IEEE 802.11 DCF (Distributed Coordination Function) was used as a physical and MAC layer for the system.

The model was used to evaluate some parameters and assess proposed method of wireless network organization. The interest was concentrating on routing layer and above. The details of the physical layer were therefore limited. The decision was to use model that combines free space propagation model and two-ray ground reflection model. When a transmitter is within the reference distance of receiver, the free space model is used. Outside of this distance, the ground reflection model is used (Broch, 1998). A noiseless channel was considered. In MAC layer implementation (IEEE 802.11 DCF) all error correction functions has been therefore omitted.

Figure 3a depicts relationship between power (power ratio referred to the idle state is shown) used by the system and number of nodes used for communication (from 2 to 24). It can be observed that the number of

nodes in the system causes almost no influence to power required from each node to maintain the system. That feature supports possibility to scale system to large networks.

As shown on fig. 3b almost no packet traffic is noticed when no data are to be sent. Therefore, when there is no data transmission planed a minimal power is required to keep network working. Actually, proposed protocol requires some packet transmission even if no data transmission takes place, but frequency of control packet exchange fundamentally decreases when no communication is required.

Communication overhead, for the analysed case, is the sum of MAC and routing layer overheads. It is almost unchanged for different packet traffic sceneries (fig. 3c). It can be noticed, that when packet traffic is high (small data transmission delay) communication overhead is even lower than for low packet traffic scenery. This feature is useful allowing maximal throughput of the network to be increased.

Proposed method is therefore, when proper parameters are provided, the best solution for this particular application, considering different known strategies of the network communication. It keeps power usage at low level simultaneously reducing communication overhead.

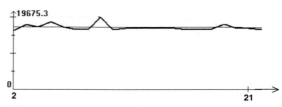

a) Power versus number of nodes

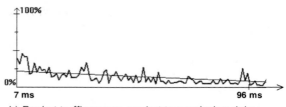

b) Packet traffic versus packet transmission delay

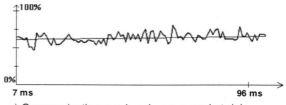

c) Communication overhead versus packet delay

Fig. 3. Selected simulation plots.

7. FUTURE DIRECTIONS

Basically, for the future work all scientific and industry advantages available should be adopted and duplication of effort already made or soon disclosed as made by others avoided. The customisation is the key of the necessity to do a lot of work anyway. There always will be a gap between standard technology available, good solutions made by others but not available (confidential or license protected) and needs postulated by customers.

The wireless network proposed here should meet some QoS requirements. The voice and video in the stream as well as packed priority of transmitted data would be then possible.

The other direction of the work includes making routing algorithms more adoptable to different conditions. The channel parameters will be included in the path finding process. For each link in the network the parameters as transmission fail probability, link throughput or security level of link will be stated. Above the routing layer some particular upper layer implementations are also planed. Especially, distributed computing is considered to make the system more resistant to failure, to reduce control traffic and increase computation capabilities.

The simulation environment is still under further development to satisfy the growing needs. The simulation of distributed computing will be included in the nearest future. The radio propagation model should be described with more details to make simulation results more complex considering QoS and routing improvements.

Complex system reconfigurability performed remotely and self-reconfigurability performed by the system itself is supposed to be the most attractive future direction to achieve the complex dynamically reconfigurable wireless network. Reconfiguration on all system levels in both hardware and software domain is the way to achieve that.

8. CONCLUSIONS

The industry technological offer on the first glance seams to fulfil all application problems possible to face in the practice. But there is no need to wait a long time to find out that it is the top of the iceberg when a real system are to be built. To overcome difficulties the economical and engineering means should be available: investment ability, organization and logistics, development methodology, technology, engineering etc. The continuity of a product is the must on the way to success. Progressive work for years to improve and develop a product has to be being accomplished step by step.

In this paper the example illustrating these thesis above was presented. Starting from the simple experience like WSB-03 through the hybrid wireless/wired system WSB-04 the quite new generation of reconfigurable wireless net is under development. This wireless network, having included standard hardware components and accumulating IEEE standardization, is the original and capable to be customized following changeable requirements coming from customers. The basic assumptions are as follows: net nodes are movable, security precautions are met, there is no base station and low power consumption is preferred over transmission speed. Other features will be exploited in the near future, features like: remote and auto-reconfiguration on system level (auto-driven movable nodes with changeable properties) or hardware/software reconfiguration applied to node resources.

The practical implementations and other more theoretical activities already made are satisfactory. The world technological progress and scientific effort gradually make available new hardware and methodology. The market all over the world absorbs more and more sophisticated wireless network technology. Wireless networks are getting to have the leadership in high technology growth dynamic. There are circumstances to think that it will be the stable tendency. It is another reason to be within this mainstream.

ACKNOWLEDGEMENTS

The authors thank people from Autocomp Electronic Ltd. where the work was founded.

REFERENCES

Anastasi, G., Lenzini, L. and E. Mingozzi (2000). HIPERLAN/1 MAC Protocol: Stability and Performance analysis. *IEEE Journal on Selected Areas in Communications*. **18/9**.

Broch, J., et al. (1998). A performance comparison of multi-hop wireless ad hoc network routing protocols. *Mobile Computing and Networking*.

IEEE (1997). Standard for Wireless LAN, Medium Access Control and Physical Layer Specification, P802.11.

Intersil Co. (1998). A Condensed Review of Spread Spectrum Techniques for ISM Band Systems. AN9820.1

Iwata, A. et al. (1999). Scalable Routing Strategies for Ad Hoc Wireless Network. *IEEE Journal on Selected Areas in Communications*, **17/8**.

Linda D.P. (2000) Exploring the Wireless LANscape. *Computer*, **33/10**.

Olech, B., and H. Poplawski (2000). ZOMBI – the example of distributed data acquisition system. *Advanced Computer Systems*, Szczecin, 363-366.

IFAC
Publications
www.elsevier.com/locate/ifac

TESTING OF TCP/IP BASED COMMUNICATION FOR CONTROL PURPOSES

Przemyslaw Plesowicz *

** Institute of Automatic Control
Silesian University of Technology,
ul. Akademicka 16 44-100
Gliwice, Poland
pleso@terminator.ia.polsl.gliwice.pl*

Abstract:
The recently fast growing popularity and functionality of Internet provides excellent possibilities in area of sophisticated remote controlled automation equipment. The aspect of real-time control over Internet/Intranet has to be measured, to answer what limits to real-time capabilities of simulator/real object imposes the network. This paper should show, how the timings and throughput of the network depends on protocol type (TCP, UDP, DataSocket, VI Server) and quantity of process variables transferred.
The measurements should be base, to make conclusion, whether it is possible to use IP networks in remote control and regulation. *Copyright © 2001 IFAC*

Keywords: automation, control, networks, protocols, real-time

1. INTRODUCTION

The recently fast growing popularity and functionality of World Wide Web creates excellent possibilities in area of sophisticated simulation equipment (Schmid 1999, Davis 1999). It is also turning into very powerful tool for education, research and industry. In education, it is possible to use Internet for remote teaching, what means allowing students to access virtual (simulated) or real instruments placed in laboratories (Metzger 2000). It should stay clear that the aim is not to eliminate real laboratories (obviously needed for hands-on experience) but creating additional laboratory support. Virtual education is rather to be treated as an additional support for didactic (real) laboratories (Stegawski, Schaumann 1998).

The second goal is to share several resources: simulators (virtual installations), real objects, experiences, or collected experimental data, with other researchers (universities) through the Internet (for some purposes also through Intranets). It can create possibility of remote experiences and research.

There are several approaches to the subject of remote experiences. The whole concept can be divided in a couple of categories, e.g. installation type, software used, hardware used, communication protocol, clients quantity, data flow direction, and the number of data transmitted. Simulator (emulator, virtual object) can be treated as a mentioned above installation type. Simulator can help learning about known, already measured and modeled object, but it can be also replaced with real object, on which it is possible to observe any new phenomena, that are not measured yet, thus they can not be already modeled. It is possible, to share (via Internet) collected (in relational database, acting as history module) installation/object data, in order to perform own object analysis (without building own simulator), or for building own simulators. Additionally researchers can take advantage from observing experiences of their colleagues from other universities (also in scope of stored historical data)

Another place to use Internet/Intranets is industrial control automation, as industrial network. In general the main idea of this paper is verifying the possibility of use TCP/IP based networks as a field-network in automatic control and its real-time capabilities.

2. REAL-TIME

The aspect of real-time control and supervisory over Internet/Intranet has to be measured, to answer what limits to real-time capabilities of simulator/object imposes the network (Chung-Ping Young et al., 2000). It is important to remember that in reality not POSIX standards make installation work in real-time. For example: in agricultural laboratories (Georgiev et al., 1999) measurement-time differences of 15 minutes can still be called real-time, and on the other side in case of very fast process even "POSIX milliseconds" do not guarantee real-time execution. Thus, it is important to use the proper definition: "real-time is ability to SYNCHRONIZE object and its model".

3. THE BASIS

Internet/Intranet idea is known to be based on 3rd layer (network layer) and 4th layer(transport layer)of ISO/OSI networking model protocols: IP (RFC774, RFC919), IPv4: (RFC791), IPv6: (RFC2460), ICMP: (RFC792) and subsequent protocols: TCP (RFC761) and UDP (RFC768). Thus, the choice of IP protocol is mandatory. Depending on installation requirement (especially real-time response) one of TCP or UDP can be chosen. In fact, there are real-time protocol standard: RTP and RTCP (RFC1889), but these are not commonly used at this time; in standard there is information describing those protocols not to provide QoS (*Quality of Service*) –the basis of control in automation. In field of automatic control, much more popular are company or consortium-specific standards. Over last years, organizations such as IAONA (*Industrial Automation Open Networking Alliance*) are trying to state one common standard for industrial automation, but not in efficient manner (in the area of IP-based networks)

Web-based installation can be built and programmed using several programming techniques. According to assumptions, it shall be client-server architecture, with simulator/object server, and students/researchers client computer, an exception from this rule can be distributed measuring-actuating equipment system. The solutions differ on the client side, and are almost identical on the server side. It can be done both, using specialized client-side software (Alfonseca et al., 1999)(LabVIEW, C++), or common tools (WWW, JAVA capable browser) (Alfonseca et al., 1999, Schmid Ch., 1999, RFC1945). Use of specialized programs (especially LabVIEW) can speed up building of simulator/object representation, and is easy to learn. The computer architecture differences does not matter since LabVIEW VI programs (Virtual Instruments), and JAVA are architecture independent. There is another one very important approach: specified programmers interface, it means publication of communications and control structures definitions (interfaces exactly), on which custom client-side programs can be built.

The data itself can be transported in one or many channels (one variable, multiple variables). Single-channel communication is necessary for two reasons: especially for disposed system, and second: test measurement of network characteristics.

The number of clients can vary from one client (peer to peer, operator, researcher), to many clients (students, or laboratories for example).

During architecture planning additional question appears: data-flow direction. Should it be unidirectional or bi-directional? Certainly during simple monitoring the unidirectional connection is fully satisfying, but problem of remote control requires bi-directional link. Another question is how to point ONE client, which is privileged to take control over simulator/object.

In area of testing various regulators using remote simulator/object, there are two approaches possible: remote regulator –regulator is running on client's computer, steering the object via Internet or local (uploadable) regulator uploaded and running on server designated for certain object, but it has to be measured, which approach is better one.

It is generally possible to divide network type to INTERNET (wide area network) and INTRANET (LAN -local area network). Use of INTERNET allows sharing data among universities, research centers. Use INTRANET can help achieving higher transfer performance, faster responses and higher security.

Table 1. mean transmission times comparison [ms]

data qty.	UDP/IP		TCP/IP		VI Server		DataSocket	
	local	remote	local	remote	local	remote	local	remote
100	1.18	19.62	0.41	17.19	1.85	20.68	1.73	0.25
200	1.14	20.39	0.80	18.03	2.55	22.84	1.08	0.23
500	1.04	19.97	26.56	19.88	2.30	29.52	2.67	1.17
1000	1.06	20.47	1.53	20.32	3.06	22.54	3.19	1.28
2000	2.07	19.98	2.34	20.03	4.32	22.39	7.82	2.15
5000	3.05	20.19	4.01	20.06	7.36	33.14	68.97	6.27
8000	5.58	20.25						
10000			7.06	40.05	13.40	69.92	143.86	20.34
20000			18.79	150.43	28.95	111.26	247.55	45.84
50000			51.27	422.16	65.25	253.84	663.36	137.06
100000			101.55	845.69	126.01	474.14	1208.39	284.51

4. EXPERIENCES PURPOSE

The main idea of experiences was to compare communication protocols for automatic control purposes. In this paper the first part is shown: comparison of IP based protocols offered by LabVIEW. For easier description, communication with VI Server over TCP is further called VI-Server protocol. During experiences, TCP, UDP, DataSocket and VI-Server transmission times has been measured in order to answer the question how long time delays on the data transmission the network imposes. All measurement data was collected on installation which idea is represented on Fig. 1.

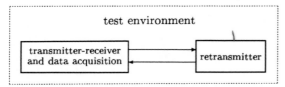

Fig. 1. transmission time measurement stand

Using different protocols (TCP, UDP, DataSocket, Vi-Server) the local „loopback" transfer times and remote (local network, based on 100Mb/s 3COM 3300 switch) data transfer times have been compared. All the test have been performed using multiplexed transmission (Fig. 2) data format used multiplexed (Fig. 3)

All the acquisited data is shown in Table (1). Transmission times in function of data quantity („double" variables) have been shown: local transmission (Fig. 4) and remote transmission (Fig. 5)

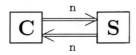

Fig. 2. multiple signals transmitted via single transmission channel

Fig. 3. multiplexed data frame format

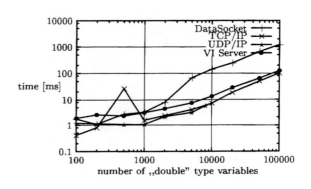

Fig. 4. local transmission times comparison

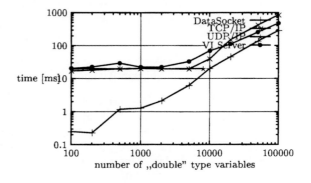

Fig. 5. remote transmission times comparison

5. IDEA

Theorem 1. (Shannon-Kotielnikow). Continuous pulse modulated signal, which modulation frequency is at least twice higher than maximal frequency of its spectrum, can be demodulated to continuous signal.

In case of using computer and computer networks for control purposes, there are pulse-modulated signal transmission (discrete-in-time signals), thus theorem (1) applies.

Using theorem (1), basing on knowledge of time constants of controlled object, and the maximal client number, it is possible to form following equations:

$$T_{sym} > 2 * T_{tr} * n \qquad (1)$$

converting:

$$\frac{2 * T_{tr} * n}{T_{sym}} > 1 \qquad (2)$$

where: T_{obj} –time constant of object, T_{tr} –total transmission time, n –client quantity

With use of above mentioned equations, it is possible to predict, whether the installation can work properly. These equations should be kind of tool for setting threshold of remote automation realization possibility.

The adjustment of parameters of an installation, to conform requirements of real-time operation can be done in two different ways: it is possible to shorten communication time (reducing transmission range, speeding up the connection, using better communication protocol) or lowering installation Time-constant. An excellent example of lowering installation Time-constant is conversion from local controller to remote controller (Fig. 6) Local controllers are more time critical, due to large amount of control data from object to controller and vice versa, on the other hand remote controllers require only very narrow bandwidth, and reduce time constants significally.

6. SUMMARY

Considering use of Internet/Intranets (IP networks in general) in automatic control area, achieving minimal transmission times is not the most important aim. The basis should be ability to SYNCHRONIZE object and its model, and assuring desired control quality coefficient.

Acknowledgements: This work was supported by the Committee of Scientific Research (KBN)

REFERENCES

Alfonseca M., de Lara J., Pulido E., 1999, *Semiautomatic Generation of Web Courses by means*

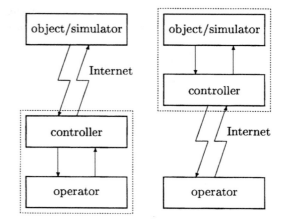

Fig. 6. local controller (server-side) versus remote controller (client side)

of an Object-Oriented Simulation Language, (SIMULATION, July 1999, 73:1, 5-12, Simulation Councils, Inc.)

Chung-Ping Young, Wei-Lun Juang, Devney M. J., 2000, *Real-Time Intranet-Controlled Virtual Instrument Multiple Circuit Power Monitoring*, (IEEE Transactions on Instrumentation and Measurement, vol. 49, no. 3, June 2000)

Davis W. J., Xu Chen, Brook A., Abu Awad F., 1999, *Implementing On-Line Simulation With the World Wide Web*, (SIMULATION, July 1999, 73:1, 40-54, Simulation Councils, Inc.)

Gerogiev G. A., Hoogenboom G., 1999, *Near Real-Time Agricultural Simulations on the Web*, (SIMULATION, July 1999, 73:1, 22-28, Simulation Councils, Inc.)

M. Metzger, *Modeling, Simulation and Control of Continuous Processes*, Edition of Jacek Skalmierski Computer Studio, Gliwice 2000

RFC 0761 *DoD standard Transmission Control Protocol*, J. Postel. Jan-01-1980. (Format: TXT=167049 bytes) (Status: UNKNOWN)

RFC 0768 *User Datagram Protocol*. J. Postel. Aug-28-1980. (Format: TXT=5896 bytes) (Also STD0006) (Status: STANDARD)

RFC 0774 *Internet Protocol Handbook: Table of contents*. J. Postel. Oct-01-1980. (Format: TXT=3452 bytes) (Obsoletes RFC0766) (Status: UNKNOWN)

RFC 0791 *Internet Protocol*. J. Postel. Sep-01-1981. (Format: TXT=97779 bytes) (Obsoletes RFC0760) (Also STD0005) (Status: STANDARD)

RFC 0792 *Internet Control Message Protocol*. J. Postel. Sep-01-1981. (Format: TXT=30404 bytes) (Obsoletes RFC0777) (Updated by RFC0950) (Also STD0005) (Status: STANDARD)

RFC 0793 *Transmission Control Protocol*. J. Postel. Sep-01-1981. (Format: TXT=177957 bytes) (Also STD0007) (Status: STANDARD)

RFC 0919 *Broadcasting Internet Datagrams.* J.C. Mogul. Oct-01-1984. (Format: TXT=16382 bytes) (Also STD0005) (Status: STANDARD)

RFC 0964 *Some problems with the specification of the Military Standard Transmission Control Protocol,* D.P. Sidhu. Nov-01-1985. (Format: TXT=20972 bytes) (Status: UNKNOWN)

RFC 1112 *Host extensions for IP multicasting.* S.E. Deering. Aug-01-1989. (Format: TXT=39904 bytes) (Obsoletes RFC0988, RFC1054) (Updated by RFC2236) (Also STD0005) (Status: STANDARD)

RFC 1945 *Hypertext Transfer Protocol – HTTP/1.0.* T. Berners-Lee, R. Fielding & H. Frystyk. May 1996. (Format: TXT=137582 bytes) (Status: INFORMATIONAL)

RFC 2460 *Internet Protocol, Version 6 (IPv6) Specification.* S. Deering, R. Hinden. December 1998. (Format: TXT=85490 bytes) (Obsoletes RFC1883) (Status: DRAFT STANDARD)

RFC 2463 *Internet Control Message Protocol (ICMPv6) for the Internet Protocol Version 6 (IPv6) Specification,* A. Conta, S. Deering. December 1998. (Format: TXT=34190 bytes) (Obsoletes RFC1885) (Status: DRAFT STANDARD)

Schmid Ch., 1999, *A Remote Laboratory Using Virtual Reality on the Web.* (SIMULATION, July 1999, 73:1, 13-21, Simulation Councils, Inc.)

Stegawski M., Schaumann R., 1998, *A New Virtual-Instrumentation-Based Experimenting Environment for Undergraduate Laboratories with Application in Research and Manufacturing,* (IEEE Transactions on Instrumentation and Measurement, vol. 47, no. 6, December 1998)

IFAC

Publications
www.elsevier.com/locate/ifac

TURBO CODES AND RADIO-DATA TRANSMISSION

Karel Vlček

Dept. of Control and Measurement
Technical University of Ostrava
Czech Republic
karel.vlcek@vsb.cz

Abstract: Using a single code requires very long codes, and consequently very complex coding system. One way around the problem to achieve very low error probabilities using coding is turbo coding (TC) application. The principle of TC is to use two codes that are not very long in a manner that reduces the complexity considerably without increasing the error rate significantly. The general model of concatenated coding system is shown an algorithm of turbo codes is given in this paper. Included are a simple derivation for the performance of turbo codes, and a straightforward presentation of the iterative decoding algorithm. The derivations of the performance is novel. The treatment is intended for further study in the field and, significantly, to provide sufficient information for the VHDL simulations of design. *Copyright © 2001 IFAC*

Key words: Concatenated coding, Code system performance, Iterative decoding, VHDL modelling

1. INTRODUCTION

The invention of turbo codes involved reviving some concepts [Unge_82] and algorithms, and combining them with some clever new ideas [Bene_97], [Pere_96], [Berr_93], [Robe_95], [Bahl_74]. Because the principles surrounding turbo codes are both uncommon and novel, it has been difficult for the initiate to enter into the study of convolutional codes [Vlce_98].

Complicating matters further is the fact that there exist now numerous papers on the topic so that there is no clear place to begin study of these codes.

This paper is addressed to this problem by including in one paper an introduction to the study of turbo codes [Aits_96], [Bene_96a].

A detailed description of the encoder is given [Reed_96] and it is presented a simple derivation of the performance in additive white Gaussian noise (AWGN). Particularly difficult for the novice has been the understanding and simulation of the iterative decoding algorithm, and so we give a thorough description of the algorithm here [Robe_95], [Berr_96].

This paper borrows from some publications, sometimes adding details that were omitted in those works. However, the general presentation and some of the derivations are novel. Our goal is self-contained, simple introduction to turbo codes for those already knowledgeable in the fields of algebraic and trellis codes.

The subsequent section then describes the iterative algorithm used to decode turbo codes. The treatment in each of these sections is meant to be sufficiently detailed so that one may with reasonable ease design a computer simulation of the encoder and decoder.

A turbo encoder is a combination of two simple encoders. The input is a block of K information bits. The two encoders generate parity symbols from two simple recursive convolutional codes, each with a small number of states. The information bits are also sent in the non-coded form.

2. ENCODER AND DECODER

The key innovation of turbo codes is an interleaver function P, which permutes the original information bits before input to the second encoder. The permutation P allows that input sequence for which one encoder produces low-weight code words will usually cause the other encoder to produce high-weight code words. Thus, even though the constituent codes are individually weak, the combination is surprisingly information bits. powerful. The resulting code has features similar to "*random*" block code with K

Random block codes are known to achieve Shannon-limit performance as *K* gets large hardware as well as the price of a prohibitively complex decoding algorithm. Turbo codes mimic to good random codes (for large *K*) using an iterative decoding algorithm based on simple individually matched to the simple constituent codes.

Each constituent decoder send information of the decoded bits to the other decoder, and uses the corresponding from the other decoder as *a priori* likelihood. The non-coded information bits are available to each decoder to initialise the *a priori* likelihood.

The decoder *"MAP"* (*Maximum a Posteriori*) bitwise decoding algorithm, which requires the states as well known Viterbi algorithm. The turbo decoder iterates between two constituent decoders to reaching satisfactory convergence. The final output is one of the likelihood estimates of either of the decoders [Dhol_94].

K BITS OF INFORMATION

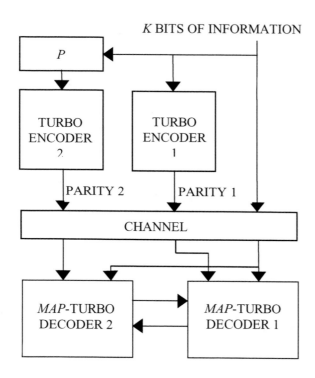

Fig. 1: Standard turbo-coding system

In addition to providing improved performance, turbo decoders [Robe_94] are lower in complexity of concatenated decoders. Decoding time is proportional to the number of iterations, unless special-purpose hardware is used to parallel-processing all of the states. Interleaver size impacts buffer requirements of decoding time: it is a primary determinant of turbo code performance is achieved with interleavers of the traditional concatenated systems.

As seen in the Fig. 1, a turbo encoder is consists of two binary rate 1:2 convolutional encoders separated by an *N*-bit interleaver or permuter, together with an optional puncturing mechanism. Clearly, without the puncture, the encoder is rate 1:3, mapping N data bits to 3*N* code bits.

The encoders are configured in a manner reminiscent of classical concatenated codes. However, instead of cascading the encoders in the usual serial fashion, the encoders are arranged in a so-called *parallel concatenation*. Observe also that the constituent convolutional encoders are of the recursive systematic variety.

Because any non-recursive (i.e., feed-forward) non-catastrophic convolutional encoder is equivalent to a recursive systematic encoder in that they possess that same set of code sequences, there was no compelling reason in the past for favouring recursive encoders.

The recursive encoders are necessary to attain the exceptional performance provided by turbo codes. Without any essential loss of generality, we assume that the constituent codes are identical.

3. SYSTEMATIC TURBO ENCODERS

Whereas the generator matrix for a rate 1:2 Non-Recursive convolutional code has the form

$$G_{N-R} = \begin{bmatrix} g_1(D) & g_2(D) \end{bmatrix},$$

the equivalent Recursive systematic encoder has the generator matrix

$$G_R = \begin{bmatrix} 1 & \dfrac{g_2(D)}{g_1(D)} \end{bmatrix}.$$

Observe that the code sequence corresponding to the encoder input $u(D)$ for the former code is

$$u(D) \cdot G_{N-R} = \begin{bmatrix} u(D) \cdot g_1(D) & u(D) \cdot g_2(D) \end{bmatrix},$$

and that the identical code sequence is produced in the recursive code by the sequence

$$u'(D) = u(D) \cdot g_1(D),$$

since in this case the code sequence is

$$u(D) \cdot g_1(D) \cdot G_R = u(D) \cdot G_{N-R}.$$

Here, we loosely call the pair of polynomials $u(D) \cdot G_{N-R}$ a code sequence, although the actual code sequence is derived from this polynomial pair in the usual way. Observe that, for the recursive encoder, the code sequence will be of finite weight if and only if the input sequence is divisible by $g_1(D)$.

The weight-one input will produce an infinite weight output for such an input is never divisible by $g_1(D)$. For any non-trivial $g_1(D)$, there exists a family of weight-two inputs of the form

$$D^j(1+D^{q-1}), \quad j \geq 0,$$

which produce finite weight outputs, i.e., which are divisible by $g_1(D)$. When $g_1(D)$ is a primitive polynomial of degree m, then $q = 2^m$; more generally, $q-1$ is the length of the pseudo-random sequence generated by $g_1(D)$.

The code word weight-one input will create a path that diverges from the all-zeros path, but never remerge, and there will always exist a trellis path that diverge and remerge later which corresponds to a weight-two data sequence. Consider the recursive code with generator matrix

$$G_R(D) = \left[1 \quad \frac{1+D^2+D^3+D^4}{1+D+D^4} \right].$$

Thus, $g_1(D) = 1+D+D^4$ and $g_2(D) = 1+D^2+D^3+D^4$ or, in octal form, $(g_1, g_2) = (31, 27)$. Observe that $g_1(x)$ is primitive so that, for example, $u(D) = 1+D^{15}$ produces the finite-length code sequence $(1+D^{15}, 1+D+D^2+D^3+D^5+D^7+D^8+D^{11})$.

Of course, any delayed version of this input, say, $D^7 \cdot (1+D^{15})$, will simply produce a delayed version of this code sequence.

The Fig. 2 gives one encoder realisation for this code. The example serves to demonstrate the conventions generally used in the literature for specifying such encoders. The function of the permute block is to take each incoming block of N

data bits and rearrange them in a pseudo-random fashion prior to encoding by the second encoder.

Unlike the classical interleaver (e.g., block or convolutional interleaver), which rearranges the bits in some systematic fashion, it is important that the permute block sort the bits in a manner that lacks any apparent order, although it might be tailored in a certain way for weight-two and weight-three inputs.

Also important is that N will be selected quite large and we shall assume N, 1000 hereafter. The importance of these two requirements will be illuminated below. We point out also that one pseudo-random permute block will perform about as well as any other provided N is large. The low-rate codes are appropriate, in other situations such as satellite communications, a rate of 1:2 or higher is preferred.

The role of the turbo code puncture is identical to that of its convolutional code counterpart, to periodically delete selected bits to reduce coding overhead.

For the case of iterative decoding to be discussed below, it is preferable to delete only parity bits, but there is no guarantee that this will maximise the minimum code-word distance.

For example, to achieve a rate of 1:2, one might delete all even parity bits from the top encoder and all odd parity bits from the bottom one.

4. TURBO DECODER

As will be elaborated upon in the next section, a *maximum-likelihood* (*ML*) sequence decoder would be far too complex for a turbo code due to the presence of the permute block. However, the sub-optimum iterative decoding algorithm to be described there offers near-ML performance. Hence, we shall now estimate the performance of an *ML*

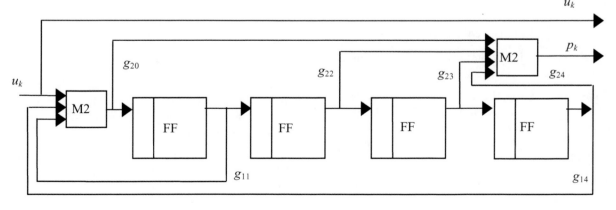

Fig. 2: Recursive encoder (g_1, g_2) = (31, 27)

25

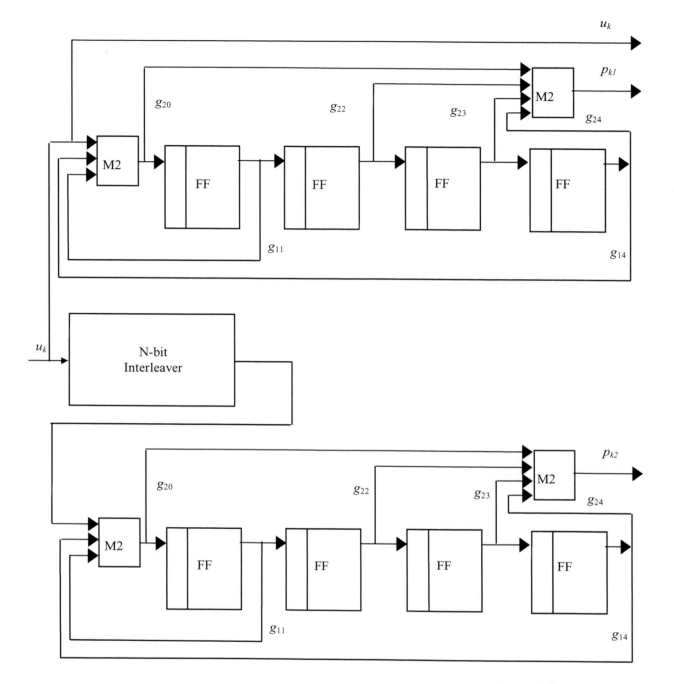

Fig. 3: Turbo code encoder

decoder (analysis of the iterative decoder is much more difficult).

The idea [Bene_96b], [Bene_97], [Berr_95] behind extrinsic information is that *MAP*-D2 provides soft information to *MAP*-D1, using only information not available to *MAP*-D1; the *MAP*-D1 does likewise for *MAP*-D2.

The Soft Decision is realised by the a posteriori LLRs (log-likelihood ratios) by the relation

$$\Lambda_i = \ln \frac{P[m_i = 1|y]}{P[m_i = 0|y]}.$$

Decoding algorithm, which processes the a priori information on its input and generates the a posteriori information on its output is called SISO (soft-input, soft-output) decoding algorithm.

The following text will introduce, that the both Viterbi, and MAP algorithms can be modified to

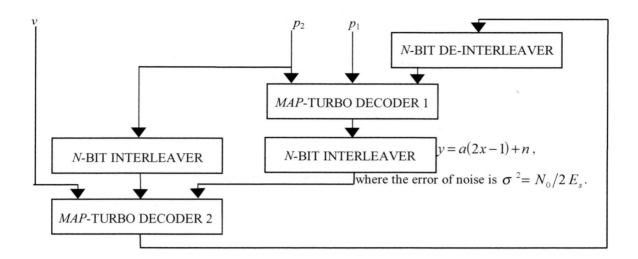

$y = a(2x - 1) + n$,

where the error of noise is $\sigma^2 = N_0 / 2 E_s$.

Fig. 4: Iterative *MAP*-turbo decoder

accept the input data with the soft decision. Similarly, the algorithms can be modified to generate output values with the soft decision.

It is clear, from the Figure 3.1, that the SISO decoder for code rate 1/2 RSC receive three inputs: systematic received sequence $y_i^{(s)}$, of parity symbols $y_i^{(p)}$, and apriori information z_i, which is generated by output of the second decoder. Output variable Λ_i of decoder is defined by previous equation.

It is usual to study the systems applying BPSK (binary phase shift keying), which do the modulation by the equation

$$y' = a\sqrt{E_s}(2x - 1) + n',$$

where a is an amplitude after feeding, the $\sqrt{E_s}(2x - 1)$ is BPSK modulation symbol, $(x \in \{0,1\})$, E_s is an energy for symbol (E_s is in correlation to energy for bit E_b, and to code ratio r, which is defined by relation $E_s = rE_b$).

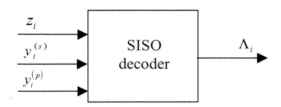

Fig. 5: Decoder with soft decision on the input and with generating of values with soft decision for convolutional code with recurrent generating matrix and ratio $1/2$.

A variable n' is the Gauss variable with error $\sigma^2 = N_0 / 2$. If it is a a constant, the information cannel is called AWGN (additive white Gaussian noise). In other cases it is called "flat-fading cannel";

the real type of fading depend on statistics a, which is usually described as Rayleigh or Rician. Other statistics is possible describe as

For soft decision LLRs (log-likelihood ratios) on the output of SISO decoder it is used the model of the information cannel, which is possible decompose into three additive components:

$$\Lambda_i = \frac{4a_i^{(s)}E_s}{N_0} y_i^{(s)} + z_i + l_i,$$

where a variable l_i is called *extrinsic information*. The first two variables on the right side of equation are in relation to received sequence $y_i^{(s)}$, on the other hand the variable, which is result of the second decoder z_i, the extrinsic information represent a new information, which depends on actual state of decoding process. To prevent problems with the positive feedback, it is important, the extrinsic information be fed from one decoder into the second one. By this way is the a priori information on the input of one decoder calculated by subtracting of the

two values of the output – the value is done by expression:

$$l_i = \Lambda_i - \frac{4a_i^{(s)}E_s}{N_0} y_i^{(s)} - z_i.$$

27

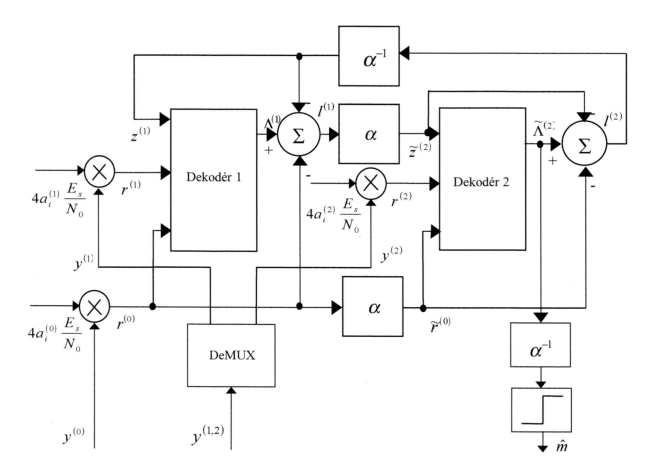

Fig. 6: Turbo Decoder Interconnection

5. TURBO DECODER INTERCONNECTION

The interconnection of the standard turbo decoder is on the fig. 6. The first decoder receives the sequence $y^{(0)}$ (multiplied by $4a_i^{(0)} E_s/N_0$), the sequence of control symbols from the first encoder $y^{(1)}$ (multiplied by $4a_i^{(1)} E_s/N_0$) and a priori information $z^{(1)}$, which is calculated by the second one.

The first decoder generates LLR (log-likelihood ratio) $\Lambda^{(1)}$. The extrinsic information of the first decoder $l^{(1)}$ is yielded as the difference of input variables of the first decoder: a priori received control sequence $r^{(1)} = y^{(1)} \cdot \left(4a^{(1)} E_s/N_0\right)$ and the received message $r^{(0)} = y^{(0)} \cdot \left(4a^{(0)} E_s/N_0\right)$. The extrinsic

information $l^{(1)}$ is used as an a priori information of the second decoder (t.j. $z_{\alpha(i)}^{(2)} = l_i^{(1)}$).

The second decoder receives an interleaved, and original message too $\tilde{r}^{(0)} = \tilde{y}^{(0)} \cdot \left(4a^{(0)} E_s/N_0\right)$, and a priori received parity sequence of the message $\tilde{r}^{(2)} = \tilde{y}^{(2)} \cdot \left(4a^{(2)} E_s/N_0\right)$. The second decoder generate LLR (log-likelihood ratio) $\Lambda^{(2)}$, from which is calculated extrinsic information $l^{(2)}$ as a result of $\tilde{r}^{(2)}$ subtraction a priori information $z^{(2)}$.

This extrinsic information $l^{(2)}$ is re-interleaved and it is used as an input sequence of the first decoder (by the equation $z_{\alpha^{-1}(i)}^{(1)} = l_i^{(2)}$) in the next iteration. After Q iterations it is calculated the final solution of the message. It is re-interleaved with hard decision by the operation:

$$\hat{m}_i = \begin{cases} 1 & \left(\Lambda_{\alpha(i)}^{(2)} \geq 0\right) \\ 0 & \left(\Lambda_{\alpha(i)}^{(2)} < 0\right) \end{cases}.$$

If it is used the code puncturing, the decoders do not receive the full parity message sequence. In this case are bits, which were through puncturing omitted, are substituted by zeros. (Linearity of the code makes it possible).

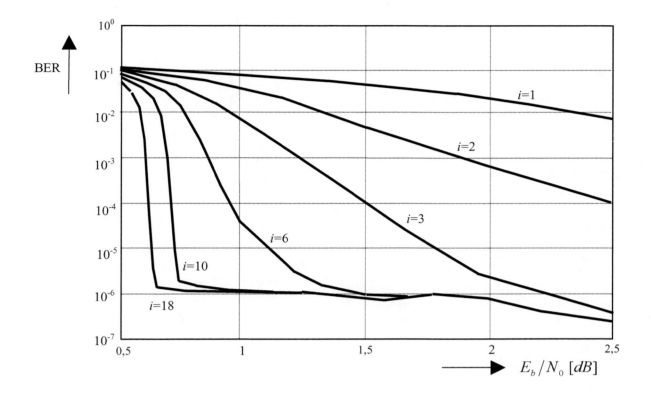

Fig. 7: The efficiency of turbo code in dependence on the number of iterations *i*.

6. EXAMPLE OF TURBO DECODING

As an illustration of iterative decoding behaviour we will consider encoder from the Fig 3. If it is calculated 18 iterations of sequential decoding, it will be fulfilled the function by the graph on Fig. 7. The variable BER is dramatically enhanced through

a bit of iterations. Later the influence become weak. This behaviour is typical not only for turbo codes, but in the other applications, in which iterative decoding is used.

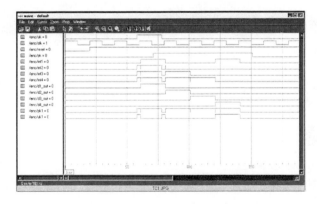

Fig. 8: Simulation of VHDL model 1st step

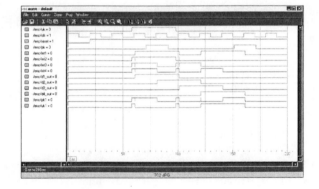

Fig. 9: Simulation of VHDL model 2nd step

7. CONCLUSION

The discovery of turbo codes, there was much interest in the coding community in sub-optimal decoding strategies for concatenated codes, involving multiple (usually two) decoders operating co-operatively and iteratively. Most of the focus was on a type of Viterbi decoder [Vite_98] which provides soft-output (or reliability) information to a companion soft-output Viterbi decoder for use in a subsequent decoding. Also receiving some attention was the symbol-by-symbol maximum a posteriori (MAP) algorithm.

8. ACKNOWLEDGEMENT

Support for GACR project 102/01/1531 „Formal methods in diagnostics of digital circuits - verification of testable design", (No. 102/00/1531) is gratefully acknowledged.

9. REFERENCES

[Aits_96] Aitsab O., Pyndiah R.: "Performance of Reed-Solomon Block Turbo Code," IEEE Global Telecomm. Conf. (Nov. 18-22 1996 London, UK), pp. 121-125.

[Arno_95] Arnold D. and Meyerhans G.: "The realization of the turbo-coding system," Semester Project Report, Swiss Fed. Inst. of Tech., Zurich, Switzerland, July, 1995.

[Bahl_74] L. Bahl, J. Cocke, F. Jelinek, and J. Raviv, "Optimal decoding of linear codes for minimising symbol error rate," IEEE Trans. Inf. Theory, pp. 284-287, Mar. 1974.

[Bene_96a] Benedetto S. and Montorsi G.: "Unveiling turbo codes: Some results on parallel concatenated coding schemes," IEEE Trans. Inf. Theory, pp. 409-428, Mar. 1996.

[Bene_96b] Benedetto S. and Montorsi G.: "Design of parallel concatenated codes," IEEE Trans. Comm., pp. 591-600, May 1996.

[Bene_97] Benedetto S., Divsalar D., Montorsi G., Pollara F.: "Serial Concatenation of Interleaved Codes: Performance Analysis, Design, and Iterative Decoding," IEEE Trans. On Inf. Theory (March 97 accepted).

[Berr_93] Berrou C., Glavieux A., and Thitimajshima P.: "Near Shannon limit error- correcting coding and decoding: Turbo codes," Proc. 1993 Int. Conf. Comm., pp. 1064- 1070.

[Berr_95] Berrou, C., Combelles, P., Penard, P., Talibart, B.: "IC for Turbo-Codes Encoding and Decoding," IEEE Solid-State Circuits Conf. (Feb 15-17 1995 San Francisco, CA, USA), pp. 90-91.

[Berr_96] Berrou C. and Glavieux A.: "Near optimum error correcting coding and decoding: turbo-codes," IEEE Trans. Comm., pp. 1261-1271, Oct. 1996.

[Dhol_94] Dholakia, A.: "Introduction to Convolutional Codes with Applications." Kluwer Academic Publishers (1994), pp. 207-208.

[Divs_95] Divsalar D. and Pollara F.: "Turbo codes for PCS applications," Proc. 1995 Int. Conf. Comm., pp. 54-59.

[Eyub_93] Eyuboglu M., Forney G. D., Dong P., Long G.: "Advanced modulation techniques." Eur. Trans. on Telecom., pp. 243-256, May 1993.

[Hage_89] Hagenauer J. and Hoeher P.: "A Viterbi algorithm with soft-decision outputs and its applications," Proc. GlobeCom 1989, pp. 1680-1686.

[Hage_96] Hagenauer J., Offer E., and Papke L.: "Iterative decoding of binary block and convolutional codes," IEEE Trans. Inf. Theory, pp. 429-445, Mar. 1996.

[Pere_96] Perez L., Seghers J., and Costello D.: "A distance spectrum interpretation of turbo codes," IEEE Trans. Inf. Theory, pp. 1698-1709, Nov. 1996.

[Reed_96] Reed M., Pietrobon S.: "Turbo-Code Termination Schemes and a Novel Alternative for Short Frames," IEEE Internat. Symp. on Personal Radio Comm. (Oct 15-18 1996 Taipei, Taiwan), pp. 354-358.

[Robe_94] Robertson P.: "Illuminating the structure of code and decoder of parallel concatenated recursive systematic (turbo) codes," Proc. GlobeCom 1994, pp. 1298-1303.

[Robe_95] Robertson P., Villebrun E., and Hoeher P.: "A comparison of optimal and sub-optimal MAP decoding algorithms operating in the log domain," Proc. 1995 Int. Conf. on Comm., pp. 1009-1013.

[Unge_82] Ungerboeck G.: "Channel coding with multilevel/phase signals," IEEE Trans. Inf. Theory, pp. 55-67, Jan. 1982.

[Vite_98] Viterbi A.: "An intuitive justification and a simplified implementation of the MAP decoder for convolutional codes," IEEE JSAC, pp. 260-264, Feb. 1998.

[Vlce_98] Vlček, K.: "The VHDL Model of Wyner-Ash Channel Coding for Medical Applications." Proc. of DDECS '98 Workshop, (Sept. 2-4 1998, Szczyrk, Poland), ISBN 83-908409-6-0, pp. 145-151.

[Vlce_00] Vlček, K., Miklík, D., Kovalský, J., Mitrych, J.: "Turbo Coding Performance and Implementation" Proc. of the IEEE DDECS 2000 Workshop, (5-7 April 2000, Smolenice, Slovakia), ISBN 80-968320-3-4, p. 71

[Zhan_96] Zhang, L., Zhang, W., Ball, J.T., Gill, M.C.: "Extremely Robust Turbo Coded HF Modem," Proc. of Conf. MILCOM 96, (Oct 21-24 1996, Washington, DC, USA), pp. 691-695.

IFAC

Publications
www.elsevier.com/locate/ifac

APPLICATION OF TMS320C31 DSP
IN ACTIVE NOISE CONTROL

Małgorzata Błażej

Institute of Automatic Control, Silesian University of Technology,
Akademicka 16, 44-100 Gliwice, Poland, mblazej@ia.polsl.gliwice.pl

Abstract. The paper presents a digital signal processing technique applied for active noise control in a laboratory system. An adaptive control algorithm is implemented using TMS320C31-40 digital signal processor from *Texas Instruments* built in the DS1102 *dSPACE* DSP Controller Board. The usefulness of LMS and RLS-based adaptive control algorithms in the system has been evaluated and the most important aspects of the optimised implementations of algorithms using C and assembler code for laboratory purpose have been discussed. *Copyright © 2001 IFAC*

Key Words. Active noise control, adaptive algorithms, digital signal processors, LMS algorithm, recursive least squares.

1. INTRODUCTION

The idea of active noise control (ANC) is an attenuation of unwanted, disturbing acoustic wave (noise) by an acoustic wave in opposing phase emitted by a control loudspeaker driven from computer. This idea has been known for many years, but it has been possible to implement since the progress in electronics and computer science gave processors with enough computational power, e.g. digital signal processors. ANC systems are very complicated both from the acoustic and control viewpoints. As it is almost impossible to describe an acoustic enclosure by means of phenomenological equations, the automatic control methods using digital signal processing procedures have to be applied. Most of ANC systems require adaptation, because: acoustic plants change due to temperature or humidity conditions, geometrical configuration of an enclosure may change, there may be persons moving inside, characteristics of the noise may change, etc. Therefore the adaptive digital signal processing algorithms are necessary for modelling and controlling acoustic plants in real-world ANC systems.

It follows from the above remarks that ANC systems are difficult to implement. Besides, there are few real-world applications. The paper presents an active

noise control system, describes the digital signal processing algorithms used to attenuate unwanted noise and implementation aspects using TMS320C31 DSP. The effectiveness of DS1102 board depends on system assumptions: system's configuration, control algorithms used, implementation code optimisation, etc. These issues are discussed further in this paper with reference to the hardware used in the described application. It is worth noticing that the system has been created for laboratory purposes what implies some additional assumptions that are not met in commercial applications.

2. DESCRIPTION OF THE ANC SYSTEM

The ANC system is configured for low-frequency noise cancellation in a small reverbant laboratory enclosure of about 23 m^3 of cubature. The frequency range is determined as 50 Hz – 110 Hz, thus sampling frequency has been chosen as 500 Hz. The hardware structure of the system and a scheme of the enclosure are shown in Figure 1. The feedforward control structure is used, as it is the most effective in this case (Niederliński, 1999). A reference microphone M_x is situated close to a primary noise source, being a loudspeaker G_h driven by a noise generator. M_x measures so-called reference signal

$x(i)$, i.e. disturbance signal. After amplification and prefiltration the reference signal is processed by a control algorithm implemented using the DSP. The control signal $u(i)$ is generated to drive loudspeaker G_s. A microphone M_e measures the error signal $e(i)$ used for adaptation purposes.

An average attenuation rate of the noise level using presented system is about 25 dB for pure tones (single sines), and up to 15 dB for broadband noises.

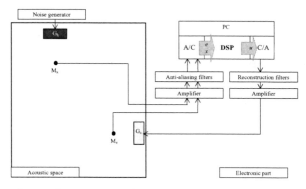

Fig. 1. The structure of the ANC system.

3. DIGITAL SIGNAL PROCESSING HARDWARE

The DS1102 DSP Controller Board from dSPACE has been used in discussed implementation. This board consists of *Texas Instruments'* TMS320C31 floating-point digital signal processor, 2K x 32-bit on-chip RAM, 128K x 32-bit RAM, 2 16-bit and 2 12-bit analog input channels and 4 12-bit analog output channels and 16 digital I/O lines. The processor is timed with 40 MHz clock that gives 50 ns single processor cycle. It allows 40 million floating-point operations per second (40 MFLOPS) that is 20 million instructions per second (20 MIPS). However, the instruction execution time may be longer that single cycle if conflicts occur on the pipeline. An example of such conflict may be observed during three-operand instruction memory read, if first of the source operands is in external memory and the second is in internal memory. In such case instruction is executed in two processor cycles (Texas Instruments, 1994). If the operands are switched, this instruction takes only one cycle. The same relationships hold for parallel instructions that are executed in one cycle if data is located in internal memory. Unfortunately the size of the on-chip RAM, for which the instruction execution time is the shortest, is very small. Thus memory location of the processed data should be carefully planned, what allows writing optimal program code despite pipeline conflicts.

The independent ALU and multiplier units using two independent auxiliary register arithmetic units that may generate two addresses per cycle give possibilities for use of parallel instructions. These instructions may operate on 4 floating-point or integer values in one cycle. For instance the parallel multiply and add instructions enable fast FIR filtering.

4. CONTROL ALGORITHMS

The block diagram of a feedforward ANC system is shown in Fig. 2. The weights of the controller FIR filter $W(z^{-1})$ are updated in every time instant i according to an adaptive algorithm. One of them, the most frequently used LMS algorithm (Widrow and Sterns, 1985; Kuo and Morgan, 1996; Niederliński, 1999) is very simple and requires low computational power. It is fast but unfortunately not very accurate, thus large controller FIR filter orders are required for appropriate system performance. The RLS algorithm is accurate but computationally complicated, that causes limitations in real-time applications implementation. However, application of orthogonal lattice filters allows to simplify the RLS algorithm, by reducing its computational complexity from square to linear (Michalczyk, 2000, Figwer and Michalczyk, 2001).

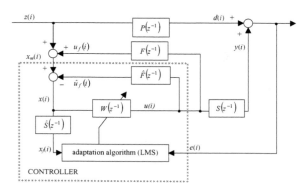

Fig. 2. Block diagram of the ANC system with the modifications for the system stability.

Control algorithms usually used are based on the LMS or RLS algorithms, that in ANC applications need additional modifications for stability to be introduced – so-called FX (Filtered X) modifications (Kuo and Morgan, 1996). These are additional filtrations of the $x(i)$ signal. While noise is controlled in a small reverbant room and a microphone works as reference sensor, there is a need for another modification for system stability, i.e. neutralisation of an acoustic feedback effect. This means additional control signal $u(i)$ filtration (Kuo and Morgan, 1996). The algorithms used in presented ANC systems are: FX-LMS algorithm with acoustic feedback neutralisation (Błażej and Ogonowski, 2001a, 2001b) and FX-SRLS (FX Schur RLS) algorithm with acoustic feedback neutralisation (Michalczyk, 2000; Figwer and Michalczyk, 2001).

FIR filters modelling the acoustic plant have to be of large orders because of the plant complexity. Thus FIR filter used for FX modification should be of order at least equal 120 (Błażej and Ogonowski, 2001b), however, the system performance is better if order 250 is applied. The order of the FIR filter used for an acoustic feedback neutralisation is 300. The order of controller FIR filter depends on kind of attenuated noise. If it is a narrowband, periodic noise, i.e. pure tone (single sine) or sum of single tones (sum of a few sines), filter order need not to be large. Thus, in this case there are no limitations for the discussed algorithms using TMS320C31-40 DSP. However, if the noise is broadband and random, the controller filter order has to be as large as possible. In that case the controller filter models the disturbance path transfer function (transfer function of an acoustic space between the reference and error microphones) that is very complicated (Michalczyk, 2000; Błażej and Ogonowski, 2001b). The controller filter order in FX-LMS algorithm is 500. FX-SRLS algorithm using controller filter order 100 gives similar results, because RLS-based algorithm is much more accurate.

5. SOFTWARE IMPLEMENTATION CONSIDERATIONS

The control program is written using C++ programming language with use of TMS assembler subroutines. Use of C++ enables simple program controlling and modification, while assembler subroutines for time critical signal processing minimise program execution time.

5.1 C code

Table 1. Approx. execution times of LMS procedures written in different programming languages.

programming language	cycles for N=100	cycles per each filter tap
C++	8101	80
C++ (optimised)	2591	25
TMS assembler	268	2

The Table 1 shows differences in execution time of LMS procedure (adaptation of the controller FIR filter). The second subroutine has been written in C++, but compiled using optimisation option of the compiler. The execution time of the assembler subroutine is different for different operand locations in memory – 2 cycles for each filter tap may be obtained if the updated filter weights are located in internal memory. The time in column 2 for the assembler example includes time for procedure call from C level, in fact it equals about 214 cycles. These differences justify the choice of the assembler programming language for signal processing routines. The presented application has been created for laboratory purposes and is often modified for

various experiments. Thus the program body is written in C++ because it allows for easier control of the program. Some short, little time consuming procedures are left in C language, while the most important ones are optimised using assembler.

5.2 Assembler routines

Table 2. Execution times of subroutines – cycles per filter tap

Subroutine	signal in memory:	
	internal	external
transversal FIR filtration	1	2
LMS algorithm for FIR filter	2	2

The execution times of subroutines used in FX-LMS control algorithm are presented in Table 2. These are the most important numbers while the filter orders and vectors sizes are large and thus processing in loops take most time of the processor. If the FIR filter order is 500, the time of FIR filtration result calculation is much longer (1 or 2 cycles per each filter tap) than time for procedure initialisation (approx. 14 cycles). Thus approximate minimal time for FX-LMS control algorithm using decoupling of the acoustic feedback for all FIR filters orders 500 equals to: 3 FIR filtrations plus 1 LMS subroutine i.e. about 2500 cycles plus 611 cycles of other algorithm operations. It gives 3111 cycles – less than 8% of the sample time.

Table 3. Comparison of FX-LMS and FX-SRLS execution times

Algorithm	controller FIR order	init. time	per filter tap	sum
FX-LMS	500	1161	3	2 661
FX-SRLS	100	1847	214	23 247

The Table 3 compares execution times of FX-LMS and FX-SRLS algorithms, both including FX modification (FIR filter order 250) and acoustic feedback neutralisation (FIR filter order 300). The algorithm implementations are built in C++ environment using assembler subroutines. The cost of accuracy of FX–SRLS algorithm is constraint by the controller filter order that should not exceed 178.

5.3 Sampling frequency choice

Sampling frequency chosen as 500 Hz gives much time for signal processing. The frequency range of attenuated broadband noise, i.e. 50 Hz – 110 Hz justifies the choice. In some applications it could be better if the sampling frequency was higher. It would cause double problems – firstly, the time for signal processing would be shorter, secondly – FIR filters orders wo uld have to be higher to model object properly. Thus the control algorithm would have more information to process in shorter time.

5.4 System structure and DSP limitations

The presented algorithms parameterised as in Table 3 give similar results in case of the ANC system with one reference sensor, one error sensor and one control unit – loudspeaker (although the transient states are different – FX-SRLS algorithm is faster and more accurate). However, in case of multipath systems with more than one sensor or loudspeaker, time for signal processing may be too short. Let consider the ANC system with one reference microphone, one error microphone and two loudspeakers. In this configuration the system works better than the system described in point 3 (Błażej and Ogonowski, 2001a). A control algorithm in this case includes 6 FIR filtrations and 2 LMS subroutines and is executed in 5621 cycles (for FIR filters orders: NS=250, NF=300 and NW=500 – see Figure 2). This time is still very short in comparison with 40 000 cycles in sampling time. Although in this case the controller filters' orders in FX-SRLS algorithm must be smaller (about 70) both algorithms may still work similarly.

However, further complication of the ANC system structure – for example 2x2x2 system (with two reference and two error microphones and two loudspeakers) – implies decreasing of the controller filter order in RLS based algorithm. In case of very complicated system structure the quality of attenuation using FX-SRLS algorithm will be unsatisfactory. The FX-LMS algorithm may be used in 2x2x2 ANC feedforward system with no change of filters' orders. The LMS-based algorithm may still include some auxiliary procedures, like on-line secondary path identification or LMS modifications for improving algorithm speed of convergence.

There is no point in analysing all possible system structures that may be used in described application, because there are many various combinations. The LMS-based algorithms seems to be the most comfortable to be implemented in case of multipath broadband ANC systems, in spite of their inaccuracy and difficult parameterising. Using RLS-based algorithms the controller filter order must be decreased along with increase of system structure complication. Thus in case of complicated acoustic phenomena the accuracy of RLS-based algorithms is insufficient due to inappropriate plant modelling.

6. CONCLUSIONS

It has been shown that for some complicated control problems, like ANC, the TMS320C31-40 DSP may be used if an appropriate control algorithm is applied. However, if a multipath control system is concerned, the simplicity of the LMS algorithm, and hence its small computational complexity, makes it the most suitable to implement. The choice between the algorithm simplicity and accuracy is imposed by the DSP limitations. However, even complicated multipath broadband ANC feedforward system using LMS-based algorithms may be implemented with TMS320C31 DSP with good results.

The described system has been created for laboratory evaluation, thus the program body has been written in C++ language. The application may be optimised if it is of commercial purpose. In that case the whole program may be written in TMS320C31 assembler, that will reduce program execution time. However, in case of FX-SRLS algorithm these savings are insignificant due to algorithm complication.

ACKNOWLEDGEMENT

The partial financial support of this research by The State Committee of Scientific Research (KBN) under grant BW-494/Rau-1/2001 is gratefully acknowledged.

REFERENCES

Błażej, M. and Ogonowski, Z. (2001a). Adaptacyjne algorytmy aktywnego tłumienia hałasu dla przestrzennych stref ciszy. *Materiały V Szkoły "Metody Aktywne Tłumienia Drgań i Hałasu"*, Krynica.

Błażej, M. and Ogonowski, Z. (2001b). Adaptive 3-D Zone of Quiet in a Reverbant Enclosure using LMS-type Algorithms, *Proceedings of the 7th International Conference on „Methods and Models in Automation and Robotics"*. Międzyzdroje.

Figwer, J. and Michalczyk, M. (2001). Active Noise Control Using Orthogonal Filters, *Proceedings of the 7th International Conference on „Methods and Models in Automation and Robotics"*, Międzyzdroje.

Kuo, S. M. and Morgan, D. R. (1996). *Active Noise Control Systems. Algorithms and DSP Implementations*. J. Wiley & Sons, N. York.

Michalczyk, M. (2000). *Zastosowanie filtrów ortogonalnych do aktywnego tłumienia hałasu*, Master Thesis, Institute of Automation, Silesian University of Technology, Gliwice.

Niederliński, A. (1999). Identyfikacja i adaptacja dla aktywnego tłumienia hałasu. *XIII Krajowa Konferencja Automatyki*. Opole.

Texas Instruments, (1994). *TMS320C3x User's Guide*, 2558539-9761 revision J.

Widrow, B., Stearns, S. D. (1985). *Adaptive Signal Processing*. Englewood Cliffs, Prentice Hall, NJ.

www.elsevier.com/locate/ifac

LP-DERIVED FEATURES FOR SPEAKER RECOGNITION AND VERIFICATION

Adam Dustor

The Institute of Electronics, Silesian University of Technology, Gliwice, Poland

Abstract: This paper contains an overview of LP-derived features that are the most commonly used feature parameters for extracting speaker specific information from speech waves. The paper focuses especially on LP cepstral techniques like cepstral mean subtraction (CMS), cepstral weighting, pole filtered cepstral mean subtraction (PFCMS) and adaptive component weighted cepstrum (ACW). Additionally the linear predictive analysis is briefly discussed as well as properties and methods for cepstrum calculation. *Copyright © 2001 IFAC*

Keywords: linear prediction; speech analysis; speaker verification; speaker identification; signal processing;

1. INTRODUCTION

Speaker recognition is the process of automatically recognizing who is speaking by analysis speaker-specific information included in spoken utterances (Furui, 1997). This process encompasses identification and verification. The purpose of speaker identification is to determine the identity of an individual from a sample of his or her voice. It can be divided into two main categories: closed set and open set. In a closed-set identification there is an assumption that only registered speakers have an access to the system which makes a decision 1 from N, where N is the number of previously registered speakers. In an open-set identification there is no such an assumption so the identification system has to determine whether the testing utterance comes from a registered speaker or not and if yes it should determine his or her identity. The purpose of speaker verification is to decide whether a speaker is whom he claims to be. Most of the applications in which voice is used to confirm the identity claim of a speaker are classified as a speaker verification.

Speaker recognition systems can also be divided into text-dependent and text-independent. The former require the speaker to provide the same utterance for both training and testing, whereas the latter do not impose such constraints. The text-dependent systems are usually based on template matching techniques in which the time axes of an input speech sample and each reference template are aligned and the similarity between them is accumulated from the beginning to the end of the utterance. Because these systems can directly exploit voice individuality associated with each phoneme or syllable, they usually achieve higher recognition performances than text-independent systems. Since there are many applications in which predetermined key words cannot be used and because human beings can recognize speakers irrespective of the content of their utterance text-independent systems have recently attracted more attention.

The main elements of a speaker recognition system are feature extraction, pattern matching, making an accept/reject decision (in verification) and enrolment to generate speaker reference models. This paper focuses on the feature extraction and presents several of the most commonly used feature parameters derived from linear predictive coding (LPC) analysis. It is organised in the following way. Section 2 presents briefly the basis for the feature extraction that is the LPC analysis. Section 3 discusses

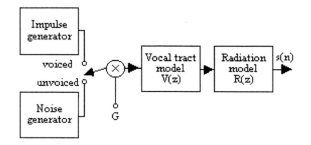

Fig. 1. The linear model for speech production.

properties and methods for cepstrum calculation, and section 4 introduces cepstral liftering. The following sections present more sophisticated cepstral techniques like CMS, PFCMS, and ACW. Section 8 includes summary and conclusion of the paper.

2. LINEAR PREDICTION OF SPEECH

Speech sounds can be classified into three categories: voiced sounds, plosive sounds and fricative or unvoiced sounds (Rabiner and Yuang, 1993). Voiced sounds are produced when the vocal cords are tensed and vibrate periodically when air flows from the lungs. In that case speech waveform is quasi-periodic. When the vocal cords are relaxed the airflow passes through a constriction in the vocal tract and becomes turbulent producing unvoiced sounds. The vocal tract is then excited by a broadband noise source.

Although there are a lot of techniques for extracting speaker specific information from the speech signal, probably the most important is a linear predictive coding analysis (LPC). It is based on the linear model for speech production (Fig. 1) where the glottal pulse, vocal tract and radiation are individually modelled as linear filters (Mammone et al., 1996). The source is either a random sequence for unvoiced sounds or a quasi-periodic impulse sequence for voiced sounds. The gain factor G controls the intensity of the excitation. The vocal tract is modelled by transfer function V(z) whereas the radiation model R(z) describes the air pressure at the lips. Combining these parts of the vocal tract yields an all-pole transfer function

$$H(z) = G(z)V(z)R(z) = \frac{G}{A(z)} \quad (1)$$

$$H(z) = \frac{G}{1 - \sum_{k=1}^{p} a_k z^{-k}} \quad (2)$$

where p is the prediction order and a_k are predictor coefficients (PC's). The transfer function also can be rewritten as

$$H(z) = \prod_{k=1}^{p} \frac{1}{1 - z_k z^{-1}} = \sum_{k=1}^{p} \frac{r_k}{1 - z_k z^{-1}} \quad (3)$$

where r_k represents the residues and z_k the poles of H(z), which are further described by

$$z_k = \sigma_k e^{j\omega_k} \quad (4)$$

where σ_k is the magnitude and ω_k the frequency of the pole z_k. The LPC model of the speech signal specifies that a speech sample s(n) can be represented as a linear sum of the p previous samples plus an excitation term

$$s(n) = \sum_{k=1}^{p} a_k s(n-k) + Gu(n) \quad (5)$$

As in speech applications the excitation term is usually unknown it is ignored and the LP approximation $\hat{s}(n)$ depends only on past output samples

$$\hat{s}(n) = \sum_{k=1}^{p} a_k s(n-k) \quad (6)$$

Unfortunately some speaker specific information is included in the excitation term (e.g. fundamental frequency) which affects on the performance of the LPC based speaker recognition systems (Campbell, 1997). Since vocal-tract changes its configuration over time, in order to model it the predictor coefficients a_k must be computed adaptively over short intervals (10 ms to 30 ms) called frames during which time-invariance is assumed. There are two standard methods of solving for the PC's: autocorrelation and covariance method. Both of them are based on minimizing the mean-square value E of the prediction error e(n) which is the difference between the actual and the predicted value of the speech sample

$$E = \sum_{n} e(n)^2 = \sum_{n} [s(n) - \sum_{k=1}^{p} a_k s(n-k)]^2 \quad (7)$$

The PC's can be found after solving the linear equations resulting from Eq. 8.

$$\frac{\partial E}{\partial a_i} = 0 \qquad i = 1, 2, \ldots p \qquad (8)$$

Assuming that speech samples outside the frame of interest are zero and defining the autocorrelation function as:

$$r(\tau) = \sum_{i=0}^{N-1-\tau} s(i)s(i+\tau) \qquad (9)$$

where N is the number of samples in a frame, the autocorrelation method yields the Yule-Walker equations given by

$$\begin{bmatrix} r(0) & r(1) & r(2) & \cdots & r(p-1) \\ r(1) & r(0) & r(1) & \cdots & r(p-2) \\ r(2) & r(1) & r(0) & \cdots & r(p-3) \\ \vdots & \vdots & \vdots & \ddots & \vdots \\ r(p-1) & r(p-2) & r(p-3) & \cdots & r(0) \end{bmatrix} \begin{bmatrix} a_1 \\ a_2 \\ a_3 \\ \vdots \\ a_p \end{bmatrix} = \begin{bmatrix} r(1) \\ r(2) \\ r(3) \\ \vdots \\ r(p) \end{bmatrix} \qquad (10)$$

Since the matrix is Toeplitz, a computationally efficient algorithm known as the Levinson-Durbin recursion (Rabiner and Juang, 1993) can be used to find the predictor coefficients a_k. After finding the PC's the magnitude response $|H(e^{i\omega})|$, which represents the spectral envelope of the speech, can be found. An example of the 12th order 11-kHz LP analysis of the vowel "a" (as in "mama") is shown in Fig. 2. Although the PC's are not usually used as the feature parameters, they are converted to other types of spectral representations like cepstral parameters, which yield better results in speaker recognition.

3. CEPSTRUM

It is an important spectral representation for speech and speaker recognition defined as the inverse Fourier transform of the log of the signal spectrum. Thus, the log spectrum can be represented as a Fourier series expansion in terms of a set of spectral

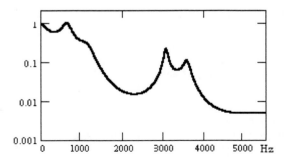

Fig. 2. Frequency response for the vowel 'a'.

coefficients c(n)

$$\ln H(\omega) = \sum_{n=-\infty}^{+\infty} c(n)e^{-jn\omega} \qquad (11)$$

The cepstrum can be calculated from the filter-bank spectrum or from LPC coefficients. In the latter case it is known as the LP cepstrum and can be calculated from the transfer function of the vocal tract H(z) in Eq. 3. Since H(z) obtained by the autocorrelation method is a minimum phase system (Rabiner and Juang, 1993), the LP cepstrum $c_{lp}(n)$ can be calculated from the equation

$$\begin{aligned} c_{lp}(n) &= \frac{1}{n}\sum_{k=1}^{p} z_k^n & n > 0 \\ c_{lp}(n) &= 0 & n \leq 0 \end{aligned} \qquad (12)$$

As Eq. 12 requires calculating poles of the H(z), another more computationally efficient recursion formula is used

$$\begin{aligned} c_{lp}(n) &= a_n + \sum_{k=1}^{n-1} \frac{k}{n} c_{lp}(k) a_{n-k} & 1 \leq n \leq p \\ c_{lp}(n) &= \sum_{k=n-p}^{n-1} \frac{k}{n} c_{lp}(k) a_{n-k} & n > p \end{aligned} \qquad (13)$$

The LP cepstrum has many interesting properties. It is causal for the minimum phase H(z) and of infinite duration.. As the cepstrum represents the log of the signal spectrum, signals represented as the cascade of two effects which are products in the spectral domain are additive in the cepstral domain. This property of separability of pitch excitation and vocal tract is considered as one of the reasons that cepstral parameters are more effective for speaker recognition than other representations of speech signal (Furui and Rosenberg, 1999). Another interesting property is the fact that $c_{lp}(n)$ decays as fast as $1/n$ as n approaches $+\infty$ so the feature vector consists of the finite number, most significant components $c_{lp}(1)$ to $c_{lp}(x)$, where x is usually approximately equal to the prediction order.

4. CEPSTRAL LIFTERING

As it can be shown that high-order cepstral coefficients are very sensitive to noise, suppression of these coefficients should lead to a more reliable calculation of spectral differences between reference and test feature vector. On the other hand low-order cepstral coefficients are mainly influenced by variations in transmission conditions (changes of telephone channel response) or speaker characteristics, so to improve speaker recognition it seems necessary to introduce some type of cepstral weighting known also as a cepstral liftering (Rabiner and Juang, 1993). It is accomplished by multiplying $c_{lp}(n)$ by a window $w(n)$ and using the weighted LP cepstrum as the feature vector. The simplest form of weighting is known as a rectangular liftering (Eq. 14) where

$$w(n) = 1 \quad n = 1,2...L$$
$$w(n) = 0 \quad n \le 0, \quad n > L \tag{14}$$

and L is the size of the window. It keeps the most significant initial coefficients. The effect of rectangular liftering on a log LPC spectrum as a function of window size is shown in Fig. 3. Other forms of weighting are known as a linear or quefrency liftering where $w(n)$ is given by

$$w(n) = n \quad n = 1,2...L$$
$$w(n) = 0 \quad n \le 0, \quad n > L \tag{15}$$

and bandpass liftering (BPL) of the form

$$w(n) = 1 + \frac{L}{2}\sin(\frac{n\pi}{L}) \quad n = 1,2...L$$
$$w(n) = 0 \quad n \le 0, \quad n > L \tag{16}$$

which de-emphasizes the lower- and higher-order components. The techniques shown above multiply all the feature vectors extracted from the utterance by

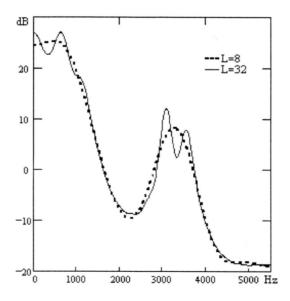

Fig. 3. The effect of rectangular weighting on a log LPC spectrum as a function of window size L.

the same fixed weights assuming that every feature vector undergoes the same distortion. Unfortunately in practical applications e.g. verification over telephone lines, an assumption that distortion introduced by telephone channel is constant with time is definitely not true. As a result of this some adaptive techniques that could overcome this problem are strongly desired. Some of them are described in the following sections.

5. CEPSTRAL MEAN SUBTRACTION (CMS)

In speaker recognition over telephone lines transmitted speech often encounters linear distortions due to the filtering effect of the channel. In the z domain it is expressed as a multiplication of H(z) corresponding to the clean, original speech and G(z) corresponding to the transfer function of the channel. As in the log domain H(z) and G(z) are additive, it is observed that the LP cepstrum of the filtered speech is a sum of the LP cepstrum of the clean speech H(z) and a component introduced by the channel. In the CMS method cepstral coefficients are averaged over the duration of an entire utterance and the averaged values are subtracted from the cepstral coefficients of each frame

$$c_{cms}(n) = c_{lp}(n) - E[c_{lp}(n)] \tag{17}$$

This method is especially effective for text-dependent applications using long utterances where testing and training were done on different channel conditions and assumes that the mean of the LP cepstrum of the clean speech is zero (Furui, 1981). Unfortunately in situations where testing and training are done on the same channel, the CMS technique degrades the performance. This is due to the implicit assumption that the mean of the original speech is zero, which is only true for the phonetically balanced voice segments, i.e. including the same amount of voiced, unvoiced and plosive sounds.

6. POLE-FILTERED CEPSTRAL MEAN SUBTRACTION (PFCMS)

This is another normalization technique improving the performance of a speaker recognition system that works on mismatched conditions i.e. the training and testing were done on different channel conditions. The concept of PFCMS is based on getting a channel estimate given by $E[c_{lp}(n)]$. It can be observed that the LP poles that lie close to the unit circle represent formants and are less sensitive to channel and noise effects. On the contrary the poles lying far from the unit circle have less speech information and model the channel effects and spectral tilt. In order to get a better estimate of the channel Naik (1995) proposed to modify the LP poles. The idea of PFCMS is shown

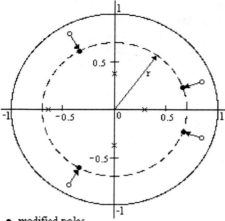

● modified poles
○ old poles
× poles within the threshold radius
r threshold radius

Fig. 4. The idea of PFCMS.

in Fig. 4. The formant poles which magnitude is bigger than selected threshold radius r are moved radially away from the unit circle whereas their frequency is left intact. The cepstrum $c_{mlp}(n)$ calculated from such modified poles contains more channel information and less speech information. The final feature vector is given by

$$c_{pflcms}(n) = c_{lp}(n) - E[c_{mlp}(n)] \qquad (18)$$

where $E[c_{mlp}(n)]$ is the channel estimate over all frames.

7. ADAPTIVE COMPONENT WEIGHTING (ACW)

Unlike the methods discussed in the previous sections, the ACW is based on intraframe processing i.e. there is no need to average cepstral coefficients over many frames. This technique modifies the LP spectrum in order to emphasize the formant structure, which makes the feature parameters more robust to channel distortions varying with time. It was observed (Assaleh and Mammone, 1994) that from the parameters describing the transfer function H(z) in Eq. 3 i.e. the residues r_k and the poles z_k given by Eq. 4 the residues are least robust to channel variability. The ACW technique exploits this property normalizing the residues so that the narrowband components are emphasized and the broadband components are de-emphasized which results in the transfer function of the form

$$H_{acw}(z) = \sum_{k=1}^{p} \frac{1}{1 - z_k z^{-1}} = \frac{N(z)}{A(z)} \qquad (19)$$

where

$$N(z) = \sum_{k=1}^{p} \prod_{i=1 \neq k}^{p} (1 - z_i z^{-1}) \qquad (20)$$

wchich can be written as

$$N(z) = p(1 - \sum_{k=1}^{p-1} b_k z^{-k}) \qquad (21)$$

The final feature vector is given by

$$c_{acw}(n) = c_{lp}(n) - c_{nn}(n) \qquad (22)$$

where $c_{lp}(n)$ is the LP cepstrum calculated from original H(z) whereas $c_{nn}(n)$ can be found using Eq. 13 involving the coefficients b_k. There is a simple formula (Zilovic et al., 1997) for coefficients b_k

$$b_k = \frac{p-k}{p} a_k \qquad 1 \leq k \leq p-1 \qquad (23)$$

which greatly simplifies the calculation of $c_{acw}(n)$.

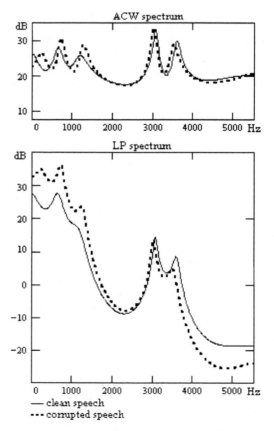

— clean speech
··· corrupted speech

Fig. 5. Comparison of magnitude spectra for clean and corrupted speech.

The ACW technique considerably improves the performance of speaker recognition in comparison with the LP cepstrum. An example of the magnitude responses of H(z) and H_{acw}(z) for a frame of clean speech and for the same frame of speech corrupted by the channel given by the transfer function of the form

$$H(z) = 1 + z^{-1} + 0.5z^{-2} \qquad (24)$$

is shown in Fig. 5. It can be seen that the mismatch in the magnitude spectrum for ACW method is reduced in comparison with H(z) which is an evidence that the ACW technique is more robust to channel variations.

8. SUMMARY

Although there have been many successes in speaker recognition technology, there are still many problems for which good solutions remain to be found. One of them is a proper selection of feature parameters that are robust to changes in speaker's voice. It is also very important to develop methods to cope with distortions due to telephone channels. This paper presents a review of the techniques used in speaker recognition with an emphasis on feature extraction. The parameters described are based on LPC analysis. It is shown that methods like CMS, PFCMS and particularly ACW can increase the performance of speaker recognition systems. It is worth noticing that ACW is based on intraframe processing whereas the other methods are interframe techniques, which require processing many frames of speech and as a result are less robust to rapid changes of channel conditions.

REFERENCES

Assaleh K.T. and R.J. Mammone (1994). New LP-Derived Features for Speaker Identification. *IEEE Trans. Speech Audio Processing*, **2**, 630-638.

Campbell J.P. (1997). Speaker Recognition: A Tutorial. *Proc. IEEE*, **85**, 1437-1462.

Furui S. (1981). Cepstral Analysis Techniques for Automatic Speaker Verification. *IEEE Trans. Acoust., Speech, Signal Processing*, **29**, 254-272.

Furui S. (1997). Recent advances in speaker recognition. *Pattern Recognition Letters*, **18**, 859-872.

Furui S. and A.E. Rosenberg (1999). Speaker Verification. In: *Digital Signal Processing Handbook* (V.K. Madisetti, D.B. Williams, Ed.). Chapman & Hall/CRCnetBase.

Mammone R.J., X. Zhang and R.P. Ramachandran (1996). Robust Speaker Recognition. *IEEE Signal Processing Magazine*, **13**, 58-71.

Naik D. (1995). Pole-Filtered Cepstral Mean Subtraction. *Proc. IEEE Int. Conf. Acoust., Speech, Signal Processing*, **1**, 157-160.

Rabiner L.R., B.H. Juang (1993). *Fundamentals of Speech Recognition*. Prentice Hall, Englewood Cliffs, NJ.

Zilovic, M.S., R.P. Ramachandran and R.J. Mammone (1997). A Fast Algorithm for Finding the Adaptive Component Weighted Cepstrum for Speaker Recognition. *IEEE Trans. Speech Audio Processing*, **5**, 84-86.

IFAC
Publications
www.elsevier.com/locate/ifac

SOFTWARE ENVIRONMENT FOR CUSTOM TV SIGNAL GENERATION

Aurel Gontean, Dan Balan, Liviu Lucaciu

*Department of Applied Electronics, "Politehnica" University Timisoara,
B-dul Vasile Parvan Nr. 2, RO-1900, Romania, gontean@ee.utt.ro*

Abstract: The goal of this paper is to introduce an efficient and yet easy to use, user friendly graphical interface for programming custom TV signals. Such waveforms are used to test the behaviour of chroma and deflection stages in TV sets and their response to disturbances. The software computes the samples for the TV lines to be generated with hardware-related equipment. It also provides communication with the board for downloading the samples or to stop a generation process. In order to customise the software according to the user's preferences, most of the TV signal parameters are programmable. *Copyright © 2001 IFAC*

Keywords: software, signal synthesis, microprocessor, programming environment.

1. INTRODUCTION

The software is designed to work together with a hardware TV signal generator. However, due to its high degree of versatility it may also be used to graphically evaluate the impact of parameter changes of the TV signal. The design simulates the distortions that appear in the output video signal of VCR due the fact that the VCR head is switching. The heads are normally correct mounted to the head drum and switch between fields. But as the tape can stretch due to ageing or to other tolerances in the VCR, the timing for the switching of the VCR heads is no longer in the vertical synchronisation period, but shifting.

2. SYSTEM OVERVIEW

In the early stages of the project, an analogue approach was thought. The rise and fall times of the synchronisation pulses could be modified with digitally controlled capacitors. A digital controlled amplifier could solve the luminance signal amplitude.

The burst signal had to be generated using a block of oscillators with frequencies corresponding with the TV standards (at least two oscillators and a PLL should have been used). For obtaining the needed burst amplitude another digital controlled amplifier had to be used. This implementation (Mitrofan, 1993) had some obvious disadvantages:

- a difficult precise control of the amplitude, rise and fall times of the signal;
- a large number of ICs needed on the board;
- low level of versatility;
- difficult future extensions.

In order to solve these inconveniences, a new approach of signals generation was needed. A fully digital design was used, with an inexpensive PC or notebook linked with a board equipped with a microcontroller and appropriate RAM used to store and output the TV signal. For generating a standard 64 µs TV line, 2560 8-bit data samples are used, so the user can have a very good control of line length and signal's amplitude. The digital control of signal parameters allows:

- timing parameters set with a 25 ns step;
- amplitude parameters set with 4 mV step;
- reduced number of ICs on the board.

In the second stage of the design, the microcontroller was intended to use for computing and sending the samples to the DAC (Gontean, 1996). In this case the role of the PC would have been only to send short messages (including TV standard, pattern, etc.) to the hardware equipment. For this type of communication and due to the small amount of data to be sent, a serial communication protocol would have been appropriated. This approach meant very fast computing and response times for the microcontroller. Due to the long interrupt responses times of the microcontroller (Infineon, 2000) and the latencies occurred at parallel ports when outputting new values, this approach was aborted.

The new approach transferred all the computing tasks to the PC. So, an inexpensive PC or notebook is used to compute the TV signal waveform samples and a much simpler board to playback these samples. In this case a big amount of data needs to be sent from the computer to the hardware device. In order to keep the transfer time within reasonable limits, a parallel communication is used. The microcontroller's role is to implement this protocol, to save the sent samples into the data memory and to control the other ICs on the board. This approach:
- allows samples fast computing time;
- allows samples fast transfer time;
- eliminates the need for additional peripheral devices and the associated software overhead;
- allows short time for any project development (by modifying only software functions a high versatility may be achieved).

All the microcontroller programming was performed in C (Tasking, 2000).

3. SOFTWARE OVERVIEW

The communication with the board is accomplished by using the parallel port (LPT1). Only 8 data lines and two control lines (STROBE and /BUSY) are used to implement the communication protocol. The DATA lines are used:
1. to transmit the samples to the board;
2. to send the end of transmission code (0FFH);
3. to send the end of process code to the board.

The STROBE line is used to send a Data Ready signal to the board. This line is used for generating an external interrupt for the microcontroller, in order to prompt that valid information is available at the Data lines. The /BUSY line is used for sending an acknowledgement signal from the board to the PC:
1 - the controller is ready to read another sample;
0 - sample transfer to the SRAM in progress.

3.1. Setting up the parallel port

The software permits to set-up the parallel port. Communication parameters may be entered, although the default values generally fit. After the user decides the configuration, a set-up file may be saved for later use (figure 1).

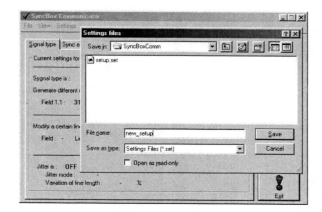

Fig. 1. Saving the set-up file.

3.2. Setting up the signal type and lines length

In case of "Interlaced scanning" selection, the program will send to the board 4 fields (representing two frames), allowing the user to modify for each field the number of lines to be send. In this mode the user can select that only one of the two of the frame's fields to be send, the other field being sent as being composed only from black lines. The user should select at least one of the two fields; otherwise the program will generate an error message. In case of an "Progressive scanning" selection the program will send to the board 2 frames; for each the user can modify the number of lines (figure 2).

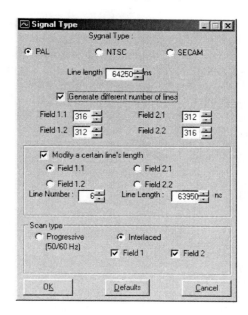

Fig. 2. Choosing the signal type.

The user can choose the TV standard to be used (PAL, SECAM or NTSC), if the fields/frames will have different number of lines than usual (+/-10% of the standard lines number), the type of scanning (progressive/ interlaced). One special feature is that the equipment can generate a certain line with a different length. The user can choose in which of the fields / frames this line is integrated and the line's number in the field/frame. The modifiable lines include the black lines (used for TXT), but don't include the vertical synchronisation lines.

3.3. Setting up the synchronisation signal parameters

The system allows the user to set the synchronisation signals, rise and fall times with a 25 ns increment. The rise / fall time can be set between 200 ns up to 1 μs, and also the synchronisation signals length can be modified between 3 μs up to 6.4 μs with a 25ns step. The synchronisation signals and burst signal amplitude can be also modified between 0~140% with a 5% step.

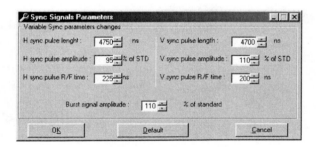

Fig. 3. Synchronisation Signal Parameters.

3.4. Jitter settings

In the "Jitter mode" the line length can be modified between 1~5% of the user selected line length. If jitter is to be generated, two choices are available (figure 4):

- Slow jitter of the line length. It is adjustable from 0 to +/-5% max. of the standard length, the jitter changes with a frequency of 200 Hz;
- A fast jitter is changing over 5 lines.

3.5. Patterns settings

The user can select in this window which pattern should be generated, and up the luminance signal amplitude (from 80% ~ 140% with a 5% step). The following patterns are available:

- 2T B/W lines
- 2T wider B/W lines
- greyscale
- white on black.

The first pattern contains an alternating 20 B/W stripes. The second type contains 10 B/W stripes. The greyscale contain 50 levels of grey from black to

white. White on black contains one white vertical stripe at 20 μs from the left side of the screen, modifying the length of the video line only the distance between the white vertical stripe of right side of the screen is modified.

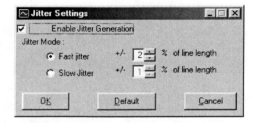

Fig. 4. Jitter Settings.

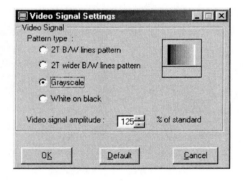

Fig. 5. Video Signal Settings.

4. VIEWING THE WAVEFORMS

The user can visualise the generated waveform for both horizontal and the vertical synchronisation by choosing one of the items in the "View" menu. The user has the possibility to visualise the standard line (blue in graphic) or the modified line (red in graphic).

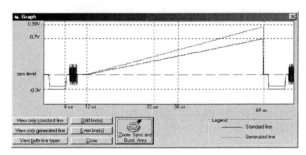

Figure 6. Comparing the standard line with the generated line.

Zooming the synchronisation and burst areas is also implemented. Using 2 cursors, time and voltage difference between two points can be measured. It is possible to view both odd and even lines. In the "Zoom on V Sync" window the user can view the five vertical synchronisation pulses and the first two pulses from post-synchronisation pulses.

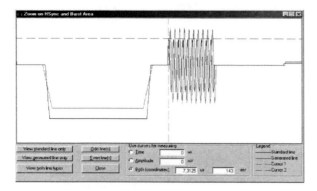

Figure 7. Zooming the burst area.

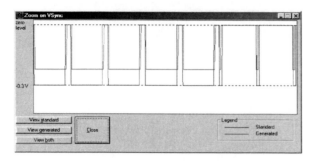

Figure 8. Zooming the vertical synchronisation area.

5. CONCLUSIONS

This paper explains a new way of computing and generating TV signals. With the continue growth in PC's computing power, more and more tasks are moved to the PC, leaving the hardware external units with simple architecture. The overall cost of both the design and the equipment significantly drops, while:

1. there is no need for a keypad or an LCD panel attached to the hardware;
2. the user has a friendly interface with a wide range of parameter to be modified;
3. each modification in the signal parameter can be seen on the screen;
4. time and voltage measurements can be accomplished with two cursors;
5. it is easy to compare the generated signal waveform with the standard signal waveform;
6. it is simple to have zoom captures from the critical areas of the signals;
7. further changes and development are easy to implement – only extra software must be written;
8. last but not least, although the design was first intended for TV signals, a random waveform may be generated with just a small programming effort, *implying no hardware changing*.

REFERENCES

Gontean, A., Babaita, M. (1996). *Programmable Logic Devices. Applications*, West Publisher Company, Timisoara, Romania (in romanian).

Infineon Technologies (2000), *C167CR 16-bit CMOS Microcontroller User's Manual,* München, Germany.

Infineon Technologies (www.infineon.com), *Data Sheet for SDA6000 microcontroller.*

Mitrofan, G. (1993). *Television Fundamentals*, Teora, Bucharest, Romania (in romanian).

Tasking (2000), *C-166 C Cross-Compiler User's Guide.*

IFAC
Publications
www.elsevier.com/locate/ifac

CUSTOM TV SIGNAL DIGITAL GENERATOR

Aurel Gontean, Liviu Lucaciu

*Department of Applied Electronics, "Politehnica" University Timisoara,
B-dul Vasile Parvan Nr. 2, RO-1900, Romania, gontean@ee.utt.ro*

Abstract: The aim of this paper is to present a microcontroller based equipment for generating custom TV signals. Such waveforms are used to test the behaviour of various stages in TV sets and their response to disturbances. The board works with dedicated software located on a PC or notebook. The communication with the PC is ensured via the parallel printer port. *Copyright © 2001 IFAC*

Keywords: signal synthesis, microprocessor, samples, D/A converter, static RAM.

1. INTRODUCTION

The main task of the board is to test and evaluate the behaviour of chroma and deflection stages in TV sets and their response to disturbances. These distortions make that in the output signal lines can appear shorter than the standard. It is common that the affected line is up to 20μs shorter than all other lines. This involves problems for the deflection circuit, which are investigated using this combined (hardware and software) equipment. The differences in the length of affected line is adjustable, also the line number where this line is to be inserted. This characteristic makes it possible to observe the locking of different PLLs which are founded in the TV set, chroma decoder or fronted PLL and the deflection or backend PLL. Another important mode is the insertion of synthetic V pulses into the video signal in front of the original V pulse without pre / equalization pulses.

2. SYSTEM REQUIREMENTS

This video generator provides multistandard video signals (PAL, NTSC and SECAM). The following parameters are programmable:

- the number of lines per frame;
- the number of lines from field to field;
- two different jitter modes;
- adjustable rise time;
- adjustable amplitude for H-synchronisation and V-synchronisation signals;
- adjustable burst amplitude;
- adjustable video amplitude.

The microcontroller used to perform the tasks of this project must have:

- enough number of I/O parallel ports needed for data transfer and control the other circuits on the board;
- good external interrupts response times for a fast data transfer from the PC to the board;
- in system programming capabilities for end-of-line programming or testing.

In the specific design, the SDA6000 (with a C166 core) was used (Infineon, 2000). For testing the software, the board was provided with a socket for EPROM/SRAM modules. For further software developments, the board has also a JTAG connector which allows a program to be downloaded into the microcontroller's program memory.

3. HARDWARE OVERVIEW

The main element of the block diagram is the SDA 6000 (figure 1). The microcontroller communicates with the PC via the parallel port, controls reading or writing data from / to the SRAM memory, enabling the DAC at appropriate moments. The firmware is loaded into a flash memory. The program was written in ANSI C (Tasking, 2000). A 22 bits address counter is used to generate the addresses for SRAM. The write clock frequency is 6 MHz, while the read frequency is 40 MHz. A NAND gate produces a reset signal for counters at a FFH byte. An 8 bit D/A converter (MAXIM 5187) is used. The DAC is disabled when data is written in the SRAM or the read data is FFH.

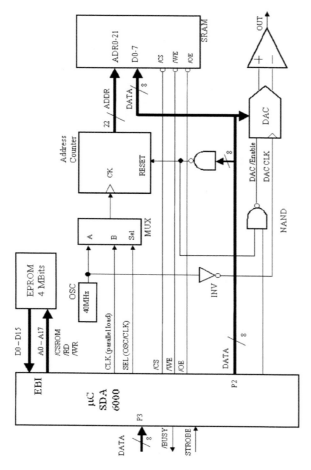

Figure 1. Block diagram of the TV custom signal generator.

The final linear stage has been designed with special care. Several operational amplifiers are used to provide the correct voltage levels, the proper output resistance and a good filtering.

4. FUNCTIONAL OVERVIEW

The desired samples are generated using the software located on the PC. A level shifter is required to shift the 5V data (from the parallel port) into the 3.3V data necessary the microcontroller. The STROBE line is used to send a Data Ready signal to the board, to interrupt the SDA6000 - for signalling that valid data is available at the Data lines. The /BUSY line is used for sending an acknowledgement signal from the board to the PC. The DATA lines are used for transmitting the computed samples to the board, to send the end of transmission code and to send the end of process code to the board.

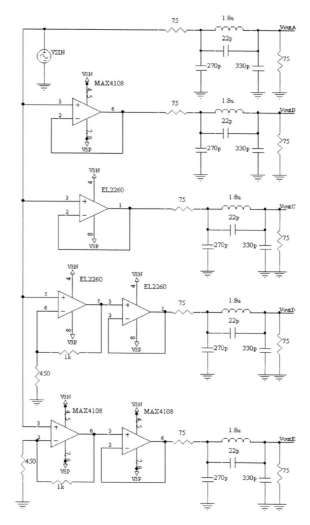

Figure 2. Searching the optimal final linear stage.

For generating a 64 μs TV line, 2560 samples are used. The microcontroller reads every sent sample and generates a strobe clock for the address counters, writing samples into the SRAM until the value of the sample is FFH. In this case it stops writing the memory, selects the 40 MHz clock for reading and selects the D/A converter. The clock frequency for the D/A converter is the inverted clock from the 40MHz oscillator. The output filter contains two identical circuits, with 6 MHz cut-off frequency. The TV signal is obtained at the filer's outputs. The output signal is available at two BNC connectors. Fast SRAMs have been chosen in order to work at 40 MHz properly.

5. DESIGN SIMULATION

As stated before, the output analogue stage has been carefully designed. The first simulation schematic (figure 2) is used for testing the frequency characteristics of two types of differential amplifiers: MAX4108 (from Maxim) and EL2260 (from Elantec), along with afferent filters. In figure 3, various frequency responses are plotted:

- $V_{OUT}A$ for the passive filter;
- $V_{OUT}B$ for the MAX4108 repeater;
- $V_{OUT}C$ for the EL2260 repeater;
- $V_{OUT}D$ for the MAX4108 amplifiers;
- $V_{OUT}E$ for the EL2260 amplifiers.

The results show little differences among the above choices. Further simulation must be performed to test the behaviour of this stage.

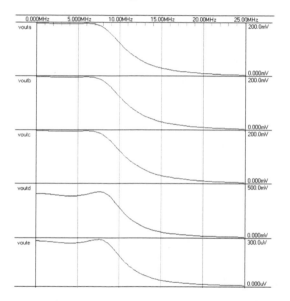

Figure 3. Simulating the final linear stage.

Good results have been obtained simulating the D/A converter and the final stage driven by a binary counter. While the TV signal is considerably different from this pattern, a random number generator implemented with a linear shift register was preferred. The schematic shown in figure 4 contains a counter multiplexed with the linear shift register, acting like a random number generator. The final simulating results are:

- V_{3DAC} – the output of the D/A converter;
- V2qo and V2q7 are LSb and MSb bits at DAC output;
- V_{4AO} – the signal amplified and shifted;
- V_{outB} – the signal amplified and filtered using a MAX4108;
- V_{outA} – the signal amplified and filtered using an EL2260.

The simulation results (figure 5) clearly show this time that MAX4108 is a better choice, and therefore it was chosen in the design.

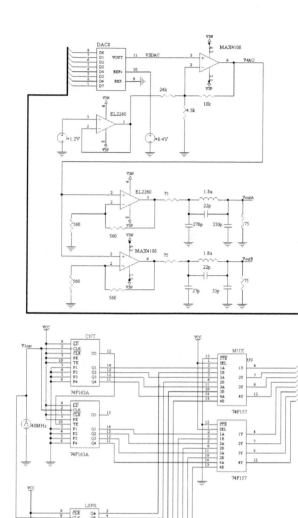

Figure 4. DAC simulating schematics.

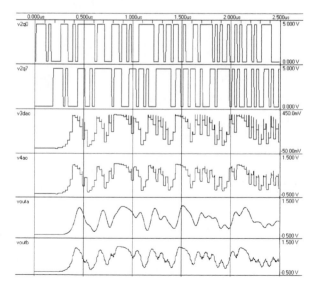

Figure 5. Simulating the DAC and the output stage.

47

6. CONCLUSIONS

This paper presents a digital way for generating custom TV waveforms. All the computation is achieved on a common PC, leaving little tasks to the external hardware equipment. This arrangement simplifies the board design, ensures a high degree of versatility and low cost future extensions. Due to the low activity of the microcontroller, a wide range of microcontrollers can be used. The digital approach allowed a dramatic cut in the number of circuits used: only 12 integrated circuits are needed for a full programmable custom TV signal generator. A possible future extension is migrating the design to the USB communication alternative.

REFERENCES

Infineon Technologies (2000), *C167CR 16-bit CMOS Microcontroller User's Manual*, München, Germany.

Infineon Technologies (2000), *SDA6000 Teletext Decoder with Embedded 16-bit Controller M2, User's Manual*, München, Germany.

Tasking (2000), *C-166 C Cross-Compiler User's Guide*.

IFAC
Publications
www.elsevier.com/locate/ifac

FPGA-BASED VIDEO PROCESSOR FOR LOG-POLAR RE-MAPPING IN THE RETINA HETEROGENEOUS IMAGE PROCESSING SYSTEM

Marek Gorgon, Miroslaw Jablonski

University of Mining and Metallurgy
Biocybernetic Laboratory, Institute of Automation,
Al. Mickiewicza 30, 30-059 Krakow, Poland

In the present publication an implementation of image re-mapping from Cartesian to Log-Polar space is presented. Realization of the algorithm has been distributed between the FPGA and DSP devices. The re-mapping process is performed in real time. In the paper the hardware architecture, realization of the Video Processor and results of the algorithm implementation are presented. *Copyright © 2001 IFAC*

Keywords: image processing, image analysis, hardware-software co-design, FPGA, digital signal processors, multiprocessing system.

1. INTRODUCTION

Technological progress in the field of production of FPGA devices in the last two years, oriented towards tailoring of the resources of FPGA devices for the needs of digital image processing, makes possible realizations of hardware or software-hardware vision systems. Solution earning a steadily growing popularity is the implementation of processing algorithms, and recently also image analysis algorithms, in reprogrammable devices (Assensi et al., 2001; Gorgon and Tadeusiewicz, 2000; Lesser et al., 2001; Wiatr, 1996; Xue, et al. 2000)

The popularity of image processing and analysis in the Life Like Perception Systems is still growing (Dorffner et al., 2001). Therefore the creation of video signal sources with characteristics close to the images originating in the human eye is so important (Sandini et al. 2000).

The first version of retina-like re-mapping board, based on TTL - logic elements and ISA-bus, was constructed in the Biocybernetic Laboratory in the 90-ies (Mikrut and Stanek, 1999). Although it is a fully functioning re-mapping system, its architecture, based on fixed-point 56001 DSP processor, does not meet the real-time speed requirements nowadays.

In the present paper heterogeneous architecture is shown, in which a single FPGA device of high density has been used for both realization of the Video Processor and controlling the resources of the

system. The paper particularly contains discussion of the Video Processor.

1.1 Log-Polar Transform

The idea of image processing and recognition in the Log-Polar space is based on human vision model. Human retina consists of thousands of receptors. A simplified model of distribution of receptors field is represented by the Log-Polar space (Schwartz 1997). The receptors of visible radiation are distributed radially in the human eye, and their density decreases with increasing distance from the central point. Such a receptor topology allows reduction of the information being processed, with local preservation of high image resolution.

The transformation of the (x,y) point from the Cartesian to Log-Polar coordinate system (u,v) is described by the equations (1), (2) and (3) below:

$$u = \ln(n) \qquad (1)$$

$$v = arctan(y/x) \qquad (2)$$

$$r = \sqrt{x^2 + y^2} \qquad (3)$$

The points of the resulting image (u_i, u_j) form a finite and denumerable set, which can be processed in a sequential way in finite time. It is assumed that the input image is homogeneously sampled in the space of (x, y) coordinates.

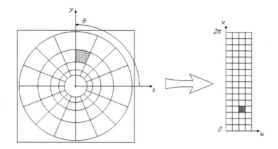

Fig.1. Re-mapping of points from the Cartesian coordinate system to the Log-Polar space.

The value $L(u_i, u_j)$ for the point in the Log-Polar space is given by equation:

$$L(u_i, v_j) = \frac{1}{A_{ij}} \sum_{(x,y) \in P_{ij}} S(x, y) \qquad (4)$$

where, P_{ij} is the set of points (further on called a segment of the source image), which form the vicinity (x_{ij}, y_{ij}) in the Cartesian coordinate system. $S(x, y)$ denote the values of points of the source image in the Cartesian coordinate system. A_{ij} denotes the size of the P_{ij} set. The L_{ij} value is therefore the arithmetic average over points belonging to the P_{ij} set. The distribution of the transformation grid points (x_{ij}, y_{ij}) and the determination of their vicinities present a separate problem, which will be studied in future, using the constructed system. As can be seen from Fig.1 the segment size increases with the increasing distance from the central point. If the digital nature of the source image is taken into account, it turns out that the volume of information has been irreversibly reduced during the transformation. Therefore the process realizes a lossy compression. The image obtained as a result of the transformation is resampled with variable spatial resolution.

1.2 Outline of the transformation algorithm

The equation (4) sets the stages of realization of the image transformation algorithm. Some of them comprise the preliminary operations, performed once before the initiation of the processing cycle (phase A), and the rest of operations have to be executed every time, in real time regime, for the images processed (phase B and C).

Phase A: preparation of the re-mapping matrix. For each (x,y) point the respective (u_i, u_j) coordinates are attributed. Thus the shape and distribution of the Log-Polar segments are determined. The obtained matrix will be further on called the Transcoder Matrix. It is also necessary to determine the number of points A_{ij} of the source image, belonging to a given segment. The above operation is executed before the start of the image acquisition.

Phase B: accumulation of values for the source image points. The values for points belonging to a given segment are summed up, providing an intermediate result in the Retina Matrix. The operation is performed in real time, for every image point.

Phase C: normalization of the contents of Retina Matrix. Because the segments contain different number of points, depending on the location, the elements of the Retina Matrix have to undergo normalization. The matrix elements, if they are used, are divided by the A_{ij} coefficient. The operation is performed between the acquisition of consecutive image frames, during the occurrence of frame synchronization pulses in the video signal.

1.3 The project assumptions

The aim of the study is a realization of the Log-Polar transform in a way enabling efficient data acquisition, with preservation of the system's flexibility. The possibility of dynamic change of the re-mapping grid topology and its orientation with respect to the input image is particularly essential, because of the planned studies of pattern recognition algorithms. This feature provides wider possibilities of algorithms implementation, in comparison to the use of specialized CCD sensors with Retina topology (Sandini *et al.* 2000). The general-purpose nature of the system assumes the use of video signal in the PAL or NTSC standards. It has been also assumed that colour pictures (with R,G,B 8-bit components), 512x512 pixels in size will be processed, with frame frequency specific for a given standard. The processing will be realized with various precessions of the input signal and in various formats. The transformation result will be provided for further elaboration by the software of the DSP processor and the Host PC. Fulfilment of all these assumptions requires a very effective distribution of the realization of processing algorithm among the available resources.

2. THE IMPLENTATION ENVIRONMENT

The implementation platform is provided by the RETINA Module, constructed in the Biocybernetics Laboratory of the University of Mining and Metallurgy (Gorgon and Tadeusiewicz 2000, Gorgon and Przybylo 2001).

2.1 The module's structure

The module's construction has been based on the Virtex 300 series programmable FPGA device, 32-bit floating point DSP96002 signal processor, BT812 video signal processor and a PCIS5933 bridge. Essential elements of the system are three blocks of fast SRAM memory, assigned for image data storage. Their independent operation enables the paralelization of some stages of the transformation algorithm, what increases the efficiency of the system. The module is provided with some extra peripheral devices, essential for the realization of the system's auxiliary functions. The devices are: the real time clock, the RS232 transceiver, the sensor of the FPGA device's temperature and the ROM memory storing the data and program for the DSP processor. Thus the platform is a 32-bit microprocessor system, enhanced by the flexibility and computational resources of the FPGA device.

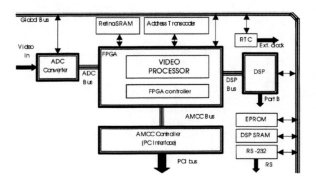

Fig. 2. The RETINA Module: block diagram of the hardware platform for realization of the Video Processor.

The heterogeneous structure of the module requires application of specialized software. For the routing of FPGA devices standard software has been used, provided by the manufacturer of the Virtex device. The software provided by the manufacturer of the signal processor enabled preparation of a program application. Additionally an application for hardware emulator is also available, which is being intensively used during the device startup and the testing of algorithm parts distributed in various resources of the module.

The RETINA Module is a device dedicated to realization of the Log-Polar transform. The selected hardware elements and their architecture have been designed in a way, which enables the partitioning of particular phases of the Log-Polar re-mapping process among the module's resources. However it is worth mentioning, that the placement of the FPGA device in the centre of the module's architecture allows its deep reconfiguration. Because of that the Retina Module can be treated as highly universal hardware-software platform for implementation of a wide variety of algorithms for Digital Image and Signal Processing.

2.2 The video channel

Programmable ADC is the source of Digital Data. Initialized by the DSP it forms the data stream, taken from the requested area of the video camera image window.

The FPGA device takes part in the video data exchange between the memories of the Retina board, Transcoder and DSP processor. Inside it integrates the Video Processor Unit (VPU), which is intended for image preprocessing during the phase B, described in section 1.2.

The role of the FPGA device exceeds beyond the range determined by the image processing algorithm. It implements the environment necessary for a proper functioning of the DSP processor and memory service, and it also creates a path of data exchange with the supervising computer via the PCI bridge, by implementing a Cross-bar Switch for external buses. The listed functions, realized for external devices and functional blocks implemented in the FPGA, have been integrated in the FPGA controller module.

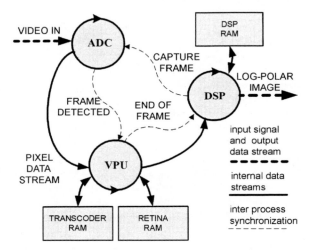

Fig. 3. The block diagram of the image processing in the Retina 2 board, for the application of Log-Polar transform, with specification of particular functional elements, data streams and communication paths between processes.

The DSP device realizes further stages of the post-processing. The particular utility of this unit in the signal processing manifests itself during the realization of floating-point arithmetical operations. In the process of Log-Polar transform the unit executes the normalization operation for the contents of RETINA matrix (see 1.2, Phase C).
In addition the DSP unit is also able to execute the system program, controlling the board. Due to that the module can also operate as a standalone unit, without a supervising computer.

3. THE VIDEO PROCESSOR REALIZATION

An integral part of the RETINA Module's video channel is the Video Processor module, implemented in the FPGA device. It works under particularly heavy load conditions, because of large volume of the data provided by the video converter and the requirement sharing of the calculation results to the DSP processor.

3.1 External memories

The work of VPU is closely related to the Transcoder and Retina Memories, which are used for data storage, according to the description in section 1.2.

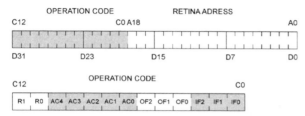

Fig. 4. Data format in the Transcoder Memory

Transcoder Memory is used as an operation code memory. It contains information about the transcoding procedure for the image point of given coordinates. The 32-bit long word of the Transcoder Memory (see Fig.4) contains the operation code for the Video Processor - equivalent of the microprocessor instruction code. The word in Transcoder Memory consists of the address field and the VPU instruction field, attributed to the given point (A18-A0 bits). The C12-C0 bits form the VPU operation code i.e. bits IF2-IF0 specify the precision of the input signal processing, while OF2-OF0 determine the format of the data storage in the Retina Memory.

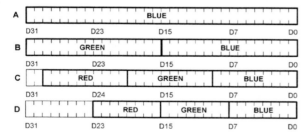

Fig. 5. Available data formats of the Retina Memory word, specified by the OF field in the Video Processor operation code.

Retina Memory is a data storage memory. In that memory the results are stored of pixel accumulation for particular sectors of the Log-Polar space. Four data formats are defined (Fig. 5). Depending on the processing mode, the calculation higher accuracy

level is selected or the parallel and simultaneous processing for all RGB components.

3.2 Functional blocks of the Video Processor

Functionally the whole VPU, except the memory block mentioned above, has been realized inside the FPGA device.

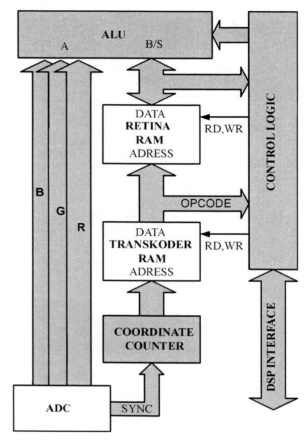

Fig. 6. The Video processor block diagram. The elements represented by shadowed units have been realized in the Virtex FPGA device.

The most important units include:
- Coordinates Counter: using the synchronization signals it calculates the address of the Transcoder Memory cell and generates the "frame end" signal for DSP,
- Control Block: it controls the memory access, including the memory blocks accesses by the DSP processor, and during the system initialization it loads the Video Processor program to the Transcoder Memory,
- Arithmetic Unit: the core of the arithmetic unit consists of 3 independent 32-bit adders, which realize the data accumulation for each colour component.

The adder's environment includes the Retina Memory data packing and unpacking system. This enables changes of the calculation precision without the necessity of adding unit modification.

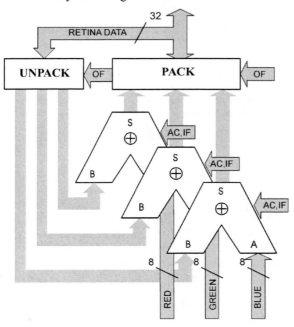

Fig. 7. Block diagram of the Video Processor arithmetic unit. In particular the input and output data and instruction buses have been identified.

Inside every adder element in Fig.7 an input data quantization unit has been located, which reduces the number of bits in the representation of the input signal to 1, 2, 4 or 8 (no quantization).

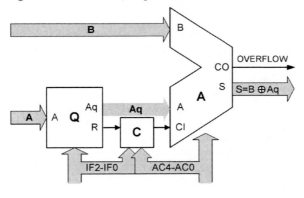

Fig. 8. Structure of a single adder channel, including the input data quantization block Q and the system of input carry-bit generation C.

Interior of the A module in Fig. 8 contains a chain of 32 segments of 1-bit adder, constructed of dedicated hardware elements of the Virtex device. In one CLB four adder segments are contained. The designed 1-bit adder is an element, which is configurable by the AC field of operation code, contained in the Transcoder Memory. It allows the selection of the operation performed by the adder during the calculations for every image pixel, without the necessity of applying the FPGA device dynamic

reconfiguration procedures, which are too slow. The AC0 and AC1 signals determine the type of operation performed by the system.

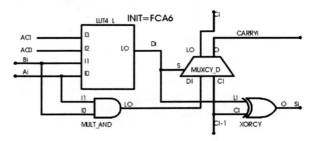

Fig. 9. Schematic of one cell of the full 1-bit adder.

Table 1 Description of the operation performed by the adder for various values of the AC1-AC0 bits of the operation code. C0 is the value of the input carry bit, provided by the C unit (see Fig.8)

AC1	AC0	equation:
0	0	S=A+B+C0
0	1	S=A+C0
1	0	S=B+C0
1	1	Si=~C0

3.3 Pipeline task execution in VP

The data provided by the ADC form a stream of bits, of a determined order. The processing rate of the image points is therefore strictly determined by the time parameters of the video standard and the converter. Particular phases of the algorithm are realized in a following sequence:

- calculation of the point coordinates: in the Coordinates Counter the address of the Transcoder Memory cell is determined,
- fetching Transcoder and Retina data,
- unpacking of the memory data, and converter data quantization,
- accumulation,
- encoding and writing the calculated value to the memory,

All the tasks listed above should be executed in one video converter cycle, i.e. 66ns. It sets very challenging requirements for the designed system.

4. IMPLEMENTATION

The design of Video Processor system configuration has been created in an iterative process. Every stage proceeded by functional and time-regime simulations of the FPGA structure, was completed by a hardware test. Predominant part of the Video Processor system project has been realized in a schematic editor. Part of the system logic has been implemented in the

VHDL language, what increases the design clarity and allows its easy parametrization. Detailed knowledge of the FPGA device architecture and the appropriate application of the directives for timing and placement constraints lead to minimization of the system implementation time and the indeterminate elements in the FPGA configuration.

Results of the Video Processor implementation. The greatest problem in the time optimization of the system is the fact that a signal change initiated by the ADC propagates outside the FPGA device via the Retina and Transcoder Memories. So finally the criterion for the Video Processor time evaluation has been set by the time interval between the appearing of the arithmetic unit operands on the buses and setting out the result. The obtained implementation results are shown in Table 2.

Table 2 The signal propagation times obtained in the Video Processor arithmetic unit

Input	Out	Delay	Logic levels
ADC	RET. RAM	18ns	8
RET. RAM	RET. RAM	15ns	6
TRA RAM	RET.RAM	9ns	3

5. SUMMARY

The presented Video Processor implementation is an example of hardware-software co-design. The computational task has been resolved by applying the architecture specific for the group of Reconfigurable Computing type devices. The FPGA device works in cooperation with two fast SRAM memories, realizing the functions of instruction and data memory respectively. The instruction memory contains the microcodes providing the possibility of hardware ALU reconfiguration in real time, for every image pixel. The above way of reconfiguration eliminates the necessity of applying dynamic reconfiguration procedures, which are too slow for the requirements of the realized application.

The constructed arithmetic unit offers the work in various processor modes: depending on the required calculation precision or the used colour components. Quantization procedures, as well as packing and unpacking of the accumulated data have been proposed, which allow the preservation of integrity and uniformity of the arithmetic unit ALU in various modes of VP operation. Reconfigurable adder unit has been designed, which chooses the type of operation performed for the given instruction specified in algorithm code. The system has been designed for real time re-mapping of the image data stream, 512x512 pixels in size, 24 bits/pixel, obtained from a typical video camera, conforming to the PAL or NTSC standard. The implementation of the Video Processor has been realized in the Xilinx Virtex XCV300BGA432-6C device.

ACKNOWLEDGEMENTS

The authors would like to thank the Xilinx University Program for software donation. The research has been performed under University of Mining and Metallurgy Research Grant no.: 11.11.120.249

REFERENCES

Assensi C.S., Pico I.F., Alvarez R. (2001) Reconfigurable Frame-Grabber for Real-Time Automated Visual Inspection (RT-AVI), *Proceedings of FPL2001*, Springer Verlag.

Dorfner G. at al. (2001). *Life like perception systems,* http://www.cordis.lu/ist/fetbi.htm

Gorgon, M. and Przybyło, J. (2001). FPGA Based Controller for Heterogeneous Image Processing System, *Proceedings of Euromicro Symposium DSD*, IEEE# PR01239, pp. 453-457, IEEE CS Press, Los Alamos.

Gorgon, M. and Tadeusiewicz R.(2000). Hardware-based image processing library for Virtex FPGA, *SPIE Proceedings*, **vol 4212**, pp.1-10, Boston, USA.

Lesser, F (2001). A MIMD-Based Multi Threaded Real-Time Processor for Pattern Recognition, *Proceedings of Euromicro Symposium DSD*, IEEE# PR01239, pp.372-375, IEEE CS Press, Los Alamos.

Mikrut, Z. and Z. Stanek. 1999. Low-Cost Board for Image Digitization, Remmaping and Processing, *University of Mining and Metallurgy, Scientific Bulletin Automatics*, **No 1/99**.

Sandini, G., P.Questa, D. Scheffer and A. Mannucci (2000). A Retina-like CMOS sensor and its applications, Proc. IEEE Sensor Array and Multichannel Signal Processing Workshop (SAM 2000), Cambridge, USA.

Schwartz, E.L. 1997. Spatial mapping in the primate sensory projection: Analytic structure and relevance to perception. *Biological Cybernetics*, **25**, pp 181-194.

Wiatr, K. (1996). Pipelined architecture of Reconfigurable Specialised Processors for a Real-Time Image Data Pre-Processing, *Proc. of ICIP*, 96TH8116, **vol.1**, pp. 649-652, Bejing, China, IEEE Press.

Xue, J., N. Zheng, X. Wang and Y. Zhang (2000). An approach to constructing reconfigurable computer vision system, *SPIE Proceedings*, **vol 4212**, pp. 11-21, Boston, USA.

IFAC

Publications

www.elsevier.com/locate/ifac

AN IMPLEMENTATION OF HUMAN VOCAL TRACT MODEL ON CARMEL DSP

Jacek Izydorczyk

Silesian University of Technology
Akademicka 16
44-100 Gliwice
Poland

Abstract: The article is about hydrodynamic model of the human speech production and its applications in the recommendations of International Telecommunication Union (ITU-T) concerned with speech compression. The model consists of tiny, compared to the length of the acoustic waves, cylinders connected in cascade. Apertures of cylinders are all the time traced by compression algorithm. The algorithm choose the most appropriate excitation for acoustic model to produce samples of speech similar to original samples. The model has been implemented as part of ITU-T G.723.1 recommendation on Carmel DSP from Infineon. *Copyright © 2001 IFAC*

Keywords: speech analysis, autoregressive models, digital signal processors, digital filters, lattice filters

Human beings hear sounds which have frequency up to 22kHz. But for human communication, by the means of the speech, sounds with spectrum nibbled to 3,5kHz are sufficient. This fact is known in telecommunication community for over one hundred years. Length of the acoustic waves which carry most of the information of the human speech is greater than about 10cm and it is greater than diameter of the human vocal system.

The human vocal system (or vocal tract) begins with vocal cords and ends at the lips. In the average male, the total length of the vocal tract is about 17.5cm (Rabiner, Schafer, 1978). It forms the acoustic tube with varying shape and aperture. The tube is excited by vibrations of the vocal cords (voiced sounds) or by turbulences of the air which flows by vocal tract contradictions (unvoiced sounds). Some sounds are produced by abrupt release of the air stopped by complete closure of the vocal tract (plosive sounds). Generally information is coded into speech by the means of excitation (voiced, unvoiced, plosive) and acoustic properties of the vocal tract.

1. A SIMPLE HELMHOLTZ MODEL

The first model of the human vocal tract origins from XIX century German physicist H.Helmholtz (Helmholtz, 1895). It consist of tiny tube of length $L = 17.5$cm with constant cross section area A opened on the end into air atmosphere. The other end of the tube is closed by vibrating membrane. The

membrane models vibrating vocal cords. Tiny tube means that the length of the acoustic waves is lower than diameter of the tube and waves propagate only in the one dimension – along the tube. The waves propagation inside tube is described by three laws. The conservation of mass law:

$$\frac{\partial \rho}{\partial t} = -\frac{\partial (\rho v)}{\partial x} \tag{1}$$

The Euler law (second Newton law formulated for fluids):

$$\frac{\partial v}{\partial t} + v \frac{\partial v}{\partial x} = -\frac{1}{\rho} \cdot \frac{\partial p}{\partial x} \tag{2}$$

The equation of the adiabatic process:

$$\left(\frac{\partial p}{\partial \rho} \right)_a = \frac{\kappa \cdot R \cdot T}{\mu} \tag{3}$$

Where ρ stands for density of the air, p is the air pressure, v is the air velocity, R is the molar gas constant (8.314510 J/mol/K), μ is molar mass of the air, T is absolute temperature of the air and κ is a ratio of the molar heat capacity of the air in the constant pressure to molar heat capacity of the air in the constant volume. Equations (1)-(3) are highly nonlinear but sounds are only small wrinkles (or distortions) on velocity field of the air. So equations are linearized for purpose of description of sound propagation. As a result we obtain telegraph equations where unknown functions are distortion of the pressure p and distortion of the velocity of the air v:

$$\frac{\partial p}{\partial t}=-c^2\rho_0\frac{\partial v}{\partial x},\quad \frac{\partial v}{\partial t}=-\frac{1}{\rho_0}\frac{\partial p}{\partial x}\qquad(4)$$

Symbol c stands for velocity of the sound in the air $c=\sqrt{\kappa RT/\mu}\approx350\,\mathrm{m/s}$ (at 36.6°C), and ρ_0 is a static value of the air density. Equations (4) are the same as equations of loseless transmission line where p is analog of voltage and the air flow Av (A stands for the tube cross section area) is analog of electrical current (Javid, Brenner, 1963). Hydrodynamic analog of characteristic impedance is $Z_0=c\rho_0/A$.

At the one end of the tube velocity of the air is determined by velocity of membrane. At the opened end of the tube distortion of the pressure must be zero (pressure in the opened air is constant). As a result transmittance function from velocity of the membrane to velocity of the air on the opened end of the tube takes the form (Landau, Lifszyc 1994):

$$H(j\omega)=\frac{1}{\cos(\omega\cdot L/c)}\qquad(5)$$

We observe resonance at frequencies $f=500\,\mathrm{Hz}$, 1500Hz, 2500Hz... i.e. when length of the tube is the odd multiple of $\lambda/4$. Resonance observed in speech signal is named formant (Kondoz 1994)0. So the first formant in the human speech is located at 500Hz the next formant is located 1000Hz higher and so on.

2. A FLEXIBLE MODEL

Helmholtz model suffers lack of mechanism of energy dissipation. Used equations does not counts the air viscosity and thermal conduction. Moreover pressure wave is completely reflected from open end of the tube. As result we observe singularities of the transmittance function at the formant frequencies. A simple method to model energy dissipation is connect the tube with the other one, probably with the greater cross section area. The second tube has infinite length. Sound wave which propagates from vocal cords to lips are reflected only partially from tubes junction (mouth). The rest propagates in the second tube and carry energy outside the system. In a such a way singularities can be eliminated, but stiff connection between formant frequencies is still a problem. In the real human speech formants are really observed round 500Hz, 1500Hz and 2500Hz and so on. But precise position of formants can be changed independently. It is especially true for two first formants. Both lie around 500Hz when a male articulate / a / vowel but can be both shifted toward low frequencies when / u / is articulated or can be exploded toward 100Hz and 1000Hz when vowel / i / is articulated.

Flexibility of the model can be obtained by partitioning the tube into N small pieces. Each two

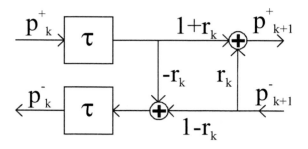

Fig. 1. The model of sound propagation inside a tube and reflection from the end of tube.

pieces has the same length but not cross section area. In the such a way the model can capture cross section area profile of the vocal tract. Sound is spitted on each junction into two parts forming wave which propagates from glottis toward lips and wave which propagates in reverse direction. Reflection coefficient of the junction r_k for the wave which propagate from k tube to k+1 tube is the same as in electrical case:

$$r_k=\frac{Z_k-Z_{k+1}}{Z_k+Z_{k+1}}=\frac{A_{k+1}-A_k}{A_{k+1}+A_k}\qquad(6)$$

The model of sounds propagation inside a tube and reflection from a junction is depicted on Fig.1. There are two channels – in the top channel waves propagate in the opposite direction then in the bottom channel. The model can be expressed in matrix form:

$$\begin{bmatrix}P_k^+\\P_k^-\end{bmatrix}=\begin{bmatrix}\dfrac{e^{j\omega\tau}}{1+r_k} & -\dfrac{r_k\cdot e^{j\omega\tau}}{1+r_k}\\[2mm] -\dfrac{r_k\cdot e^{-j\omega\tau}}{1+r_k} & \dfrac{e^{-j\omega\tau}}{1+r_k}\end{bmatrix}\begin{bmatrix}P_{k+1}^+\\P_{k+1}^-\end{bmatrix}\qquad(7)$$

where P_k^+ is a Fourier transform of pressure wave traveling in the top channel and P_k^- is a Fourier transform of pressure wave traveling in the bottom channel. In the our model tubes are chained to form system with flexibility similar to flexibility of the human vocal tract. Pressure wave is generated by vibrations of glottis on the one end of the system which on the other end has very long (infinite) tube where waves propagate only in the top channel (this tube models propagation outside the human vocal tract). This last tube can be exchanged by impedance (analog of impedance) of the tube. Transmitance function from the wave generated by glottis to the wave in the last tube has the following form (Rabiner, Schafer 1978):

$$H(j\omega)=\frac{e^{-j\omega\tau N}\cdot\prod_{k=1}^{N}(1+r_k)}{D_N(e^{-2j\omega\tau})}\qquad(8)$$

where $D(x)$ is a polynomial in the form:

$$D_N(x)=1-\sum_{k=1}^{N}\alpha_k x^k\qquad(9)$$

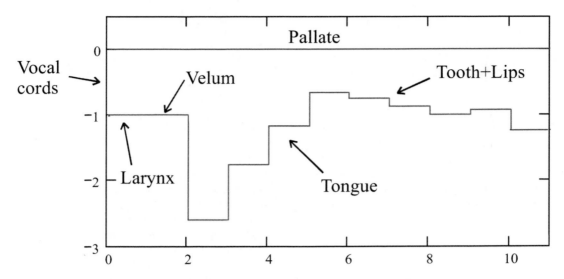

Fig. 2. Profile of author's vocal tract during articulation of the consonant / ∫ /.

It can be shown that the polynomial can be computed by recurrence formula:

$$D_0(x)=1$$
$$D_k(x)=D_{k-1}(x)+r_k x^{-1}D_{k-1}(x^{-1}) \qquad (10)$$

This formula suggest that reflection coefficients can be easily extracted from the human speech signal by the means of the Levinson-Durbin algorithm (Kondoz 1994). From reflection coefficients relative area of cross-section of each tube in the model can be obtained and shape of the vocal tract can be plotted in real time.

The human vocal tract model can be easily implemented on digital computer where delay line is simulated by the array in the memory. The most desirable situation arise when delay line can be reduced to exactly one memory cell which holds one sample of the wave. Such discrete system is equivalent to continuous one under assumption that delay time 2τ (see polynomial $D_N(x)$ in the equation (8)) is not greater then $1/(2 \cdot f_0)$ where f_0 is the greatest frequency in the wave spectrum (Shannon Sampling Theorem). For telecommunication purposes modeling of the waves with spectrum restricted to 4kHz range is sufficient. It translates to delay $2\tau =125\mu s$ or tube length $\Delta\ell \approx 2.2cm$. It means that number of pieces should be no less then $N = \ell/\Delta\ell = 17.5cm/2.2cm \approx 8$.

3. ITU G.723.1 RECOMMENDATION

The model of the human vocal tract have been implemented by the author on Carmel DSP processor from Infineon as a part of G.723.1 international ITU recommendation. The recommendation describes dual rate 6.3 kb/s / 5.3 kb/s codec (ITU-T Rec.

G.723.1 1996). The input to the codec is formed by 13-bit samples of speech signal taken at 8kHz rate. Samples are groped into 240 element frames. Each frame consist of four subframes. Frames are windowed by Hamming window and ten LPC coefficients are computed to all subframes by the means of the Levinson-Durbin algorithm. But only LPC parameters of the speech for the last subframe of the frame are transmitted to the decoder. Before it LPC parameters are transformed and quantized. Transformation consist of conversion to reflection coefficients following equation (10) and after it transformation to the form of the spectral line pairs (LSP) (Kondoz 1994). Reflection coefficients was used in my implementation of G.723.1 codec to obtain relative cross-section area of tubes of the human vocal tract model. Fig. 2. shows an example of results – author's vocal tract profile during articulation of the consonant / ∫ /. Table 1 shows numerical values of reflection coefficients used to compute profile.

From LSP parameters DC bias is subtracted and differences are vector quantized in 3-3-4 groups (ten LSP parameters). For quantization 256 entry static tables are used so resultant code describing the shape of the vocal tract is 24-bit.

Coded samples of the speech are used to compute estimation of the pitch period – period of the glottis vibrations for voiced speech or period of self similarity for unvoiced speech. Speech is filtered to flatten spectrum in the formant regions (perceptual filtering) and eroded by earlier samples of the speech using long time predictor (LTP) based on estimation of the pitch period. Residual signal obtained in the output of LTP predictor is approximated by the output of the synthesis (LPC) filter excited by the CELP (5.3 kb/s) or MPLPC (6.3 kb/s) pulse train.

Table 1. Reflection coefficients for author's consonant / ∫ /

ρ_0	ρ_1	ρ_2	ρ_3	ρ_4	ρ_5	ρ_6	ρ_7	ρ_8	ρ_9
$6,2\times10^{-4}$	0,743	-0,373	-0,393	-0,499	0,104	0,158	0,130	-0,064	0,276

Excitations are computed by the means of the analysis by synthesis (AbS) technique.

The following parameters are transmitted to the decoder: LSP parameters (vocal tract profile), parameters of LTP predictor and parameters of pulse train excitation. Bit allocation in the output stream is detailed in ITU recommendation (ITU-T Rec. G.723.1 1996).

4. AN IMPLEMENTATION ON CARMEL DSP

The main challenge of the implementation of the G.723.1 recommendation have been a strange architecture of Carmel DSP from Infineon Technologies AG (Infineon 2001). This architecture combine Very Long Instruction Word paradigm for high speed execution with low cost and low energy consumption. As a result Carmel DSP is fixed point core which can be integrated with memory and application specific logic on single ASIC chip.

Carmel core consist of two execution units which can work in parallel – see Fig.3. The first (the left) execution unit EU1 consist of MAC unit, ALU and shifter. The second (the right) execution unit (EU2) consist of MAC unit and ALU. Instructions are fetched and executed in 48-bit words. The word can be one 48-bit instruction executed by EU1 or two 24-bit instructions executed serially by EU1 or two 24-bit instructions executed in parallel – one by EU1 and second by EU2. In this mode of operation each instruction can use only one element of the execution unit. As a result at most two MAC operations or two ALU operations or one MAC and one ALU/shifter operation per clock cycle can be executed.

The most impressive speed of execution is achieved when fetched word is 48-bit header of CLIW – configurable long instruction word. This header picks a word from special 96-bit CLIW memory where up to 1024 user-defined instructions can be stored. CLIW instruction can use both MAC units and both ALU units (or ALU and shifter) in parallel. Header feeds CLIW instruction with up to four arguments. As a result one CLIW instruction can substitute four to six conventional 24/48-bit instructions and can be executed effectively in the one clock cycle.

One clock execution time is achieved by the mean of eight stage pipeline processing. First two stages fetch instructions; next two stages decode instructions; next stage compute addresses of arguments; next fetch arguments from memory; next execute instructions and the last (eighth) writes results to memory. Four memory buses are used to feed this machine with data. But such deep pipelining – essential for high steady state speed of processing – is very troublesome during conditional jumps and interrupts. As a result Carmel can be heavily slow down during execution loops with conditional and random termination condition.

A small example of the Carmel code is presented in the table 2. There is code used to vector quantization of a group of three LSP parameters. Weighted

Table 2. Carmel code for vector quantization of LSP parameters.

```
clr(a1,a2,a3,a4,a5)   || nop;
rep(256)block{
    cliw LspQnt010(r1++,r1++,r5++,r5++)
    {a01=round(*ma1**ma3)||a4+=a3||a11=round(*ma2**ma4)||nop||ff1=*ma3||ff2=*ma4;}
    cliw LspQnt020(r4++,r4++)
    {a3=a01**ma1||max(a5,a4)||a4=a11**ma2||nop;}
    cliw LspQnt030()
    {a3-=a01*ff1||ifexe a1h=dec(a21)||a4-=a11*ff2||nop;}
    cliw LspQnt040(r1++,r5++)
    {a01=round(*ma1**ma2)||nop||nop||inc(a21)||ff1=*ma2;}
    cliw LspQnt050(r4++)
    {a3+=a01**ma1||nop||a4-=a01*ff1||nop;}
    }
a4+=a3;
max(a5,a4);
ifexe a1h=dec(a21);
pacfg=0;
*(r7+d_indx)=a1h;
```

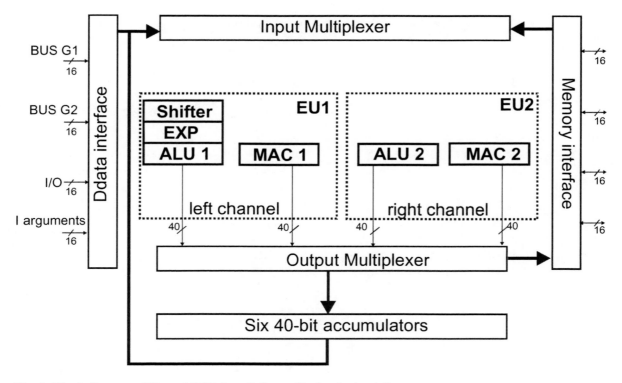

Fig. 3. Block diagram of Carmel DSP from Infineon Technologies AG.

Euclidian distance between LSP parameters $\underline{p}$ and quantized ones $\hat{p}$ is computed during first loop execution:

$$\varepsilon_n^2 = \sum_{k=0}^{2} \left(2 \cdot w_k \cdot p_k \cdot \hat{p}_{k,n} - w_k \cdot \hat{p}_{k,n}^2\right) \quad (11)$$
$$n = 0 \ldots 255$$

where w_k is weighting function. The computations needs nine MAC operations which are executed in five clock cycles by MAC units of EU1 and EU2. In the next loop execution ALU portion of EU1 and EU2 is used for all necessary comparisons to find index n which maximize Euclidian distance ε_n^2.

Value of ε_{n+1}^2 is computed in parallel. In the example we can observe all main techniques used by superscalars to accelerate code execution and explore instruction level parallelism. It is loop unrolling, zero overhead looping, software pipelining and conditional execution of instructions.

5. THE END REMARKS

Described implementation of ITU G.723.1 recommendation consist of circa 11000 lines of assembler code written under Tasking tools for Carmel (Tasking 2000)0. About one thousand lines of the code are macros. The coder portion of the implementation needs in average about 7.7 MIPS and maximally about 9 MIPS (such situation is almost impossible) versus 16 MIPS of ITU estimation (Cox, Kroon 1996). Such impressive

results are possible because computations are made by two execution units in parallel. Computations connected with vocal tract modeling needs about 1.1 MIPS.

REFERENCES

Cox R.V., Kroon P. (1996). Low Bit-Rate Speech Coders for Multimedia Communications, *IEEE Communications Magazine*, Vol.34, № 12, pp.34-41.

Helmholtz von H.L.F. (1895). *Die Lehre von den Tonempfindungen als physiologische Grundlage der Theorie der Music*, F.Vieweg und Sohn, Braunschweig.

Infineon Technologies AG (2001). *Carmel 10xx User's Manual*.

ITU-T Rec. G.723.1 (1996). *Dual Rate Speech Coder for Multimedia Communications Transmitting at 5.3 and 6.3 kbit/s*.

Javid M., Brenner E. (1963): *Analysis, Transmission and Filtering of Signals*, McGraw-Hill Book Company.

Kondoz A.M. (1994). *Digital Speech. Coding for Low Bit Rate Communication Systems*, John Wiley & Sons, Inc.

Landau L.D., Lifszyc E.M. (1994). *Hydrodynamics*, Państwowe Wydawnictwo Naukowe PWN, Warsaw, (in Polish).

Rabiner L.R., Schafer R.W. (1978). *Digital Processing of the Speech Signals*, Prentice Hall.

Tasking Inc. (2000). *Carmel v.3.0 Cross-Assembler, Linker/Locator, Utilities User's Guide*, MA523-00-00, Doc.ver. 1.9.

IFAC
Publications
www.elsevier.com/locate/ifac

BASE ACOUSTIC PROPERTIES OF POLISH SPEECH

Jacek Izydorczyk, Piotr Kłosowski

Department of Electronics
Silesian University of Technology, Gliwice

Abstract: The article presents physical basics of human speech production. There are
considered the most fundamental phonemes - vowels. Acoustic properties of vowels are
extracted. It is observed that the most important are positions and shapes of first two formants.
The rest of the spectral attributes of vowels, exception is voiced excitation, can be cancelled.
Vowels are still recognized by human. In the article acoustic model of vowels production is
proposed and tested. Results are presented in graphic manner. *Copyright © 2001 IFAC*

Keywords: speech, speech analysis, speech recognition, speech synthesis

1. HUMAN SPEECH PRODUCTION

Human speech is produced by the vocal cords in the
larynx, the trachea, the nasal cavity, the oral cavity, the
tongue, and the lips. Figure 1 shows the human speech
organs. The glottis is the space between the vocal cords.
For voiced sounds such as vowels, the vocal cords
produce a series of pulses of air. The pulse repetition
frequency is called the glottal pitch. The pulse train is
referred to as the glottal waveform. The rest of the
articulatory organs filter this waveform. The trachea, in
conjunction with the oral cavity, the tongue, and the
lips, acts like a cascade of resonant tubes of varying
widths. The pulse energy reflects backward and forward
in these organs, which causes energy to propagate best
at certain frequencies. These are called the formant
frequencies [6].

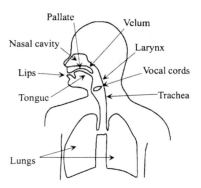

Figure 1:The speech organs

Phonemes are the smallest units of speech that
distinguish one utterance from another in a particular

language. An allophone is an acoustic manifestation of a
phoneme. A particular phoneme may have many
allophones, but each allophone (in context) will sound
like the same phoneme to a speaker of the language that
defines the phoneme. Another way of saying this is, if
two sounds have different acoustic manifestations, but
the use of either one does not change the meaning of an
utterance, then by definition, they are the same
phoneme. Phonemes are classified by linguists into
following sets:

- Vowels - which are considered in detail later.
- Diphthongs are sounds that change smoothly from
  one vowel to another.
- Nasals [m, n]. To produce nasals, a person opens the
  velar flap, which connects the throat to the nasal
  cavity.
- Liquids are the vowel-like sounds [l] and [r].
- Glides (semivowels) are the sounds [j] and [ł].
- Fricatives [f] and [s]. Breath passing through a
  constriction creates turbulence and produces such
  unvoiced sounds.
- A stop (also called a plosive) is a momentary
  blocking of the breath stream followed by a sudden
  release. The consonants [p,b, t,d,k] and g are stop
  consonants.
- Aspirates. Opening the mouth and exhaling rapidly
  produces the aspirate [h]. Other consonants such as
  [p,t], and [k] frequently end in aspiration, especially
  when they start a word.

Affricative is a stop immediately followed by a
fricative. The Polish sounds [cz] (as in *czołg*) and [dż]
(as in *dżem*) are affricates.

Linguists have developed an International Phonetic Alphabet (IPA) that has symbols for almost all phones [7]. This alphabet uses many Greek letters that are difficult to represent on a computer. Polish linguists have developed the phoneme alphabet to represent Polish phonemes using normal ASCII characters. Extra symbols are provided that either combine certain phonemes or specify certain allophones to allow the control of fine speech features. Speech researchers often use the short-term spectrum to represent the acoustic manifestation of a sound. The short-term spectrum is a measure of the frequency content of a windowed (time-limited) portion of a signal. For speech, the time window is typically between 5 milliseconds and 25 milliseconds, and the pitch frequency of voiced sounds varies from 80 Hz to 280 Hz. As a result, the time window ranges from slightly less than one pitch period to several pitch periods. The glottal pitch frequency changes very little in this interval. The other articulatory organs move so little over this time that their filtering effects do not change appreciably. A speech signal is said to be stationary over this interval. The spectrum has two components for each frequency measured a magnitude and a phase shift. Empirical tests show that sounds that have identical spectral magnitudes sound similar. The relative phase of the individual frequency components plays a lesser role in perception. Typically, we perceive phase differences only at the start of low frequencies and only occasionally at the end of a sound. Matching the spectral magnitude of a synthesized phoneme (allophone) with the spectral magnitude of the desired phoneme (taken from human speech recordings) always improves intelligibility. A spectrogram is a plot of spectral magnitude slices, with frequency on the y-axis and time on the x-axis. The spectral magnitudes are specified either by colour or by saturation for two-colour plots. Depending on the time interval of the spectrum window, either the pitch frequency harmonics or the formant structure of speech may be viewed. It is even possible to ascertain what is said from a spectrogram. Figure 2 shows spectrogram of human speech. The formant frequencies are the upper regions that move up and down as the speech organs change position. Fricatives and aspiration are characterized by the presence of high frequencies and usually have much less energy than the formants.

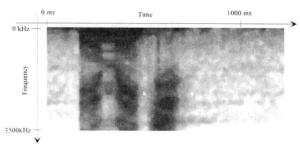

Figure 2: A speech spectrogram

2. ACOUSTIC PROPERTIES OF POLISH VOWELS. AN ATTEMPT TO IDENTIFY THEM INSIDE WORDS.

It is believed that vowels are the most fundamental phonemes. Vowels are classified by the place of the articulation, particularly by the position of the tongue. It is because the rest of the distinctive articulatory features are the same for all of them. All vowels are:
- Voiced. During articulation of the vowel vocal cords vibrates.
- Opened. Airflow inside oral cavity is frictionless and aperture of the lips are relatively high.

Vowel phonemes are described by two features *height* and *position*. Feature height refers to the vertical position of the tongue in the mouth. Vowels articulated with the body of the tongue close to the palate are high vowels. There are also mid and low vowels articulated with lowered tongue. Feature position refers to the horizontal position of the tongue in the mouth. There are front, central and back vowels. Tytus Benni used five vertical positions of the tongue to create famous Polish vowels triangle [1]:

Table 1:Triangle of the Polish vowels by Tytus Benni

Height – vertical position of the tongue	Position – hor. pos. of the tongue				
	front				back
	1	2	3	4	5
1	i	y			u
2		e		o	
3			a		

Such classification, made by articulation properties, is not useable for synthesis and recognition purposes. Acoustic properties of vowels is needed.

Vowels are periodic or near periodic signals produced by excitation of oral cavity by vibration of vocal cords. Figure 3 shows 25 [ms] of vowel [a] produced by a male.

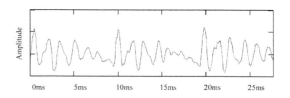

Figure 3: Vowel [a] produced by a male

There are showed about three "periods" of the signal or three vibrations of vocal cords. All of them are very similar but not the same. The amplitude spectrum of the signal is presented on the Figure 4.

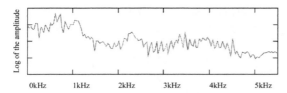

Figure 4: Amplitude spectrum of vowel [a]

The spectrum poses high, sharp peaks and deep and sharp holes. We observe interference of spectrums of three similar periods. We assume that spectrum of the signal produced by vocal cords is flat in 0 kHz to 5 kHz range [2],[3]. So the envelope of spectrum from the Figure 4 is amplitude characteristic of oral cavity of the male.

Spectrum envelope can be extracted effectively by the means of Linear Predictive Coding (LPC) [3]. There is created linear predictor of speech sample

$$s_k \uparrow \hat{s}_k (a_1,a_2,...,a_N) = \sum_{n=1}^{N} a_n \, s_{k-n} \qquad (1)$$

Predictor is described by N coefficients $a_1, a_2, ..., a_n$ which are computed to minimize mean prediction error over analyzed time range:

$$\varepsilon^2 (a_1,a_2,...,a_N) = \frac{1}{K} \sum_{k=0}^{K-1} [s_k - \hat{s}_k (a_1,a_2,...,a_N)]^2 \qquad (2)$$

Amplitude characteristic of the IIR filter described by the transmittance:

$$H(z) = \frac{\varepsilon}{1 - \sum_{n=1}^{N} a_n \, z^{-n}} \qquad (3)$$

is envelope of amplitude spectrum of speech signal - see Figure 5.

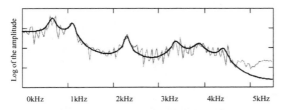

Figure 5: Envelope of the amplitude spectrum of the vowel [a]

Constant ε is a minimal mean prediction error. In our example voice of the male has been sampled with frequency $f_0 = 11.025 \ kHz$, so signal from the Figure 3 consist of $K=306$ samples. There has been used predictor of order $N=14$. Value of the order is sufficient to capture all formants. Moreover in the most cases there are two conjugate poles of transmittance $H(z)$ per

formant. Transmittance $H(z)$ can be expanded into partial fractions:

$$H(z) = \sum_n H_n(z) \qquad (4)$$

where each $H_n(z)$ is second-order or first-order factor. It was checked by experiments that first order-factors can be omitted. As a result there is peer-to-peer relation between second-order factors and formants:

$$H(z) = \sum_n H_n(z) = \sum_n \frac{\alpha_n + \beta_n z^{-1}}{1 + 2 \cdot \cos(\omega_n) \cdot e^{-\gamma_n} \cdot z^{-1} + e^{-2\gamma_n} \cdot z^{-2}} \qquad (5)$$

where $\omega_n/2\pi$ is n-th formant frequency and parameter γ_n controls shape of the amplitude characteristic of $H_n(z)$ factor near formant frequency. Factors $H_n(z)$ nicknamed formants by authors, are sorted from lowest formant frequency to the highest formant frequency and named F1, F2, F3, ... Using second-order IIR filters $H_n(z)$ connected in parallel it is possible to synthesize vowels and freely add and discard any formant - see Figure 6 and 7.

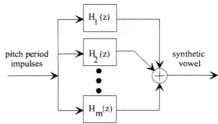

Figure 6: Production of the synthetic vowels

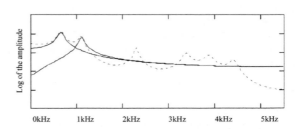

Figure 7: Amplitude spectrum of the vowel [a] and formants F1 and F2 (lines)

Vowels have been synthesized and formants has been discarded from the highest frequency to the lowest. Voice without high formants is somewhat artificial but absolutely recognizable for a human as we preserve two lowest formants. So we concluded that for recognition of vowels only two formants are needed - formants on two lowest frequencies named F1 and F2. If we take amplitude characteristics of vowels with only two lowest formants and arrange then into vowels triangle we obtain Figure 8.

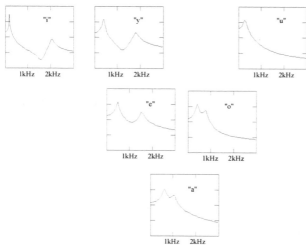

Figure 8: Logarithm of amplitude spectrum of the vowels arranged into vowels triangle

Tongue of the speaker is in most "neutral" position during articulation of vowel [*a*]. It is in the low vertical position and mid horizontal position. When tongue is moved toward lips (horizontal position front) and palate (vertical position high) formants F1 and F2 repels. F1 moves toward low frequencies and F2 moves toward high frequencies. It is transition [*a-e-y-i*]. When tongue is moved toward velum (horizontal position back) and toward palate (vertical position high) formants F1 and F2 together moves toward low frequencies. It is transition [*a-o-u*]. Sometimes, as in our example, formants F1 and F2 during articulation of [*u*] are so close that amplitude spectrum has only one maximum.

Let us assume we have a speaker. Each vowel articulated by her/him occupy small area in a two dimensional space which consist of frequency of formant F1 and F2 - see Figure 9.

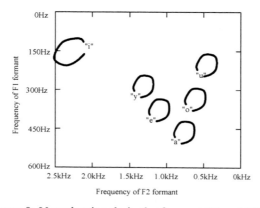

Figure 9: Vowels triangle in the formant F1 and F2 frequency space

Even more, vowels are again arranged into vowel triangle. Similar figures can be obtained for allophonic forms of vowels.

Vowels can be recognized by two first formants not only in separately articulated phonemes. It is possible extract them from continuous speech.

Figure 10 shows time trajectories of formants F1 and F2 in the word *zawiesić*. Phonetic transcription of the word is *zawiësić*.

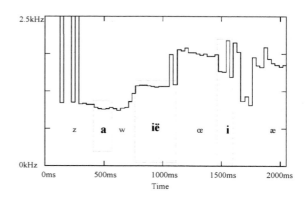

Figure 10: Time trajectories of F1 and F2 formant frequencies in the word "zawiesić"

We can easily identify position of vowel [*a*], next is transition [*ië*] and very short vowel [*i*] before affricate [*ć*]. It is possible that all human speech (all phonemes) are "constructed" with two formant frequencies in mind. How an idea can be extended for e.g. stop consonants is not clear for authors. We know that beside positions of two formant frequencies other acoustic properties of the signal are important eg. voiced/voiceless property. But as in classification by articulation properties quantity of features of phonemes should be restricted to three or four.

During our investigations we have found some peculiarities about acoustic properties of vowels.

- Some persons has first two formants in vowel [*a*] so close that it forms one maximum in the amplitude spectrum - see Figure 11. Discrete structure of the maximum is showed during articulation of other vowels.

- If we used to synthesize vowels only one formant then vowels can be recognized by the human but vowel [*e*] has been recognized as [*o*] and vowels [*y*] and [*i*] has been recognized as [*u*]. It probably means that mid-back vowels [*a-o-u*] are recognized by one formant.

- We tried to synthesize vowels using filter with linear logarithmic amplitude characteristic in the region of the first two formants (about 0-3kHz). Amplitude characteristic of the filter has been obtained from spectrum of natural vowel using least squares method. Again vowels [*a-o-u*] can be recognized by the human.

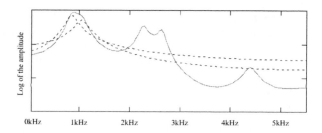

Figure 11: Amplitude spectrum of vowel [a] (line) can posses one extremum created by two formants (dots).

3. MATHEMATICAL MODEL OF VOWELS ARTICULATION

It is believed that the simplest model of articulation of vowel [a] is a tiny cylinder 17.5 cm long, opened in the one end (mouth) and excited by vibrations of a membrane on the second one (glottal cords) [2]. Tiny cylinder means that aperture of the cylinder is small compared to length of acoustic waves.
Acoustic waves inside cylinder are described by [5]:

Euler law, specific version of the second Newton law formulated for continuous medium:

$$\frac{\partial v}{\partial t} + v \cdot \frac{\partial v}{\partial x} = -\frac{1}{\rho} \cdot \frac{\partial p}{\partial x} \qquad (6)$$

where v is velocity of the air, p is a pressure of the air and ρ is a density of the air.
Continuity law:

$$-\frac{\partial (S\rho v)}{\partial x} = S \cdot \frac{\partial \rho}{\partial t} \qquad (7)$$

where S is aperture of the cylinder.

Adiabatic process law:

$$\frac{p}{\rho^{\kappa}} = const. \qquad (8)$$

where κ is ratio of molal specific heat of the air on the isobaric process.

We take into account only small changes of air pressure p' around static value of the atmospheric pressure p_0:

$$p = p_0 + p' \qquad (9)$$

and small changes of air density ρ' around static value ρ_0:

$$\rho = \rho_0 + \rho' \qquad (10)$$

We omit small values of the second order and obtain equation which describes sounds in cylinder:

$$\frac{1}{S} \cdot \frac{\partial}{\partial x}\left(S \cdot \frac{\partial p'}{\partial x}\right) + \frac{1}{c^2} \cdot \frac{\partial^2 p'}{\partial t^2} = 0 \qquad (11)$$

where c is acoustic waves velocity:

$$c = \sqrt{\frac{\kappa \cdot R \cdot T}{\mu}} \qquad (12)$$

For cylinder with constant aperture it can be simplified to classical form:

$$\frac{\partial^2 p'}{\partial x^2} + \frac{1}{c^2} \cdot \frac{\partial^2 p'}{\partial t^2} = 0 \qquad (13)$$

It can be shown [5] that amplitude characteristic of the cylinder length l, excited in the one end and opened in the other end takes the form:

$$\left(\frac{v_o}{v_m}\right)^2 = \frac{1}{\cos^2\left(\frac{\omega}{c} l\right)} \qquad (14)$$

where ω is frequency of the wave, v_m is amplitude of the air velocity near the membrane and v_0 is amplitude of the air velocity in the opened end of the cylinder. Such result is not very realistic. Sounds are mirrored from the opened end of the cylinder and energy is not convected outside the system. Pulse response of the system takes the form of the sinusoidal signal. Human speech is not such spectrally sharp. For example DTMF symbols used in telephony for number signaling consist of two sinusoidal signals which frequencies are close to frequencies of first two formants of vowels. Signal is near periodic. But we never mismatch DTMF symbols with a human speech. So we have created models of oral cavity which always are opened to cylinder with infinite length. This cylinder models opened space. Energy of acoustic vowels generated by glottal cords are convected into the cylinder. The model consist of constant number of cylinders with various length and aperture connected in series. We have changed length of cylinders and their aperture to minimize square root error between amplitude characteristic of oral cavity of human speaker and amplitude characteristic of system of cylinders. This "optimization" have been done in the range of frequencies characteristic to first two formants. E.g. for vowel [a] it is 0-1500 Hz range. Results for vowel [a] are presented on Figure 12.

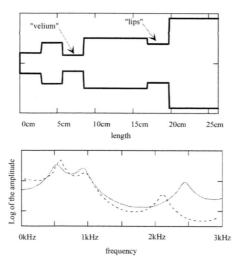

Figure 12: High – cross section of the model of the oral cavity during vowel [a] articulation,
Low – Amplitude spectrum of the synthetic [a] (line) and human speaker produced [a] (dots).

There is showed cross section of the model which consist of six cylinders with variable length and aperture. An exception is the last (sixth) cylinder which length equals to infinity. We can easily to see lips and velum contraction. Tongue is in low position and there is resonance space between velum and lips. Amplitude characteristic of the model, from glottal cords to the edge of the lips, resemble two first formants of the vowel [a]. There is even the third formant in the position quite near the position of the third formant of human speaker. But we have not considered position of the third formant in our optimization!

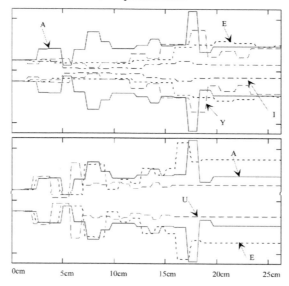

Figure 13: Cross section of the model of the human oral cavity during vowels articulation

It is obvious for us that to obtain flexibility of the model comparable with flexibility of human oral cavity we need lot of cylinders with variable length and aperture. It leads to great computational complexity of "optimization" problem. We stopped our investigations with model of vowel production which consist of:

- Sixteen cylinders with equal length and variable aperture. Sum of the lengths of cylinders are variable about 17.5 cm.
- Ending cylinder with infinite length and variable aperture. An model of acoustic waves radiation into open space.

The model possess 18 parameters - apertures of 17 cylinders and length of the oral cavity. Optimal values of the parameters for Polish vowels are presented on Figure 13. There are cross sections of the model which produce vowels from mid-low to front-high [a-e-y-i] and mid-low to back-high [a-o-u].

4. CONCLUSIONS

We believe that attributes of the speech used by human to recognize the meaning of the words are simple. We believe that the number of the attributes which must be considered are bounded to small number. The most important of them are voiced/unvoiced attribute and position and shape of first two formants. We believe that a human can control position only of two first formants. Other formants are not (but can be) controlled. We used model of cascaded cylinders (tubes) to simulate vowels production. We observed that the shape of amplitude spectrum of produced vowels depends heavily on acoustic waves radiation outside oral cavity. Observed shapes of tubes are the similar to the shapes of the human oral cavity during vowels production. For other phonemes production the model must be supplemented by the model of nasal cavity and more sophisticated excitations.

REFERENCES

[1] Benni T., *Fonetyka opisowa języka polskiego*, Ossoliński Publishing, Wrocław 1959 (in polish).
[2] Bristow G.: *Electronic Speech Systems*, McGraw-Hill Book Company, 1984.
[3] A. M.Kondoz:, *Digital Speech. Coding for Low Bit Rate Communication Systems*, John Wiley & Sons, Inc., 1994.
[4] Ostaszewska D., Tambor J., *Podstawowe wiadomości z fonetyki i fonologii współczesnego języka polskiego*, Uniwercity of Silesia nr 488, Katowice 1993, (in polish).
[5] Landau L.D., Lifszyc E.M., *Hydrodynamika*, PWN Publishing, 1994.
[6] Jassem W., *Podstawy fonetyki akustycznej*, PWN Publishing, Warszawa 1973, (in polish).
[7] Flanagan J., *Speech Analysis, Synthesis, and Perception*, 2nd ed.,New York: Springer-Verlag, 1972.

IFAC
Publications
www.elsevier.com/locate/ifac

COMB FILTER APPLIED TO TEXAS INSTRUMENTS-TMS320C3X

Juliusz Maćkowiak

Institute of Electronics, Silesian University of Technology
mackow@boss.iele.polsl.gliwice.pl

Abstract: This article deals with the problem of applying of the adaptive comb filter to the Digital Signal Processing. This algorithm was dedicated to speech filtration, which was corrupted by white additive noise. This practical realization was made for floating point DSP, Texas Instruments, TMS320C31. *Copyright © 2001 IFAC*

Keywords: filter, digital signal processors, adaptive digital filters, white noise

1. INTRODUCTION

The focus of this paper is the enhancement of speech degraded by additive noise by exploiting the periodic nature of a voiced speech. The idea of the comb filter (Lim, 1978) is to average signal by using information about the periodicity. This information is contained at the fundamental frequency.

The speech signal is not fully periodic but it is concerned to be quasi periodic- to have some fluctuations.

Period of the signal varies from 5 ms up to 15 ms (Kodoz, 1994), and depends on sex, age, kind of voiced speech, or other qualities of the person.

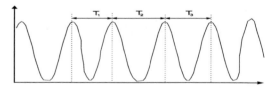

Fig. 1 Example of pseudoperiodical signal.

In the Fig. 1 is presented pseudoperiodical signal at the moment period satisfies the equation:

$$\underset{k \neq n}{\exists} \, T_n \neq T_k \qquad (1)$$

when $-L \leq k \leq L \wedge -L \leq n \leq L$ and L is the boundary of the signal.

Variations in the pitch periods diminish the intelligibility of filtered speech. Possible manner of solving this problem, can be applying time warping method (Maćkowiak, 2001).

2. COMB FILTER

The operation of an adaptive comb filter (Lim, 1978) can be explained by considering its unit sample response for one period:

$$h(n) = \sum_{k=-L}^{L} a_k \cdot \delta(n - N_k)$$

$h(n)$ -is denoted as unit sample response, and $\delta(n)$ as unit sample function. The length of the filter is 2L+1 pitch periods, a_k is the coefficient and N_k is given by the following equations:

$$N_k = -\sum_{l=k}^{-1} T_l \quad \text{for } k < 0$$
$$(2)$$
$$0, \qquad \text{for } k = 0$$
$$\sum_{l=0}^{k-1} T_l \quad \text{for } k > 0$$

where pitch period is denoted as T_k. Period of the speech waveform was detected by the autocorrelation algorithm. It was realized by finding maximum value of the function (3) and next by finding j which corresponds to this value.

3. METHOD[1]

Scheme of the algorithm is shown in the Fig. 2. The autocorrelation is calculated over the frames of speech that overlaps. The size of the overlapping depends on the computational burden. After filtering and sampling there is a block, which responds for decision: voiced, or unvoiced. This decision is made using the autocorrelation analysis that is presented as:

$$y(j) = \sum_i x(i+j) \bullet x(i) \qquad (3)$$

where: x-input sample, y- output sample. It is calculated for each frame having 200 samples at 8kHz sampling rate.

If the signal is periodic the autocorrelation function reaches maximum at the pitch period $y_{max}(j)$. It is important to notice that comparing this value to $y(0)$ can be inferred whether this frame is voiced or not.

Block, which checks whether pitch period is in the boundary, is used because it is indispensable to avoid estimating noise period as pitch period.

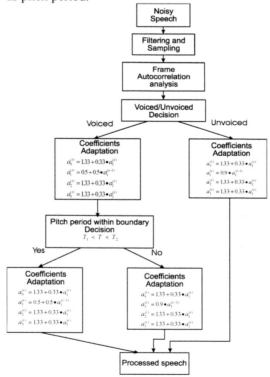

Fig. 2 Adaptive comb filtering algorithm.

There are four filter coefficients the sum of which equals four. When these are calculated for unvoiced speech three coefficients are

---

[1] All references

attenuated but one -gained. The same action is made on the testing of the pitch boundary. If speech is voiced four coefficients have the same value 1, and when it is unvoiced speech, only one coefficient is nonzero. The listing of the program is presented in Appendix.

4. DISCUSSION

Algorithm improves intelligibility of the voiced speech. Subjective assessment of this method is positive. For some noisy words it is easier to infer what was told, but the difference between speech before and after filtration is not so great. This program was created for the Polish language. SNR for the voiced speech decreases about 27dB, and for the unvoiced one is unchanged. It yields property that for some part of the utterance voice is more audible. In the time of very high noise distortion appears, and the longer filter the more distortion in speech. This distortion appears because the speech had been pseudoperiodic and was averaged.

5. REFERENCES

Maćkowiak J. (2001). Time Warping in Application To Speech Filtration. In: *VI International Conference Proceedings SYMBIOSIS 2001(Tkacz E.),pp. 200-202. Silesian University of Technology, Szczyrk, Poland*.

Lim, J. S., et al. (1978). Evaluation of an Adaptive Comb Filtering Method for Enhancing Speech Degraded by White Noise Addition. *IEEE Transaction on Acoustics, Speech, and Signal Processing,* **VOL. ASSP-26, No. 4**, pp. 345-458.

Kondoz, A. M. (1994), *Digital Speech,* University of Survey, West Sussex.

Texas Instruments (1994), *TMS320C3X User's Guide,* Texas Instruments Incorporated, Owensville.

Texas Instruments, *TMS320C3X DSP Starter Kit, User's Guide,* Texas Instruments Incorporated, SPRU163a.

Texas Instruments (1998), *TMS320C3X General-Purpose Applications, User's Guide,* Texas Instruments Incorporated, SPRU194.

6. APPENDIX

In this listing is included main loop and two interrupt service routines from serial port.

Following code presents calculating autocorrelation for $j = 0$:

```
ldf     0.0, R2
mpyf3  *AR2++%, *AR1++%, R0
```

```
        ldi     R4, RC
        rptb    A1          ; y(j)=x(i+j)*x(i)
        mpyf3   *AR2++%,*AR1++%,R0
                    ; IR1-j, AR1-i
A1 ||   addf3   R0,R2,R2
        addf    R2, R0
        stf     R0,@Max01
```

Maximum value of this function (3) is stored at *Max01* location, and is used to verify whether the signal holds information about period.

```
        ldi     @Inde, IR1
        ...
        ...
        ...
        ldf     0.0, R2
AutoL   nop     *AR2++(IR1)%
        mpyf3   *AR2++%,*AR1++%,R0
        ldi     R4, RC
        rptb    Autok1  ;y(j)=x(i+j)*x(i)
        mpyf3   *AR2++%,*AR1++%,R0
Autok1 || addf3 R0,R2,R2
        addf    R2, R0
        cmpf    @Max, R0
        callgt  NewMax
        cmpi    IR1, R3
        bgtd    AutoL
        addi    1,IR1
        ldf     0.0, R2
        nop
```

Inde variable includes starting point of *j*. It is used because character of the autocorrelation function decreases very slowly, and it is necessary to evade detection of the false maximum. Maximum value is stored at *Max* by the subroutine *NewMax*.

```
NewMax      stf     R0, @Max
            ldi     IR1, AR3
            subi    1, AR3
            sti     AR3, @MaxI
            rets
```

j value which corresponds to the *MAX* is saved as *MaxI*. This variable is employed in the process that adjust filter coefficients:

```
        ldi     @ADC_last, AR1
        nop     *AR1++(50)%
        sti     AR1, @SYN
        ...
        ...
        ...
m       ldi     @ADC_last, AR1
        ldi     AR1, AR0
        nop     *AR0++(100)%
        cmpi    @SYN, AR0
        bne     m
```

The above code is utilized to synchronize work of the filter and the autocorrelation module. Calculation of the autocorrelation consumes a lot of time and this is why a delay is necessary.

This portion of the program represents adaptation of the coefficients. First, disable of the interrupts is needful to avoid maladapting.

```
        ldi     @MaxI1, IR1
        ldi     @FIR_coefx,AR0
        nop     *AR0++(100)
        andn    GIE, ST
        ldi     0,IE
;----------------------------------
        ldf     0, R0
        stf     R0, *-AR0(IR1)
        stf     R0, *AR0
        stf     R0, *++AR0(IR1)
        stf     R0, *AR0
                ;Removing last coefficients
;----------------------------------
        ldf     @Level, R0
                ;Load last coefficient value
        ldi     IR1, R4
        ldi     @MaxI, IR1
        cmpi    P1, IR1      ;P1<IR1 jump to
N1
        calllt  Zakres
        cmpi    P2, IR1
        callgt  Zakres
                ;Verify whether MaxI
liettttttttttts             ;between P1, and
P2
                ;P1<j<P2
                ;if not attenuates coefficient in
                ;Zakres subroutine
;------------------------------------------
        ldi     @FIR_coefx,AR0
        nop     *AR0++(100)
;------------------------------------------
        ldf     @Max01,  R2
        ldf     @Max,    R3
        mpyf    0.9, R2
        cmpf    R2, R3
                ;Verify if value of Max reaches
                ;appreciable level if so go to N1
        bgt     N1
        mpyf    @Att, R0
        b       N2
;------------------
N1      ldf     1, R2
        subf    R0, R2, R2
        mpyf    0.5, R2
        addf    R2, R0
        ;Increasing the value of coefficients
        ;------------------------------
N2      subi3   R4, IR1, R4
        ldilt   1,R4
        ldigt   -1, R4
        addi    R4, IR1
        ldf     R0, R2
        mpyf    3, R2
        subf    4, R2
        absf    R2
        ;Calculation of the coefficient which
        ;disables filter for non-voiced speech-
```

```
;only one is nonzero or all but one is         ldi     0xF4,IE
;attenuated                                    or      GIE, ST
;-------------------------------               sti     IR1, @MaxIl
stf     R0, *-AR0(IR1)                         cmpf    0.0, R0
stf     R2, *AR0                               ldflt   0.0, R0
stf     R0, *++AR0(IR1)                        cmpf    1, R0
stf     R0, *AR0                               ldfgt   1, R0
;coefficients update                           stf     R0, @Level
;-----------------------                        b       main
                                               ;------------------------------
                                               subroutine which attenuates
                                     Zakres    ldi R4, IR1
                                               mpyf @Att, R0
                                               rets
```

IFAC
Publications
www.elsevier.com/locate/ifac

AN IMPROVED APPROACH TO INTERPOLATED DFT

Wojciech Oliwa

Silesian Technical University Gliwice, Poland

Abstract: This paper presents some ways of Interpolated DFT method improvement namely a high precision algorithm that is used to measure frequency of spectral lines in DFT spectrum. An improved approach including a new measuring relation and algorithm of IDFT are proposed. That approach comprises all advantages of already known methods yet additionally is very simple and offers a standardised procedure for all windowing functions. Copyright © 2001 IFAC

Keywords: frequency measurement, spectrum analysis, discrete Fourier transform, Interpolated DFT, windowing functions

1. INTRODUCTION

One of the most widely used DSP algorithm is Discrete Fourier Transform (or its improved version FFT). DFT changes the domain in which signal is analysed from time domain into frequency domain, hence it is natural that the primary goal of applying DFT is spectrum analysis.

Unfortunately direct interpretation of spectrum obtained as a result of the DFT leads to poor frequency measurement precision and in consequence some methods that improve precision of frequency estimation have to be used.

2. FREQUENCY OF DFT SPECTRAL LINE; CLASICAL APROACH

A simplest, classical, method of "measuring" frequency via DFT is to find in discrete spectrum a point that has local maximum amplitude. The precision of such measurement is not high due to a discrete nature of spectrum obtained from DFT. This precision depends on N - number of points in time

series (transformed to spectrum that consists of N/2+1 different points) and f_s – sampling frequency at which series of samples was acquired:

$$e = \pm f_s \cdot \frac{1}{2N} \qquad (1)$$

Sampling frequency is selected to fit an input signal bandwidth and could not be decreased to satisfy expected frequency measurement precision. Hence the only way of increasing precision of such measurement is to increase N. This is very time consuming solution because DFT calculation time is proportional to factor $N*\log_2 N$. Of coarse, unless desired frequency precision improvement is not great that drawback has no significant meaning.

Increasing N means that more samples should be transformed. It could be done simply by acquiring more samples. However, often, instead of acquiring more samples the time series are enlarged by adding zero valued samples after sampling and windowing. This method is called "zero padding".

3. SOME IMPROVEMENTS OF INTERPOLATED DFT METHOD

Another way of improving frequency measurement precision, completely different from classical approach is an Interpolated DFT method. The IDFT method takes advantage of the spectral leakage in DFT spectrum.

3.1 Spectral leakage and windowing functions.

Spectral leakage is considered as the worst drawback of DFT. It appears whenever acquisition time of time series is not an integer multiple of input signal period and it is a consequence of discontinuity at the ends of time series. To cope with that problem windowing functions are used.

There are a lot of windowing functions (Harris, 1978), however only four types became very popular (Lyons, 1998):

- Blackman-Harris with minimum sidelobe level,
- Blackmann-Harris with maximum sidelobe decay,
- Keiser-Bessel,
- Dolph-Chebyshew.

All these windowing function have some common properties that are important to spectral analysis and IDFT:

- they considerably decrease spectral leakage when acquisition time of time series **is not** an integer multiple of input signal period;
- they cause spectral leakage when acquisition time of time series **is** an integer multiple of input signal period;
- values of spectral points (amplitudes) of spectral line vary when input signal frequency changes.

In conclusion, if windowing function is used, spectral leakage always take place and every spectral line consists of at least three DFT spectrum points even though average leakage is much decreased. The values of that points change even if input signal frequency change less than e from equation (1).

Properties mentioned above result from the fact that, from mathematical point of view, DFT is discrete convolution of true input signal spectrum and windowing function spectrum. So if input signal is pure sinusoidal signal the points of DFT spectrum lies on displaced continuous windowing function spectrum. This is the reason why even small variation of input signal frequency cause changes of the values of DFT spectrum points. Relative changes are independent of input signal amplitude and could be used to determine true frequency from points of discrete spectrum. The first step is to recover true

position n_p of the spectral line. Then true frequency f_{in} may be easy obtained from equation:

$$f_{in} = \frac{n_p}{N} \cdot f_s \qquad (2)$$

Consider that n_p could be any real number between 0 and $N/2+1$.

3.2 The idea of Interpolated DFT method.

As mentioned above the phenomenon of spectral leakage is used to increase frequency measurement precision via DFT by a method called Interpolated DFT. Originally the method was proposed by Rife and Vincent (1970) and after that some modifications of Interpolated DFT were proposed by other authors (Jain, et al., 1979; Grandke, 1983; Renders, et al., 1984; Andria, et al., 1989; Offelli, et al., 1990).

All these variations of IDFT method use quotient of values of two greatest amplitude points that consist on spectral line to compute input signal frequency.

Only method developed by Render et al. (1984) uses amplitudes of real and imaginary part of two greatest points. In example in Figure 1 it would be points with index 5 and 6.

3.3 Classical frequency measuring relation.

Generally indexes of points L_1 and L_2 are defined as follows:

$$L_1 = L \leq n_p = L + \delta_p < L_2 = L + 1 \qquad (3)$$

for $A(L_1) > A(L_2)$, and

$$L_1 = L - 1 \leq n_p = L - \delta_p < L_2 = L \qquad (4)$$

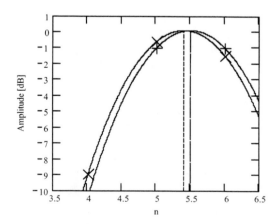

Fig. 1. Influence of input signal frequency change from 5.4 to 5.5 on DFT spectrums: the values of DFT points lie on the continuous curves of Blackman windowing function spectrums.

for $A(L_1) < A(L_2)$. In equations (3) and (4) δ_p is the real number within range 0 to 0.5, δ_p is called here frequency correction factor. L is the greatest point index, and $A(i)$ is amplitude of point with index i.

Let us define a coefficient α – a quotient of values of two greatest amplitude points:

$$\alpha = \begin{cases} \dfrac{A(L+1)}{A(L)} & for \quad A(L+1) > A(\{L-1) \\ \dfrac{A(L-1)}{A(L)} & for \quad A(L+1) < A(\{L-1) \end{cases} \tag{5}$$

The α depends on frequency correction factor δ_p. For chosen windowing functions that dependence is monotonic and unequivocal within range of δ_p that changes from –0.5 to 0.5, hence it could be used as true frequency measuring relation. It is also an even function therefore the considerate range could be narrowed to 0 to 0.5. In order to be useful measuring relation it should have a zero error frequency measurement for δ_p equal to 0 and 0.5. This step could be called normalisation:

$$\alpha_{norm}(\delta_p) = \frac{1}{2} \cdot \frac{\alpha(\delta_p) - \alpha(0)}{\alpha(0.5) - \alpha(0)} \tag{6}$$

where $\alpha(x)$ is a value of α coefficient for δ_p equal to x. Since the spectrum of all windowing functions is symmetrical, the coefficient $\alpha(0.5)$ is always equal to 1. Hence, only one coefficient, $\alpha(0)$, per windowing function has to be known.

Normalised measuring relation are non-linear for all windowing functions, therefore it causes frequency measurement errors (see Figure 2 and Table 1). The best solution of that problem is to find analytical solution. It was found only for one family of windowing functions: Blackman-Harris with maximum sidelobe decay (including very popular Hanning window) by Rife and Vincent (1970):

$$\delta_p = \frac{(M+1)\alpha - M}{\alpha + 1} \tag{7}$$

where M is order of windowing functions.

To improve precision of frequency measurement for other windowing functions a linearization of normalised measuring relation should be done. One of possible ways is to compensate an error using trigonometric series:

$$\alpha_{lin}(\delta_p) = \alpha_{norm}(\delta_p) - \sum_{i=1}^{P} e_i \cdot \sin(2 \cdot i \cdot \pi \cdot \delta_p) \tag{8}$$

where P is number of compensation elements and e_i

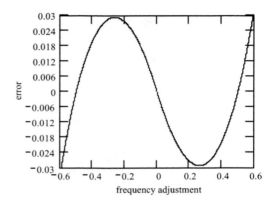

Fig. 2. A frequency recovery error vs. δ_p frequency correction factor for Blackman window, classical measuring relation.

Table 1 Maximum error of classical frequency measuring relation for some windows

Window	$\alpha_{norm}\ (\delta_p)$	$\alpha_{lin}\ (\delta_p)$
Hanning	0.036	7.78E-4
Blackman	0.029	5.45E-4
4 term min. sidelobe level	0.023	3.89E-4
4 term max. sidelobe decay	0.017	2.65E-4
Keiser-Bessel 75dB	0.026	4.62E-4
Dolph-Chebyshew 100dB	0.024	4.08E-4

coefficients are calculate from equation:

$$e_i = \frac{1}{0.5} \int_{-0.5}^{0.5} \alpha_{norm}(x) \cdot \sin(2 \cdot i \cdot \pi \cdot x) dx \tag{9}$$

The mutual independence of e_i coefficients and independence from P is the advantage of proposed linearization. The most right column in Table 1 presents the errors of measuring relation linearized with 3 compensation elements.

3.4 New frequency measuring relation.

Classical normalised measuring relation based on quotient of values of two greatest amplitude points is practically useless because of its significant non-linearity. It is used either within analytical solution for one class of windowing functions or with some kind of linearization for other windows. However it is possible to define different measuring relation that is much more linear and therefore could be useful even without linearization (see Figure 3).

Let us define new frequency measuring relation as follows:

$$\gamma(\delta_m) = \begin{cases} \dfrac{A(L)+2\cdot A(L+1)}{A(L)+A(L+1)} & for \quad A(L+1) > A(L-1) \\[2mm] \dfrac{A(L)+2\cdot A(L-1)}{A(L)+A(L-1)} & for \quad A(L+1) < A(L-1) \end{cases} \qquad (10)$$

As in the classical relation, the new frequency measuring relation still need normalisation:

$$\gamma_{norm}(\delta_p) = \frac{1}{2}\cdot\frac{\gamma(\delta_p)-\gamma(0)}{\gamma(0.5)-\gamma(0)} \qquad (11)$$

Again, since spectrum for all windowing functions is symmetrical, the coefficient $\gamma(0.5)$ is always equal to 1.5 and coefficient $\gamma(0)$ has to be known only.

It could be shown that the equations (10) and (11) leads to essentially the same solution as equation (7). Hence new normalised measuring relation for Blackman-Harris windows with maximum sidelobe decay are exact. For other windowing functions the same linearization method as proposed in equations (8) and (9) could be applied. The most right column in Table 2 presents the errors of measuring relation linearized with 3 compensation elements.

Table 2 Maximum error of new frequency measuring relation for some windows

Window	$\gamma_{norm}\ (\delta_p)$	$\gamma_{lin}\ (\delta_p)$
Hanning	2E-15	-
Blackman	1.01E-3	3.15E-5
4 term min. sidelobe level	9.5E-4	3.41E-5
4 term max. sidelobe decay	3E-15	-
Keiser-Bessel 75dB	1.04E-3	3.26E-5
Dolph-Chebyshew 100dB	1.03E-3	3.21E-5

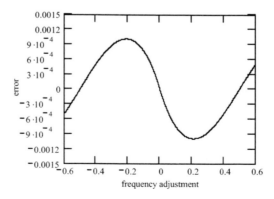

Fig. 3. A frequency recovery error vs. δ_p frequency correction factor for Blackman window, new measuring relation.

3.5 A procedure of using Interpolated DFT with any windowing function.

The results of using 6 different windowing function are presented in this paper, but other windowing function could be easily applied with proposed improved Interpolated DFT method. Windowing function should be symmetrical in frequency domain, should comprise a main lobe wider than 2 points and a second derivative within the range of main lobe must be monotonic. It seems that all known windowing function obey this limitation. Therefore proposed improvement could be used with any windows.

An algorithm of preparing for apply Interpolated DFT with some windowing function is as follows:
1. Analytically calculate a continuous spectrum function of windows.
2. Calculate values of spectrum for 0 and 0.5.
3. Use equation (10) to obtain coefficient $\gamma(0)$.
4. Optionally calculate from equation (9) compensation coefficients to obtain desired frequency measurement precision.
5. Use equation (11) and optionally (8) to achieve true frequency of spectral line

First and second steps could be done using a program for which coefficient $\gamma(0)$ is required. That procedure must be applied only once per window. The values of coefficient $\gamma(0)$ for some popular windows are presented in Table 3.

Table 3 Values of coefficient $\gamma(0)$ for some windows

Window	$\gamma(0)$
Hanning	1.3(333333)
Blackman	1.373134328
4 term min. Sidelobe level	1.402171675
4 term max. Sidelobe decay	1.4(285714)
Keiser-Bessel 75dB	1.388329223
Dolph-Chebyshew 100dB	1.398615444

4. AN INFLUENCE OF NOISE

It is impossible to acquire a signal free of noise. On the other hand Interpolated DFT method uses weak changes of amplitudes to obtain true frequency with great precision. Therefore it is obvious that maximum obtainable precision of measurement frequency via Interpolated DFT strongly depends on input signal SNR.

A number of simulations were done to estimate an influence of noise level on maximum frequency measurement precision. The results of the simulations are presented in Figures 4 and 5. Each point on graphs is the worst case of 100 simulation done for different frequencies with the same SNR. The

Hanning window was chosen because a new measuring relation is the best possible solution for this window.

It could be seen in Figure 5 that if SNR level is less than about 36dB, the maximum obtainable frequency measurement precision depends only on noise level. But even for signals with greater SNR an influence of noise level is significant, for example to recover true frequency with 10^{-4} precision input signal should has SNR greater than about 52dB. To achieve such SNR a noiseless signal of peak to peak value equal to 0.7 full conversion range has to be sampled by 9 bit ADC. For that reason the precision offered by new normalised measuring relation could be in many cases sufficient.

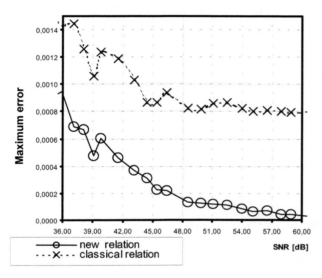

Fig. 4. Maximum error of frequency recovery vs SNR for Hanning window.

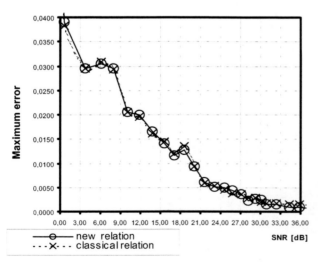

Fig. 5. An maximum error of frequency recovery vs. SNR for Hanning window.

5. CONCLUSIONS

Proposed improved and standardised method of Interpolated DFT has some advantages:
- is very simple, requires a knowledge of only one easily obtainable coefficient;
- for Blackman-Harris windows with maximum sidelobe decay is errorless;
- for other widowing functions offers sufficient precision without additional linearization for measuring signals with SNR less than 34dB.
- is applied in the same manner to all windows types.

REFERENCES

Andria G., Savino M., Trotta A. (1989)
Windows and Interpolation Algorithms to Improve Electrical Measurement Accuracy. *IEEE Transactions on Instrumentation and Measurement*, **Vol. 38**, pp.856-863

Grandke T.(1983),
Interpolation Algorithms for Discrete Fourier Transform of Weighted Signals. *IEEE Transactions on Instrumentation and Measurement* **Vol. 32**, pp. 350-355

Harris F.J. (1978),
On use of windows for harmonic analysis with the discrete Fourier transform. *Proceedings of the IEEE* **Vol. 66**, pp. 51-83

Jain V. H., Collins W. L., Davis D. C. (1979),
High-Accuracy Analog Measurement via Interpolated FFT. *IEEE Transactions on Instrumentation and Measurement*, **Vol. 28**, pp. 113-122

Lyons R.(1998),
Windowing Functions Improve FFT Results – Part I&II. *Test & Measurement World* June 1998, September 1998

Offelli C., Petri D.(1990),
Interpolation Techniques for Real-Time Multifrequency Waveform Analysis. *IEEE Transactions on Instrumentation and Measurement*, **Vol. 39**, pp. 106-111

Renders H., Schoukens J., Vilain G. (1984),
High-Accuracy Spectrum Analysis of Sampled Discrete Frequency Signals by Analytical Leakage Compensation. *IEEE Transactions on Instrumentation and Measurement*, **Vol. 33**, pp.287-292

Rife D.C., Vincent G.A .(1970),
Use of the Discrete Fourier Transform in the Measurement of Frequencies and Levels of Tones. *Bell System Technical Journal*, **Vol. 49**, pp. 197-228

IFAC

Publications

www.elsevier.com/locate/ifac

EFFICIENT ALGORITHM FOR IMPULSIVE NOISE REDUCTION

Bogdan Smolka [*,1] **Marek Szczepanski** [*,1]
Cristian Cantón-Ferrer [**]

** Department of Automatics,
Electronics and Computer Science
Silesian University of Technology
Akademicka 16 Str, 44-101 Gliwice, Poland
bsmolka@ia.polsl.gliwice.pl*

** Polytechnical University of Catalonia, Spain

Abstract: In this work a novel approach to the problem of impulsive noise reduction
for color and gray scale images is presented. The new image filtering technique
is based on the maximization of the similarities between pixels in the filtering
window. The new method removes the noise component, while adapting itself to
the local image structures. In this way, the proposed algorithm eliminates impulsive
noise, while preserving edges and fine image details. Since the algorithm can be
considered as a modification of the standard vector median filter driven by fuzzy
membership functions, it is fast, computationally efficient and very easy to implement.
Experimental results indicate that the new method is superior to the commonly used
algorithms for impulsive noise reduction. *Copyright © 2001 IFAC*

Keywords: biomedical systems, medical applications, filtering techniques, comet
assay, biomedical preprocessing, computer vision

1. BRIEF OVERVIEW OF STANDARD COLOR NOISE REDUCTION FILTERS

A number of nonlinear, multichannel filters, which
utilize correlation among multivariate vectors using various distance measures, have been proposed
to date [1-5]. The most popular nonlinear, multichannel filters are based on the ordering of vectors
in a predefined moving window. The output of
these filters is defined as the lowest ranked vector
according to a specific vector ordering technique.

Let $\mathbf{F}(x)$ represent a multichannel image and let
W be a window of finite size n (filter length).

The noisy image vectors inside the filtering window W are denoted as $\mathbf{F}_j$, $j = 0, 1, ..., n - 1$.
If the distance between two vectors $\mathbf{F}_i, \mathbf{F}_j$ is de-
noted as $\rho\{\mathbf{F}_i, \mathbf{F}_j\}$ then the scalar quantity $R_i = \sum_{j=0}^{n-1} \rho\{\mathbf{F}_i, \mathbf{F}_j\}$, is the total distance associated
with the noisy vector $\mathbf{F}_i$.

The ordering of the R_i's : $R_{(0)} \leq R_{(1)} \leq ... \leq R_{(n-1)}$, implies the same ordering to the corresponding vectors $\mathbf{F}_i$: $\mathbf{F}_{(0)} \leq \mathbf{F}_{(1)} \leq ... \leq \mathbf{F}_{(n-1)}$.
Nonlinear ranked type multichannel estimators
define the vector $\mathbf{F}_{(0)}$ as the filter output. However, the concept of input ordering, initially applied to scalar quantities is not easily extended to
multichannel data, since there is no universal way
to define ordering in vector spaces.

[1] This work has been partially supported by the KBN
grants 7 T11A 010 21, PBZ-KBN-040/P04/08 and NATO
Collaborative Linkage Grant LST.CLG.977845

To overcome this problem, distance functions are often utilized to order vectors. As an example, the *Vector Median Filter* (VMF) uses the L_1, L_2 norms to order vectors according to their relative magnitude differences [1, 3, 6].

The orientation difference between two vectors can also be used as their distance measure. This so-called *vector angle criterion* is used by the *Vector Directional Filters* (VDF) to remove vectors with atypical directions [4, 7].

The *Basic Vector Directional Filter* (BVDF) is a ranked-order, nonlinear filter which parallelizes the VMF operation. However, a distance criterion, different from the L_1, L_2 norms used in VMF is utilized to rank the input vectors. The output of the BVDF is that vector from the input set, which minimizes the sum of the angles with the other vectors. In other words, the BVDF chooses the vector most centrally located without considering the magnitudes of the input vectors.

To improve the efficiency of the directional filters, a new method called *Directional-Distance Filter* (DDF) was proposed [4]. This filter retains the structure of the BVDF but utilizes a new distance criterion to order the vectors inside the processing window.

Another efficient rank-ordered technique called Hybrid Directional Filter was presented in [8]. This filter operates on the direction and magnitude of the color vectors independently and then combines them to produce a unique final output.

All standard filters detect and replace well noisy pixels, but their property of preserving pixels which were not corrupted by the noise process is far from the ideal. In this paper we show the construction of a simple, efficient and fast filter which removes noisy pixels, but has the ability of preserving original image pixel values.

2. NEW FILTERING TECHNIQUE

Let us start from a gray scale image in order to better explain how the new algorithm is constructed. Let the gray scale image be represented by a matrix F of size $N_1 \times N_2$, $F = \{F(i,j) \in \{0,\dots,255\}, i = 1, 2, \dots N_1, j = 1, 2, \dots, N_2\}$.

Our construction starts with the introduction of the similarity function $\mu : [0; \infty) \to \mathbf{R}$. We will need the following assumptions for μ:

1. μ is decreasing in $[0; \infty)$,

2. μ is convex in $[0; \infty)$,

3. $\mu(0) = 1$, $\mu(\infty) = 0$.

In the construction of our filter, the central pixel in the window W is replaced by that one, which maximizes the sum of similarities between all

its neighbours. Our basic assumption is that a new pixel must be taken from the window W (introducing pixels which do not occur in the image is prohibited like in the VMF and BVDF).

For this purpose μ must be convex, which can be easily shown. For the gray scale images we define the following fuzzy measure of similarity between two pixels F_k and F_l [11] :

$$\rho\{F_k, F_l\} = \mu(|F_k - F_l|). \qquad (1)$$

Let us now assume that F_0 is the center pixel in the window W and the pixels $F_1, F_2, \dots, F_{n-1}$ are surrounding F_0, (Fig. 1).

The filter works as follows:
In the first step the total sum R_0 of the similarities between the central pixel F_0 (suspected to be noisy) and its neighbours $F_i, i = 1, \dots, n$ is calculated. In the second step each of the neighbours of the central pixel F_0 is moved to the center of the filtering window and the central pixel is removed from W. For each pixel F_i of the neighbourhood, which is being placed in the center of W, the total sum of similarities R_i is calculated and then compared with R_0. It has to be stressed that in the second step the total sum of similarities is calculated without taking into account the original central pixel, which is rejected from the filter window.

In this way, the central pixel F_0 is replaced by that F_i from the neighbourhood, for which the total similarity function R_i, which is a sum of all values of similarities between the central pixel and its neighbours reaches its maximum. In other words if for some i

$$R_i = \sum_{j=1}^{n-1} (1 - \delta_{i,j}) \, \rho\{F_i, F_j\}, i = 1, \cdots, n-1, (2)$$

is larger than

$$R_0 = \sum_{j=1}^{n-1} \rho\{F_0, F_j\}. \qquad (3)$$

then the center pixel is replaced by F_i.

Generally the pixel F_0 is given the value F_{i_*} where $i_* = \arg\max_i R_i$

$$R_i = \delta_{i,0} \sum_{j=1}^{n-1} \rho\{F_i, F_j\} + (1 - \delta_{i,0}) \sum_{j=1}^{n-1} (1 - \delta_{i,j}) \, \rho\{F_i, F_j\} (4)$$

This approach can be in a easily applied to color images. In this case, we use the similarity function defined by $\rho\{\mathbf{F}_k, \mathbf{F}_l\} = \mu(||\mathbf{F}_k - \mathbf{F}_l||)$, where $||\cdot||$ denotes the specific vector norm.

Now in exactly the same way we maximize the total similarity function R for the vector case.

In finding the maximum in (4), we obtain $(n - 1)$ nonzero components in R_0. If we replace the central pixel by one of its neighbourhood (by F_2 in Fig.1 a), then we obtain only $(n - 2)$ nonzero components in R, as the pixel which has been put into the center disappears from the filter window (Fig. 1 b). In this way the filter replaces the central pixel only when it is really noisy and preserves the image structures.

The *BASIC* code, which can be used for the fast computer implementation (L_1 vector norm) is presented in the APPENDIX.

3. RESULTS

The performance of the new algorithm was compared with the standard procedures of noise reduction used in color image processing.

The color image *LENA* has been contaminated by 4% of impulsive "salt & pepper" noise added independently to each RGB channel.

The root of the mean squared error (RMSE), peak signal to noise ratio (PSNR), normalized mean square error (NMSE) have been used as quantitative measures of quality for evaluation purposes.

We investigated the behaviour of the proposed filter using various convex functions in order to compare the new approach with the standard filters presented in Tab. 1, and obtained the best results when applying the following :

$$\mu_1(x) = e^{-\beta_1 x} , \qquad \beta_1 \in (0; \infty), \qquad (5)$$

$$\mu_2(x) = \frac{1}{1 + \beta_2 x} , \qquad \beta_2 \in (0; \infty), \qquad (6)$$

$$\mu_3(x) = \frac{1}{(1 + x)^{\beta_3}} , \qquad \beta_3 \in (0; \infty), \qquad (7)$$

$$\mu_4(x) = 1 - \frac{2}{\pi} \arctan(\beta_4 x) , \quad \beta_4 \in (0; \infty), \qquad (8)$$

$$\mu_5(x) = \frac{2}{1 + e^{\beta_5 x}} , \qquad \beta_5 \in (0; \infty), \qquad (9)$$

$$\mu_6(x) = \frac{1}{1 + x^{\beta_6}} , \qquad \beta_6 \in (0; 1), \qquad (10)$$

$$\mu_7(x) = \begin{cases} 1 - \beta_7 x & \text{if } x < 1/\beta_7, \\ 0 & \text{if } x \geq 1/\beta_7, \end{cases} \quad \beta_7 \in (0; \infty). \quad (11)$$

There are no special reasons to choose exactly these forms of the similarity function. One can easily find other μ fuctions, which meet the required conditions and also yield good filter results. We expect that there exists something like an "optimal shape" of the similarity function, but it depends in extremely complicated way on the statistical properties of the noise and the image structure.

Table 2 gives the best values of parameters β_i for functions μ_i and test images distorted by impulsive "salt & pepper" noise up to 10 % on each RGB channel. Figure 6 shows the graphs of these functions. According to the results depicted in Tab. 3 and extensive simulations with other noise intensities and color test images we suppose that an optimal shape of the similarity function is somewhere between μ_5 and μ_7.

Obviously, the presented functions $\mu_1, \ldots, \mu_7$ are rather plain and it is easy to propose procedures which can give us function closer to the optimal one. We did not do it for two reasons. Firstly, more complicated form of the similarity function makes the filter significantly slower. Secondly, we do not think that significant improvement of the filter efficiency is possible.

Table 3 summarizes the results obtained for the test image LENA distorted by 4 % impulsive noise. We have used the L_2 norm and the values of β_i from Tab. 2 in order to obtain results shown in Tab. 3. All proposed functions μ give very good results, although especially worth attention are μ_1, μ_5, μ_7. Table 4 shows *RMSE* values obtained using the proposed filter for four different norms. As can be seen, the best choice is as expected L_2.

The efficiency of the new filtering technique as compared with the vector median and other related filters is shown in Fig. 3.

Figure 4 depicts the result of noise reduction using the new method applied to a gray scale image *LENA* in comparison with the standard median filter. The test image was contaminated by 4% 'salt & pepper' noise and a 3×3 filtering mask was used. As can be seen the new class of filters eliminates efficiently impulsive noise, while preserving important image structures like edges, corners, lines and fine texture.

Another interesting property of the presented method of noise attenuation is shown in Fig. 5. Iterating the filtration process improves the image quality, whis is not the case when using the standard VMF. Additionally the output image converges much faster to its root than the VMF, when we repeat the filtration process.

4. CONCLUSIONS

In this letter, a new class of filters has been presented. Experimental results included in this letter, indicate that the new method of noise reduction significantly outperforms standard procedures used to restore gray scale and color images contaminated with impulsive noise. The new technique is fast and very easy to implement. The BASIC code is given in the APPENDIX, so that the filter can be easily evaluated by the image processing community.

5. REFERENCES

[1] I. Pitas, A. N. Venetsanopoulos, 'Nonlinear Digital Filters : Principles and Applications', Kluwer Academic Publishers, Boston, MA, (1990)

[2] K.N. Plataniotis, A.N. Venetsanopoulos, 'Color Image Processing and Applications', Springer Verlag, (June 2000)

[3] I. Pitas, P. Tsakalides, Multivariate ordering in color image processing, IEEE Trans. on Circuits and Systems for Video Technology, 1, 3, 247-256, (1991)

[4] D.G. Karakos, P.E. Trahanias, Generalized multichannel image filtering structures, IEEE Trans. on Image Processing, 6, 7, 879-881, (1996)

[5] I. Pitas, A.N. Venetsanopoulos, Order statistics in digital image processing, Proc. of IEEE, 80, 12, 1893-1923, (1992)

[6] J. Astola, P. Haavisto, Y. Neuovo, Vector median filters, IEEE Proc., 78, 678-689, (1990)

[7] P.E. Trahanias, A.N. Venetsanopoulos, Vector directional filters: A new class of multichannel image processing filters, IEEE Trans. on Image Processing, 2, 4, 528-534, (1993)

[8] M. Gabbouj, F.A. Cheickh, Vector Median-Vector Directional hybrid filter for color image restoration, Proc. of EUSIPCO-96, 879-881, (1996)

[9] K.N. Plataniotis, D. Androutsos, A.N. Venetsanopoulos, Color image processing using adaptive vector directional filters, IEEE Trans. on Circuits and Systems-II: Analog and Digital Signal Processing, 45, 10, 1414-1419, (1998)

[10] K.N. Plataniotis, D. Androutsos, V. Sri, A.N. Venetsanopoulos, A nearest neighbour multichannel filter, Electronics Letters, 1910-1911, (1995)

[11] B. Smolka, K. Wojciechowski, Random walk approach to image enhancement, Signal Processing, Vol. 81, 465-482, (2001)

[12] B. Smolka, K.N. Plataniotis, A. Chydzinski, A. N. Venetsanopoulos, K. W. Wojciechowski, On the Reduction of Impulsive Noise in Multichannel Image Processing, Optical Engineering, Vol. 40, No. 6, 902-908, (2001).

6. APPENDIX

```
BASIC CODE OF THE NEW ALGORITHM
'cr(N1,N2), cg(N1,N2), cb(N1,N2) - input color image,
'wr(N1,N2), wg(N1,N2), wb(N1,N2) - output color image
'beta - similarity function coefficient
'sim - total similarity between pixels in 3x3 window
For i=0 To 255
For j=0 To 255
expo(i,j)=Exp(-beta*Abs(i-j))
Next
Next
For i=2 To N1-1
For j=2 To N2-1
max=-1
For g=-1 To 1
For h=-1 To 1
w=i+g
z=j+h
sim=0
For r=-1 To 1
For s=-1 To 1
x=i+r
y=j+s
If Not w=x Or Not z=y Then
If Not r=0 Or Not s=0 Then
simr=expo(cr(x,y),cr(w,z))
simg=expo(cg(x,y),cg(w,z))
simb=expo(cb(x,y),cb(w,z))
sim=sim+simr+simg+simb
End If
End If
Next
Next
If sim>max Then
max=sim
pixr=cr(w,z)
pixg=cg(w,z)
pixb=cb(w,z)
End If
Next
Next
wr(i,j)=pixr
wg(i,j)=pixg
wb(i,j)=pixb
Next
Next
```

Table 1. Filters compared

Notation	Filter	Ref.
AMF	Arithmetic Mean Filter	[1]
VMF	Vector Median Filter	[6]
ANNF	Adaptive Nearest Neighbor Filter	[10]
BVDF	Basic Vector Directional Filter	[7]
HDF	Hybrid Directional Filter	[8]
AHDF	Adaptive Hybrid Directional Filter	[8]
DDF	Directional-Distance Filter	[4]
FVDF	Fuzzy Vector Directional Filter	[9]

Table 2. Optimal values of constans β_i $[10^{-3}]$.

β_1	β_2	β_3	β_4	β_5	β_6	β_7
5,04	6,62	192	6,97	7,90	266	3,72

Table 3. Comparison of the new filter with the standard techniques (*LENA* color image contaminated with 4% "salt & pepper" noise added independently to each RGB channel).

METHOD	NMSE $[10^{-4}]$	RMSE	PSNR [dB]
NONE	514,95	32,165	17,983
AMF	82,863	12,903	25,917
VMF	23,304	6,842	31,427
ANNF	31,271	7,926	30,149
BVDF	29,074	7,643	30,466
HDF	22,845	6,775	31,513
AHDF	22,603	6,739	31,559
DDF	24,003	6,944	31,288
FVDF	26,755	7,331	30,827
PROPOSED			
$\mu_1(x)$	**4,959**	**3,157**	**38,145**
$\mu_2(x)$	5,398	3,294	37,776
$\mu_3(x)$	9,574	4,387	35,288
$\mu_4(x)$	5,064	3,190	38,054
$\mu_5(x)$	**4,777**	**3,099**	**38,307**
$\mu_6(x)$	11,024	4,707	34,675
$\mu_7(x)$	**4,693**	**3,072**	**38,384**

Fig. 2. Similarity functions $\mu_1, \ldots, \mu_7$.

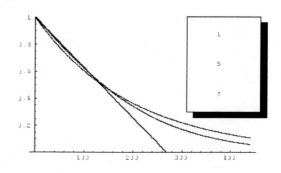

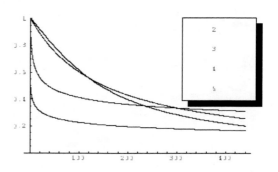

Table 4. Comparison of the new filter results (RMSE) using different norms (*LENA*).

	L_1	L_2	L_3	L_∞
$\beta_1(x)$	3,615	**3,157**	3,172	3,462
$\beta_5(x)$	3,579	**3,099**	3,167	3,694
$\beta_7(x)$	3,838	**3,072**	3,138	3,752

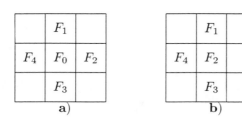

a)   b)

Fig. 1. Illustration of the construction of the new filtering technique for the 4-neighbourhood case. If the center pixel F_0 is replaced by its neighbour F_2, then the similarity measure $R_2 = \rho\{F_2, F_1\} + \rho\{F_2, F_3\} + \rho\{F_2, F_4\}$ between F_2 (new center pixel) is calculated. If the total similarity R_2 is greater than $R_0 = \rho\{F_0, F_1\} + \rho\{F_0, F_2\} + \rho\{F_0, F_3\} + \rho\{F_0, F_4\}$ then the center pixel is replaced, otherwise it is retained.

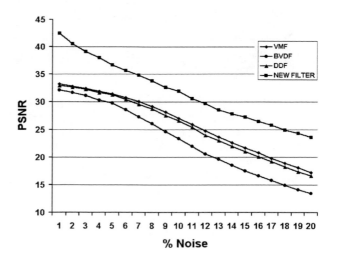

Fig. 3. Dependence of the noise reduction efficiency on the percentage of impulsive noise for the new method, VMF, BVDF and DDF, (*LENA* colour image, $\beta_1 = 5.04 \cdot 10^{-3}$).

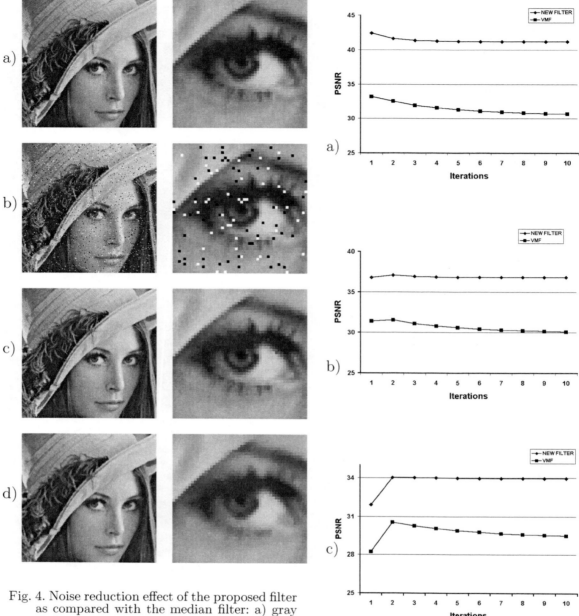

Fig. 4. Noise reduction effect of the proposed filter as compared with the median filter: a) gray scale test image *LENA*, b) image distorted by 4% 'salt & pepper' noise, c) filtered with the new method $\beta_1 = 5.04 \cdot 10^{-3}$ (PSNR=42.02), d) median filter (PSNR=34.08). To the right zoomed image portions.

Fig. 5. Dependance of the noise reduction efficiency of the proposed filter and VMF on the number of iterations for colour test image distorted by: a) 1% impulsive noise b) 5% impulsive noise c) 10% impulsive noise. (*LENA* colour image, $\beta_1 = 5.04 \cdot 10^{-3}$).

IFAC
Publications
www.elsevier.com/locate/ifac

DECOMPOSITION OF BOOLEAN FUNCTIONS IN MORPHOLOGICAL PRE-PROCESSING OF BINARY IMAGES

Urszula Stańczyk * **Konrad Wojciechowski** **
Krzysztof Cyran *

* *Institute of Computer Science*
** *Institute of Automatic Control*
Silesian University of Technology,
Gliwice, Poland

Abstract: Transformations that operate on a finite window of a binary image can be expressed as logical functions, where the central pixel and its neighbourhood (usually defined in terms of 4-, 6- or 8-connectivity) act as variables of a function. When the size of the window increases so does the processing time and the amount of data to be stored. The paper proposes to decompose such logical functions and then use sub-functions to obtain the result of the operation. This approach can simplify the process and shorten the time needed for the transformation. *Copyright © 2001 IFAC*

Keywords: switching functions, image processing, decomposition

1. INTRODUCTION

Mathematical morphology is a branch of digital image analysis and processing with origins in the sixties of the XXieth century that bases on shape and structure of objects within an image. The theoretical basis of it was provided by two researches at the Paris School of Mines in Fontainebleau, George Matheron and Jean Serra, working at the time on problems in petrography and mineralogy. Due to their fundamental work and later developments morphology has gained the status of a powerful tool in image processing with numerous applications in pattern recognition, medical imaging, microscopic imaging, computer vision and other areas.

Mathematical morphology uses concepts both from algebra and geometry. The central idea is to examine the structure of an image by matching it with relatively small patterns at various locations in the image. These patterns are called *structuring elements* and by varying their size and shape one can obtain useful information about the structure and shape of the image objects and relations among them. Such approach allows to define a transformation by declaring its type, (choosing from the set of predefined morphological operations, such as basic dilation, erosion or more complex opening, closing or Hit-or-Miss transforms) then giving the structuring element as a single shape geometrical parameter of the transformation.

As with all image processing and analysis operations, when performing a morphological transformation the time factor is crucial. The size and shape of a structuring element greatly affect the overall processing time, so from early on the main efforts to optimise (understood in terms of reducing time) went to the decomposition of the structuring element. Due to algebraic properties, the chain rule for both dilation and erosion (Haralick *et al.*, 1987) applied to a structuring element allow to decompose (Zhuang and Haralick, 1986; Park and Chin, 1994) single stage operation with the structuring element into a superposition of consecutive transformations with smaller elements, which saves time, and can also be implemented using pipeline architecture hardware that operates on the neighbourhood of a pixel.

Some examples of transformations to be performed by the pipeline processor may be found (Pratt, 1991) to execute shrinking, thinning, skeletonizing and thickening of binary image. The operations by their given definition cannot be performed in the single stage of processing with 3x3 window, which corresponds to the closest neighbourhood of a pixel. Such window does not provide enough information, which would result in loss of connectivity and total erasure of some parts of the image. Instead, the algorithm has two stages - first the conditional marking is performed, then, depending on a local value of an erasure inhibiting logical variable, the image is processed to obtain the output value. The difference of this attitude when compared with decomposition of a structuring element is that Pratt does not define these operations by any element, using rather the set of masks causing "hits", which is another way of describing a transformation.

Given the set of "hits" for an operation, one can consider yet another approach to defining and performing a morphological transformation, that is expressing it as a logical function. Since (Heijmans, 1995) every finite window operator can be written as a Boolean function depending on surrounding pixels, hits are easily translated as "on-set" for the function and as such can be used to obtain logical expression either directly in the Sum of Products (SoP) canonical normal form, or through minimisation process for some reduced or minimal form.

Treating a morphological transformation as a logical function offers another way of dealing with the problem of optimisation. Instead of a structuring element decomposition one may consider decomposition of Boolean function that performs image transformation and such approach is addressed in this paper, including some discussion of advantages and disadvantages for possible software and hardware implementations.

2. BASIC CONCEPTS OF MATHEMATICAL MORPHOLOGY

Morphology treats real images as sets of pixels in a space of some dimension, for example N-dimensional Euclidean space E^N. In the context of computer vision and graphics there is used a discrete or digitised equivalent of an Euclidean space - the set of N-tuples of integers that give coordinates of a pixel and its "value". Sets in Euclidean 2-space denote foreground regions in binary imagery.

2.1 Morphological transformations

There are two elementary morphological transformations: *dilation* and its dual - *erosion*. These fundamental operations are rarely used as single transformations, but they give basis to define some more complex ones such as opening, closing or Hit-or-Miss.

Dilation combines an image and a structuring element using vector addition of sets elements. For $A, S \subseteq E^N$, the dilation of A by S, denoted by $A \oplus S$, is defined as: $A \oplus S = \{x \in E^N \,|x = a + s$ for some $a \in A$ and $s \in S\}$.

Erosion uses the vector subtraction of set elements to combine an image and a structuring element. For $A, S \subseteq E^N$, the erosion of A by S, denoted by $A \ominus S$, is defined as: $A \ominus S = \{x \in E^N \,|x + s \in A$ for every $s \in S\}$.

As already stated, dilation and erosion are dual operations and in turn they are used to define another pair of duals: opening and closing transformations. Opening is erosion followed by dilation. For $A, S \subseteq E^N$, the opening of A by S, denoted by $A \circ S$, is defined as: $A \circ S = (A \ominus S) \oplus S$.

Closing is dilation followed by erosion. For $A, S \subseteq E^N$, the closing of A by S, denoted by $A \bullet S$, is defined as: $A \bullet S = (A \oplus S) \ominus S$.

Opening and closing transformations are idempotent - i.e. iterative application of an operation does not change the resulting image.

Another operator that is defined by erosion and dilation is Hit-or-Miss. Let $A, B, C \subseteq E^N$, where A is an image and B and C are two such structuring elements that $B \bigcap C = \emptyset$. The Hit-or-Miss operator, denoted by $A \otimes (B, C)$, is defined as: $A \otimes (B, C) = \{x \in E^N \,|B_x \subseteq A$ and $C_x \subseteq A^C\}$, where C_x denotes translation and A^C complementation.

Dilation and erosion can also be expressed as respectively union and intersection of original image translations. Their vectors are given by the definition of the structuring element used. This fact affects properties and gives different interpretation of operations, emphasising the role of image shifting in implementation, as it may be effectively performed by some pipeline processor.

2.2 Role of a structuring element

Structuring element, as a parameter of a morphological transformation, plays very important role in image processing and analysis. Its shape and size determine the processing time and computational complexity. Typical structuring elements are of some isotropic shape - that is causing the same behaviour in all directions - a square, a circle, other commonly used are a rectangle or an ellipse. The local origin usually is placed at the centre.

Dilation with 3x3 square structuring element results in growing image objects uniformly in space by a single pixel-width, while erosion with the same structuring element shrinks objects. Unconditional dilation and erosion do not prevent total erasure or loss of connectivity, nor do they check whether unconnected objects become connected or not. However, such additional conditions can be implemented using different definitions of transformations, applying the concept of Hit-or-Miss operations.

When processing objects without holes with dilation, erosion, opening and closing, the results are seen mainly on the outline of an object. In case of holes, everything depends on their size - if they are smaller than the structuring element, they may disappear completely or become even smaller (for dilation, closing) or they grow (for erosion, opening).

3. DECOMPOSITION IN MATHEMATICAL MORPHOLOGY

Given the same general conditions, the processing time of a morphological transformation depends on three factors: its structuring element, the type of the operation and the image to be processed. For general consideration this last element may be disregarded - studying and comparing different implementations of a transform can only be done providing the same class of images. That leaves the structuring element - how it is defined, how many pixels it consists of, and the type of the operation. When an image is subjected to a transformation, it is scanned over with the structuring element as a kind of mask for possible matches. It means that when generating the output value for the central pixel, each single pixel belonging to the element, or rather the point of the image corresponding to it, is considered. As a result of it, the size of the structuring element greatly affects the processing time for the single pixel and in turn the overall time of the transformation. That is why the concept of decomposition of a structuring element is so important.

The idea of decomposition bases on the property of dilation, which is often referred to as *chain rule* and can be expressed as follows:

$$A \oplus (B \oplus C) = (A \oplus B) \oplus C \qquad (1)$$

The form $(A \oplus B) \oplus C$ allows a considerable savings in the number of operations to be performed. While dilating A by D, the number of operations to be performed depends on the number of pixels the structuring element consists of. As D is

B dilated by C, this number of pixels may be equal to the result of multiplication the number of B pixels by the number of C pixels. When first dilating A by B, then by C the number of operations corresponds to the sum of the numbers of points B and C consist of, which is a reduction in computational complexity. In the first case we may have exponensial complexity in as many as N^2 operations to be performed, while in the later only linear $2N$.

With respect to structuring element decomposition, a chain rule for erosion holds when the structuring element is decomposable through dilation:

$$(A \ominus B) \ominus C = A \ominus (B \oplus C) \qquad (2)$$

This relation is as important as the chain rule for dilation, because it permits a large erosion to be computed by two successive erosions.

The structuring element can also be decomposed by using set union in the following way:

$$A \oplus (B \bigcup C) = (A \oplus B) \bigcup (A \oplus C) \qquad (3)$$

$$A \ominus (B \bigcup C) = (A \ominus B) \bigcap (A \ominus C) \qquad (4)$$

This equalities are significant, as they allow for a further decomposition of a structuring element into a union of structuring elements.

The idea of implementing a morphological transformation as a sequence of iterative operations on a pixel neighbourhood is not necessarily the most efficient or universal approach. For example, not all structuring elements can be decomposed into iterative neighbourhood dilations. What is more, such implementation may be not particularly efficient in terms of processing time or hardware requirements.

The difference between structuring element decomposition by dilation and by union reflects itself in the efficiency of computing the dilations. Take under consideration the structuring element consisting of 16 points that form a square 4x4. It is possible to decompose this element into the union of 16 structuring elements, each consisting of a single pixel which is suitably displaced from the origin. Dilation by such elementary structuring element is simply a shift of the original image and requires 15 shifts and 15 unions. By contrast, the decomposition of structuring element into the four elementary structuring elements permits dilation through the chain rule. In that case, only four shifts and four unions are required. Computationally, the difference involves a shift and union of the previously computed result when applying the chain rule, whereas decomposition by union independently accumulates the individual shifts of the original image.

4. BOOLEAN FUNCTIONS IN IMAGE PROCESSING

The transformations described above operate on the finite window of an image, which means that they can be expressed as logical functions (Heijmans, 1995). Let $A, S \subseteq E^N$ where A is an image and S a structuring element. Morphological operator ϕ_b is defined by: $\phi_b(A) = \{x \in E^N \,|\, \phi_b(A(x + s_1), \ldots, A(x + s_n)) = 1\}$

With such definition it is possible to write morphological transformations for some exemplary structuring element S consisting of n points in the following way, where $x \cdot y$ stands for x AND y (operation of logical product), $x + y$ stands for x OR y (operation of logical sum) and $\overline{x}$ stands for NOT x (operation of logical complementation):

- Dilation
  $$f_b(a_1, a_2, \ldots, a_n) = a_1 + a_2 + \ldots + a_n$$
- Erosion
  $$f_b(a_1, a_2, \ldots, a_n) = a_1 \cdot a_2 \cdot \ldots \cdot a_n$$
- Hit-or-Miss transformation
  $$f_b(a_1, a_2, \ldots, a_n) = $$
  $$a_1 \cdot a_2 \cdot \ldots \cdot a_m \cdot \overline{a_{m+1}} \cdot \overline{a_{m+2}} \cdot \ldots \cdot \overline{a_n}$$

Hit-or-Miss operations can be described not only by structuring elements but also by directly given set of masks that cause "hits". This form actually even makes it easier to express them as logical functions, as a set of masks corresponds to the "on-set" for the function. Such a function is by its definition fully specified. Next it is enough to minimise the function in order to make the processing faster and the task is completed. However, the operators may be given in the natural language as well. For some of them finding the function is fairly easy, for some it requires advanced heuristic algorithms to obtain the formula.

Taking under consideration 8-connected neighbourhood of a pixel, in which the central pixel is marked as x - the local origin, and $x_0, x_1, x_2, x_3, \ldots, x_7$ are the neighbouring points, numerated in an anti-clockwise fashion, starting from three o'clock, other examples of additive and subtractive operators are as follows:

- Interior fill - crates a black pixel if all 4-connected neighbourhood pixels are black.
  $$f_b(x_0, x_1, \ldots, x_7) = x + x_0 \cdot x_2 \cdot x_4 \cdot x_6$$
- Diagonal fill - crates a black pixel if such creation eliminates 8-connectivity of the background.
  $$f_b(x_0, x_1, \ldots, x_7) = x +$$
  $$(x_0 \cdot \overline{x_1} \cdot x_2 + x_2 \cdot \overline{x_3} \cdot x_4 + x_4 \cdot \overline{x_5} \cdot x_6 + x_6 \cdot \overline{x_7} \cdot x_0)$$
- Isolated pixel remove - erases a black pixel with eight white neighbours.
  $$f_b(x_0, x_1, \ldots, x_7) =$$
  $$x \cdot (x_0 + x_1 + x_2 + x_3 + x_4 + x_5 + x_6 + x_7)$$

- Interior pixel remove - erases a black pixel if all 4-connectes neighbours are black.
  $$f_b(x_0, x_1, \ldots, x_7) = x \cdot (\overline{x_0} + \overline{x_2} + \overline{x_4} + \overline{x_6})$$

As an example of a transformation that is given by description in the natural language one can consider widely known and used median operators. Such operator should set the output to the value of the majority of neighbouring pixels. For the exemplary window of 3 by 3 the number of pixels is nine - so the output value should be equal to the state of five (or more) surrounding points. The Boolean function that performs such operation on a central pixel and its neighbourhood, written in the Sum of Products minimal form would be equal to summed products of all possible combinations of fives of function variables. Due to Boolean algebra theorems, these terms cover the rest of terms with all possible combinations of 6, 7, 8 and 9 variables of the function. There are 126 combinations of fives of variables, so the function would have that number of terms.

More complex transformations are rarely trivial to express. Take for example shrinking of a binary image (Pratt, 1991). It should erase black pixels in such a way that an object without holes erodes to a single point at or near its centre and objects with holes erode to connected rings laying midway between each hole and its nearest outer boundary. This kind of transformation cannot be performed using single-stage 3x3 Hit-or-Miss operation, as this window does not provide sufficient information to prevent total erasure or loss of connectivity. The window 5 by 5 would be enough, but such an approach would result in increased computational complexity as there would be 2^{25} possible patterns to be examined. This can be helped by dividing the process into stages. At first there is performed a conditional marking of pixels for possible erasure, then the pixel and its marked neighbourhood are examined to determine which points can be unconditionally erased without total erasure or loss of connectivity.

5. BOOLEAN FUNCTION DECOMPOSITION

Logical functions are omnipresent in various areas of computer science such as logic synthesis (Luba, 2000), artificial intelligence, computer vision, information systems, machine learning. Complexity of tasks that are solved when using Boolean functions usually require them to be of many variables, which in turn affect their effectiveness. That is one of the reasons for popularity of the concept of logical function decomposition. Its object is to divide a function into a set of subfunctions, which in terms of logic circuit can be imagined as breaking some given circuit into a number of smaller interacting components.

Decomposability is such a property of a logic function F that expresses ability of describing the function as:

$$F(X) = H(A, G(B)) \qquad (5)$$

Here $A \bigcup B = X$ and $n_A + n_B \leq n$. Each of the functions G and H is less complex than F as they depend on fewer variables.

Over the years, several algorithms for functional decomposition have been developed, using among others decomposition charts, partition-based representation of functions, graph colouring heuristics. These strategies can be divided into two distinct groups: parallel and serial decomposition.

Parallel decomposition partinions the set of outputs (the problem is in general addressed to multiple-output functions) into two disjoint subsets in such a way that the input sets of those two subsets are smaller that the set of the original inputs. Serial decomposition is defined by the following. Let $X = A \bigcup B$ be the set of input variables, $C \subseteq A$ and Y the set of output variables. A serial decomposition of a Boolean function F exists if there is $F = H(A, G(B, C) = H(A, Z)$, where $G(B, C) = Z$ and $H(A, Z) = Y$. If $C = \emptyset$ then decomposition is called disjoined.

Due to extensive popularity in modern logic synthesis such elements as FPGA and PLD numerous techniques have been invented (Luba *et al.*, 1992) with the main view of meeting the requirements of implementation and with the goal of obtaining more optimal solutions. Particularly promising is a balanced multilevel decomposition, which combines the approaches of both serial and parallel decomposition. This method has been implemented as an academic tool - programme called DEMAIN (Rawski *et al.*, 1998). In algorithm of the multilevel synthesis serial decomposition is applied recursively to the results from the previous step with assist of parallel decomposition. The process stops when some predetermined conditions are satisfied - for example all obtained sub-functions depend on a certain number of variables. The efficiency of this approach is strongly influenced by the number of values in sub-functions (Rawski *et al.*, 1999).

Originally functional decomposition was not intended for multiple-value Boolean functions, but it has been extended to cover that area as well, when more and more circuits have appeared with included decoders as their internal parts, preventing the necessary use of additional elements.

6. IMAGE TRANSFORMATION DECOMPOSITION

When decomposing image transformation treated as logical function the goals to achieve are: re-

duced processing time and/or decreased amount of data to be stored, finally the relative simplicity for either software or hardware implementation resulting from the former.

Disadvantages may be found into following considerations. First of all, by definition all functions to be decomposed are fully specified and as such prove to be extremely difficult to optimally decompose. Furthermore, as the problem of decomposition is addressed at all only for functions operating on the window bigger than the closest neighbourhood of a pixel, the number of variables and their possible combinations increases exponentially. Even for the window 5x5 that is not relatively very big, there are 25 variables and 2^{25} combinations, which in turn means that for such window as many as $2^{2^{25}}$ could be defined. To operate on one of those functions it is necessary to store all information about it choosing one of possible representations.

The information about Boolean function can be presented in the form of a truth table, but such solution requires to store all values of the functions, ones and zeros alike. Assuming certain weights for all variables and following some order with function values it is then unnecessary to "remember" the actual combination of variables. Their values can be re-acquired from the position in the ordered set of the function values. With one bit per each value it gives the amount of data to be stored equal to the total number of function values divided by 8 to express it in bytes. But each time to check the value of the function it is necessary to access some memory where this information is stored first computing the index to it by proper interpretation of surrounding points.

Since most functions have either many more zeros than ones or more ones than zeros, it is possible to store just on-set or off-set. That cuts the number of values significantly. In such a case, however, for each value the combination of variables have to be stored as well. What is more, even with some predefined ordering it is not possible to obtain some easy way to index such a set, which in turn causes some time-consuming searching.

Instead of keeping the values of a function, its logical expression may be used. Yet not all functions are effectively minimised. If the formula is complex then the processing time of evaluating the output value increases significantly. And it is first necessary to minimise it, which is not that simple with all minimising algorithms design rather for weakly specified functions (with a large number of don't-care conditions).

Minimisation is a very serious problem when dealing with logical functions. From well-known techniques Karnaugh maps are useful only for few variables, MacClusky or Kazakov method, or even ESPRESO exploit don't-care conditions and computational complexity for fully specified function of many variables is not very encouraging.

There exists yet another approach - decomposition of function using the multiple-value subfunctions first to reduce the amount of data to be stored, and by doing it also to decrease time. The proposed algorithm in the pessimistic case has the same parameters as the processing without decomposition, on average bringing significant gains. In general it can be described by the following stages:

- First, the size of a processing window is established along with the description of an operation to be performed.
- Next, dividing this window in rows and columns, the possible combinations for them are listed with assigning to them some value of the sub-function. The number of such values should be the smallest possible, as it determines the processing time for each subfunction.
- Obtaining the formal representation for subfunctions and the original function depending on sub-functions.
- First stage of actual processing - gathering auxiliary information either horizontally or vertically scanning through an image.
- Second, final stage of transformation - using the date from the first stage producing the output value for pixels.

Of course, the above heuristics should be treated as a very simple example of more general approach, which has been implemented as software to obtain experimental results. Apart from possible application to already given functions it is also possible to use such algorithm to obtain the formal representation for more complex functions.

7. CONCLUSION

In this paper a new attitude to the problem of decomposition in mathematical morphology is proposed. This concept bases on first expressing a transformation as logical function and then decomposing it, instead of decomposing a structuring element, as required by traditional approaches. Apart from gains in the overall processing time and stored data, such algorithm simplifies the process of obtaining function representation when some operation is given only in normal language and operates on the window bigger that the closest neighbourhood of a pixel. Especially multiple-value Boolean sub-functions prove to be useful, as they store more information depending on the fewer variables.

8. REFERENCES

Haralick, Robert M., Stanley R. Sternberg and Xinhua Zhuang (1987). Image analysis using mathematical morphology. *IEEE Transactions on Pattern Analysis and Machine Intelligence.*

Heijmans, Henk J.A.M. (1995). Mathematical morphology: a modern approach in image processing based on algebra and geometry. *SIAM Review.*

Luba, Tadeusz (2000). *Synteza ukadów logicznych.* Wydawnictwo Wyzszej Szkoly Informatyki Stosowanej i Zarzadzania. Warszawa.

Luba, Tadeusz, Maciej Markowski and Bogdan Zbierzchowski (1992). Logic decomposition for Programmable Gate Arrays. *Proceedings of Euro ASIC.*

Park, Hochong and Roland T. Chin (1994). Optimal decomposition of convex morphological structuring elements for 4-connected parallel array processor. *IEEE Transactions on Pattern Analysis and Machine Intelligence.*

Pratt, William K. (1991). *Digital image processing.* John Wiley and Sons, Inc.. NY.

Rawski, Mariusz, Lech Jóźwiak and Tadeusz Luba (1999). The influence of the number of values in sub-functions on the effectiveness and efficiency of the functional decomposition. *Proceedings of the 25th EUROMICRO Conference.*

Rawski, Mariusz, Pawel Tomaszewicz and Paul Amblard (1998). Comparison of different decomposition techniques of a digital circuit - a study case. *Proceedings of the MIXDES Conference.*

Zhuang, Xinhua and Robert M. Haralick (1986). Morphological structuring element decomposition. *Computer Vision, Graphics and Image Processing.*

IFAC
Publications
www.elsevier.com/locate/ifac

Threshold Optimisation in the Wavelet Compression

Karel Vlček[1] and David Klecker[2]

[1] Dept. of Measurement and Control,
FEI, Technical Univ. of Ostrava
17. listopadu 15
Ostrava-Poruba, 708 33
Czech Republic
karel.vlcek@vsb.cz

[2] APP Czech, spol. s r.o., Prague,
Czech Republic
dklecker@appg.com

Abstract: An application of algorithm of the discrete wavelet transformation (DWT) represents the universal calculation process for the time-frequency analysis of data, similarly to discrete Fourier transform (DFT). The starting steps of the systematic research of the wavelet theory are dated into the first part of eighties. The applications of this theory are developed ten years later. The high number of application comes from the region of data analysis, the image compression, and the system of relations in the numeric mathematics. The theory of DWT is applied as a universal mathematics tool in more then sixteen-application areas, the development of which was separated in the previous years. The application regions are defined not only in mathematics, but a lot of applications of 1D signal processing, as well as an image processing. *Copyright © 2001 IFAC*

Keywords: Discrete Wavelet Transform, Image Compression, and Threshold Optimisation

1. Introduction

The actual state of the theoretical knowledge on DWT allows the user comparison of formal properties of the wavelet analysis, and harmonic analysis. The simplest and the first in history example of wavelet are Haar function. Later described Reisz basis, which is not acceptable, on one hand, in criterion of the orthonormal basis, but on the other hand, it is possible to do the complete reconstruction of the original function from the wavelet coefficients. For an evaluation of the properties the Fourier basis will be repeated first.

The properties of harmonic functions, which generate this basis, fulfil the acceptation of the complete orthonormal system. The Parseval identity is valid for the system. The theory of wavelets results, the Reisz basis is the special case of Parseval identity, with reservations fulfilled. The base wavelet, in spite of, must not exist by the way, to be interpreted in spectral modus as an unambiguous

$$c_k = c_k(x) = \frac{1}{\sqrt{T}}\rangle x, e_k \langle = \frac{1}{\sqrt{T}}\int_a^b x(t)\frac{1}{\sqrt{T}}e^{-j2\pi\frac{k}{T}t}dt = \frac{1}{T}\int_a^b x(t)e^{-j2\pi\frac{k}{T}t}dt$$

is the Fourier coefficient. The set of all $c_k(x)$ is the Fourier's frequency spectrum of function $x(t)$.

The important property of one another relation of description of time function, and the existence of

expression of properties of an analysed function on the frequency.

2. Harmonic Analysis

The spectral analysis by the DFT is in the signal processing the basic method, which allow the computation of the spectral coefficients, and by this way the expression of the signal spectrum. It is asked the whole signal description with the independent variable in the time $x(t)$.

The problem is that we must suppose the knowledge of the signal in the past as well as in the future. The spectrum calculation must be expressed by the Fourier basis of function. The Fourier basis is expressed by using Euler formula of complex sinus wave by this way, to be expressed every function in the Fourier sequence:

$$x(t) = \sum_{k=-\infty}^{\infty} x_k \frac{1}{\sqrt{T}}e^{j2\pi\frac{k}{T}t} = \sum_{k=-\infty}^{\infty}\langle x.e_k\rangle \frac{1}{\sqrt{T}}e^{j2\pi\frac{k}{T}t} = \sum_{k=-\infty}^{\infty} c_k e^{j2\pi\frac{k}{T}t},$$

Fourier coefficients is Bessel-Parseval identity. For $x_k = \langle x, e_k \rangle = \sqrt{T}c_k$ is formed Bessel-Parseval identity formula:

$$\frac{1}{T}\int_a^b x(t)\overline{y(t)}dt = \sum_{k=-\infty}^{\infty} c_k(x)\overline{c_k(y)}.$$

The special case is the Parseval identity; if it is $x = y$, in this case it is possible the identity describe as:

$$\frac{1}{T} \int_a^b |x(t)|^2 \, dt = \sum_{k=-\infty}^{\infty} |c_k(x)|^2.$$

If $x(t)$ the analogue signal, the value of a certain integral $\int_a^b |x(t)|^2 \, dt$ is in relation to the energy of the signal at the interval $[a, b]$

Every harmonic wave $e_k(t)$ is generated in the harmonic analysis process only by changing of scale of independent variable t in the function $e_k(t) = w(kt)$ for every numbers $k \in Z$, where Z is definition region of the complex numbers. The whole base system is defined only by one basic wave function $w(t)$.

For $|k|$ big $e_k(t) = w(kt)$ oscillates this wave function with the high frequency and for $|k|$ small it oscillates with the low frequency. The expression $f_k = \frac{|k|}{T}$ describes the k-th harmonic.

3. WAVELET ANALYSIS

The harmonic analysis allows find harmonic approximation functions. To be the expression calculable, we find the wave functions $w_k(t)$, which must be not important for $t \to \pm\infty$. It is advantage due the practical reasons, the attenuation be rapid in the time. In the equation $x(t) = \sum_{k=-\infty}^{\infty} x_k w_k(t)$ every approximation component $x_k w_k(t)$ will contribute to signal expression $x(t)$ only in the little region of the location, in which it is shifted. If the harmonic function used as an approximation function $w(t) = e^{\frac{j2\pi}{T}}$ fill the whole region of the time axis, the wavelet approximation function is for example $\psi(t)$, which do not cover the whole region if the time axis. Shifting of does the sufficient covering of the time axis $\psi(t - k)$. With respect to the technique of calculation it is advantage to consider this shift, which is described by the expression:

$$\psi_{j,k}(t) = 2^{j/2} \cdot \psi(2^j t - k) = 2^{j/2} \cdot \psi\left(2^j\left(t - \frac{k}{2^j}\right)\right) = 2^{j/2} \cdot \psi\left(\frac{t - k/2^j}{2^{-j}}\right)$$

The wavelet function $\psi_{j,k}(t)$ is obtained from $\psi(t)$ by the binary dilatation with the factor 2^{-j} and by the shift $\frac{k}{2^j}$, which go down with the relation to shortening at $j \to +\infty$, and it go up with relation to becoming large at $j \to -\infty$ in dependability to dilatation factor 2^{-j}. Wavelet has a compact basis

function, if their definition set is bounded. Every function can be expressed by the wavelet sequence:

$$x(t) = \sum_{j,k=-\infty}^{\infty} c_{j,k} \psi_{j,k}(t),$$

where are the wavelet coefficients.

$$c_{j,k} = c_{j,k}(x) = \rangle x, \psi_{j,k} \langle = \int_{-\infty}^{\infty} x(t) \cdot \overline{\psi_{j,k}(t)} \, dt =$$
$$= \int x(t) \cdot \overline{\psi\left(\frac{t - k/2^j}{2^{-j}}\right)} \, dt$$

The function ψ is called mother wavelet. The mother wavelet is defined on the basic functions, and it is created by the multiresolution from the father wavelet in the relation:

$$\psi(x) = \sqrt{2} \cdot \sum_n g_n \cdot \phi(2x - n),$$

the expression $g_n = (-1)^n h_{1-n}$.

The father wavelet is solved by the dilatation of relation:

$$\phi(x) = \sum_n h_n \cdot \phi(2x - n).$$

Compact basic functions ϕ and wavelet ψ demands, to be a final number of nonzero filter coefficients h_n in dilatation relation, and the demand of the orthonormality of functions:

$$\phi_{m,n}(x) = 2^{-m/2} \phi(2^{-m} x - n)$$

If it is fulfilled there are defined values of coefficients h_n. There is an example of the certain values of Daubechies wavelet:

$$h_0 = \frac{1 + \sqrt{3}}{4\sqrt{2}}, \quad h_1 = \frac{3 + \sqrt{3}}{4\sqrt{2}}, \quad h_2 = \frac{3 - \sqrt{3}}{4\sqrt{2}}, \quad h_3 = \frac{1 - \sqrt{3}}{4\sqrt{2}}.$$

The definition region of the father wavelet carrier ϕ is the interval $[0,3]$, the carrier of the mother wavelet ψ is the interval $[-1,2]$. To be applicable in time-frequency analysis the functions $\psi_{j,k}$ must be reliable located with respect to the time axis, as well as the frequency axes. This contradiction demand is fulfilled by the "quick attenuation" of $\psi(t)$. For this property is used so called window.

4. SIGNAL FILTRATION

A signal, which is corrupted by additive noise, can be processed by filtration with application of wavelet functions. The filtration is analogous to filtration based on Fourier transformation. The Fourier filtration is based on computation of coefficients c_j, which are multiplied by weight function: the transmission characteristics of filter, the condition of realisation of which is decreasing of influence of coefficients c_j with increasing $|j|$.

When is used wavelet filtration, the influence of the coefficients $c_{j,k}$ in dependence to chosen k is local only, but it is not global, as at Fourier filtration. The wavelet filter is easy adaptable by the local characteristics of signal. This is the way to removing the main disadvantage: non-stable noise. This property can cause some complications, if the signal is loaded by the very non-stable noise.

The periods of signal with the higher change of error probability can be considered as the dynamic errors, and they can be non-sufficiently removed. It is a general problem of criteria evaluation in the noise removing. For the wavelet coefficient modification are used various criteria with the determining of the thresholds.

When the "Hard Threshold" is used, the wavelet coefficients are compared with the fixed value of threshold λ. If the wavelet coefficient is less then the threshold, it is put equal to zero, in the other case it is left without changes:

$$c_{j,k}^{hard} = \begin{cases} 0 & |c_{j,k}| < \lambda \\ c_{j,k} & |c_{j,k}| \geq \lambda \end{cases}.$$

When the "Soft Threshold" is used, the wavelet coefficients are shifted to zero with value λ.

$$c_{j,k}^{soft} = sign\left(c_{j,k}\right) \max\left(0, |c_{j,k}| - \lambda\right).$$

The other way of the quantity of variable is so called "Quantile Threshold". This is method, which describes process similar to the hard threshold, but as a constant λ is used quantiles from the set of all wavelet coefficients. The result is, for example, that 30% of smallest wavelet coefficients are putted equal to zero.

The frequently used is the process of "Universal Threshold" with the global threshold value $\lambda = \sigma \cdot \sqrt{2 \cdot \log n}$. The character of data chooses the optimal criterion. The number n is an amount of data, and σ is the error from the non-correlated white noise. This method defines the number λ, and it is possible to use the both hard and soft threshold. For the correlated noise it is possible to derive method by the same way.

5. SIGNAL COMPRESSION

In the process of compression, similar as filtration, are the data expressed by the vector $a = \left(a_0, a_1, \ldots a_{M-1}\right)^T$, where $M = 2^K$, we consider the function $A = \sum_n a_n \phi_{mn}$ (coefficients a_n are calculated

as periodical prolongation of vector a), which lay in the linear space:

$$V_m = \overline{\lambda\{\phi_{mn}, n \in Z\}}.$$

The functions ϕ_{mn} are created by the dilatation and shifting, that is consequently:

$$V_{m+1} \subset V_m, \text{ pro } n \in Z.$$

Let us write:

$$W_{m+1} = \overline{\lambda\{\psi_{m+1,n}, n \in Z\}},$$

from this results also:

$$V_m = V_{m+1} \oplus W_{m+1}, \text{ pro } n \in Z,$$

where W_{m+1} is an orthogonal supplement V_{m+1} in V_m.

The basic idea used in compression (similarly as in filtration) is done by the projection of the function $A \in V_m$ into V_{m+1} and W_{m+1}. The orthogonal projection by operators $P_{m+1}: V_m \to V_{m+1}$ and $Q_{m+1}: V_m \to W_{m+1}$ is possible A decompose into:

$$A = P_{m+1}A + Q_{m+1}A,$$

where

$$P_{m+1}A = \sum_n \left\langle A, \phi_{m+1,n} \right\rangle \phi_{m+1,n},$$

$$Q_{m+1}A = \sum_n \left\langle A, \psi_{m+1,n} \right\rangle \psi_{m+1,n}.$$

The coefficients of these projections are numbers

$$a_n^1 = \left\langle A, \phi_{m+1,n} \right\rangle \text{ and } d_n^1 = \left\langle A, \psi_{m+1,n} \right\rangle.$$

With respect to, that it is valid

$$\left\langle \phi_{ml}, \phi_{m+1,k} \right\rangle = h_{k-2l} \text{ and } \left\langle \phi_{ml}, \psi_{m+1,k} \right\rangle = g_{k-2l},$$

After rearrangement

$$a_k^1 = \sum_l a_l \left\langle \phi_{ml}, \phi_{m+1,k} \right\rangle = \sum_l h_{k-2l} a_l,$$

$$d_k^1 = \sum_l a_l \left\langle \phi_{ml}, \psi_{m+1,k} \right\rangle = \sum_l g_{k-2l} a_l.$$

The coefficients a_l was created by the periodical prolongation of the vector a, and by this reason coefficients a_k^1 a d_k^1 $(k \in Z)$ will have periodical character too. As the result we will consider final vectors

$$a^1 = \left(a_0^1, a_1^1, a_2^1, \ldots, a_{\frac{M}{2}-1}^1\right)^T, \quad d^1 = \left(d_0^1, d_1^1, d_2^1, \ldots, d_{\frac{M}{2}-1}^1\right)^T.$$

The vector a^1 defines projection of function $A \in V_m$ in the space V_{m+1}. Since V_{m+1} has the identical structure as V_m but it is in "half density" (it is created by dilatation of V_m) there will be the global character of input vector a expressed by the vector a^1 with the half-length. The space W_{m+1} is an

91

orthogonal supplement V_{m+1} in V_m. Vector d^1 will contain the information supplement about local changes in the input vector a. The vector contains d^1 the numbers that are not far from zero in the practical use, if the changes of input vector are slow.

The quick changes cause the vector d^1 will contain the big values at the short interval of signal or image.

6. ALGORITHM OF CALCULATION

For the case of four non-zero coefficients h_n, which were introduced in previous text, it is possible for calculation of a_k^1 a d_k^1 use the matrix notation: $\begin{pmatrix} a^1 \\ d^1 \end{pmatrix} = \begin{pmatrix} H \\ G \end{pmatrix} a$, where matrixes H and G are in dimensions $M/2 \times M$, and they have the structure as follows:

$$H = \begin{pmatrix} h_0 & h_1 & h_2 & h_3 & & & \\ & & & \ddots & \ddots & & \\ & & & & h_0 & h_1 & h_2 & h_3 \\ h_2 & h_3 & & & & & h_0 & h_1 \end{pmatrix},$$

$$G = \begin{pmatrix} g_2 & g_3 & & & & g_0 & g_1 \\ g_0 & g_1 & g_2 & g_3 & & & \\ & & \ddots & \ddots & \ddots & & \\ & & & & g_0 & g_1 & g_2 & g_3 \end{pmatrix}.$$

The matrixes are, thanks to, compact structure of wavelet carrier ϕ, ψ, rare, what is advantage for very quick and effective implementation of matrix multiplication. Very important result of orthonormality of wavelets is orthogonality of matrix $\begin{pmatrix} H \\ G \end{pmatrix}$. The reconstruction of input a from the vectors of projection coefficients a^1, d^1. The transposed matrix calculates it:

$$a = \begin{pmatrix} H \\ G \end{pmatrix}^T \begin{pmatrix} a^1 \\ d^1 \end{pmatrix}.$$

The computation process can be repeated for the vectors $a^1, a^2, \ldots$ For repeated compression or filtration. After j steps, we will obtain:

$$A = P_{m+j}A + \sum_{i=1}^{j} Q_{m+i}A,$$

due to $P_{m+j}P_{m+j-1} = P_{m+j}$. We obtain the vectors of coefficients of separate projections:

$$a^j, d^j, \ldots, d^1,$$

where the vectors a^j, d^j have $M/2^j$ components.

The vectors d^1 are values that are not far from zero. If it will be used for data compression, the effective memory management will be based on the vectors d^1, which will not write into memory. The

reconstruction of data will be lightly different from original information. The advantage is, if the data are interpreted as a signal, the lost data will not influence global spectrum of the signal.

What advantages represent application of wavelets in signal processing? The properties of processing are similar to Fourier analysis. By this way it is possible to obtain wavelet coefficients, which are used similarly as Fourier coefficients. There are used algorithms Fast Fourier Transform (FFT) for DFT calculation, the complexity of which is expressed by $O(n \cdot \log_2(n))$, similarly in DWT the complexity is less as $O(n)$.

The optimisation of DWT calculation is more effective. The morphology of matrixes H and G allow applying (thanks to belt structure) another mechanisms for effective manipulation of data [9]. These properties are important in the case of large data files. The image processing is one of these applications in which the effective processing is important.

The effective algorithm of image compression was derived from the introduced wavelet transformation. The algorithm was used for compression of the real image, which were obtained from the coloured TV image as red (R) component of RGB image. The image is transformed into "bitmap" file. The two orthogonal bases were used in DWT: Haar, and Doubechies. With respect to 2D character of carrier, there are used for calculation the arrangement of formulas by the multiresolution as follows (Nacken, 1993):.

Let us consider matrix $M \times M$ with the image elements $c_{mn}^{(0)}$. The 2D multiresolution we will write for generator ϕ, defined on the compact carrier. Its dilatation is described by the equation:

$$\phi(x) = \sqrt{2} \sum_{0}^{2N-1} h_k \phi(2x - k),$$

and its correspond mother wavelet ψ is defined as

$$\psi(x) = \sqrt{2} \sum_{0}^{2N-1} g_k \phi(2x - k), \quad kde$$

$$g_k = (-1)^k \cdot h_{1-k}.$$

The multiresolution of signal matrix $c^{(0)}$ is described as j-th level of recursion

$$P_j \cdot c^{(0)} = \sum_{mn} c_{mn}^{(j)} \cdot \phi_{jm}(x) \cdot \phi_{jn}(y), \text{ and}$$

$$Q_j \cdot c^{(0)} = \sum_{mn} \begin{bmatrix} d_{mn}^{(x)(j)} \cdot y_{jm}(x) \cdot f_{jn}(y) + d_{mn}^{(y)(j)} \cdot f_{jm}(x) \cdot y_{jn}(y) + \\ + d_{mn}^{(xy)(j)} \cdot y_{jm}(x) \cdot y_{jn}(y) \end{bmatrix}.$$

It is important, that the calculations can be computed easy as the matrix operations by the proper implementation of hardware, which is devoted to vector and matrix operations:

$$c^{(j+1)} = \mathbf{H}\, c^{(j)} \mathbf{H}^{T}, \quad d^{(x)(j+1)} = \mathbf{H}\, c^{(j)} \mathbf{G}^{T},$$

$$d^{(y)(j+1)} = \mathbf{G}\, c^{(j)} \mathbf{H}^{T}, \quad d^{(x)(j+1)} = \mathbf{G}\, c^{(j)} \mathbf{G}^{T},$$

Where matrix operators $\mathbf{H}$, $\mathbf{G}$ have size $M/2{\times}M$ of matrixes composed from the coefficients h_k, g_k (Daalhuis, 1993). Signal $c^{(j+1)}$ is an information of pixels in coarse scale and three signals: $d^{(x)(j+1)}$, $d^{(y)(j+1)}$, $d^{(xy)(j+1)}$ are less then fix border of amplitude of brightness, which are located out of quadrant, in which is located the main image.

The difference between amplitudes of brightness is three times, in the reconstruction there are substituted by zeros. This is important for writing of data into memory and for transmission of image. If it is used orthonormal basis of wavelets (Daubechies, 1988), the reconstruction algorithms is defined similarly by the simple recursive relation.

The images, which are processed as the examples are computed by wavelet transformation with the coefficients by Haar, (values 1, and -1) and by Daubechies as values $h_0, ..., h_3$ (Daalhuis, 1993). The original image is transformed into images, which represent compression and expansion in various conditions. The main information rests in left upper corner. The other parts contain difference signals.

7. RESULTS

The original image was arranged from the colour image with RGB components. This arrangement is processed from the R component. The image was chosen due to its various parts: the marmot is sharp part in the centre, the background

Fig. 1: Original image

Fig. 2: The wavelet compression by Haar basis, threshold ration = 97.25, decomposition level = unlimited.

Fig. 3: Image wavelet compression by Haar basis, threshold ration = 93.16, decomposition level = unlimited.

Fig. 4: Image wavelet compression by Haar basis, threshold ration = 85.27, decomposition level = unlimited

Fig. 5: The wavelet compression by Daubechies basis – four coefficients, threshold ration = 98,44, decomposition level = unlimited

Fig. 6: The wavelet compression by Daubechies basis – four coefficients, threshold ration = 96.7, decomposition level = unlimited

Fig. 7: The wavelet compression by Daubechies basis – four coefficients, threshold ration = 93.76, decomposition level = unlimited

Fig. 8: The wavelet compression by Daubechies basis – four coefficients, threshold ration = 77.96, decomposition level = unlimited

The images was processed by the programme system "Wavelet Development Kit" version 1.4, the author of programme system is Ing. David Klecker, affiliation: APP Czech.

8. CONCLUSION

The short wavelets, such as Daubechies wavelets can be processed more effective. An implementation is advantage, when parallel instructions are used. An instruction set of such instructions is accessible in the signal processor TMS320C50, for example (Vlček, 2000).

The accesses to data as well as effective and sophisticate instruction set are the main advantages in implementation of the wavelet transformation algorithms in digital signal processor. The method of image compression is recommended for its simplicity and effective implementation, the indications are, that application specific integrated circuit will be an effective hardware implementation.

The next method enhancement is now in the stadium of the standard definition under the name JPEG 2000. It is expected the more effective solution in connection with image decomposition as well as duly arrangement of sequence of image processing.

9. ACKNOWLEDGEMENTS

The research is supported by the project GAÈR No. 201/00/1531.

REFERENCES

Nacken, P. (1993): Image compression using wavelets, In *Wavelets: An Elementary Treatment of Theory and Applications* (Ed. Tom H. Koornwinder), World Scientific Publ., London 1993, pp. 81-91

Daalhuis, A.B.O. (1993): Computing with Daubechies' wavelets, (*the same book,* pp. 93-105)

Daubechies, I. (1988): Orthonormal bases of compactly supported wavelets, *Comm. Pure and Appl. Math.* **41** (1988), pp. 909-996.

Kucera, R., Vlček, J., Vlček, K. (1996): Image Compression using Multiresolution Analysis. In *Analysis of Biomedical Signals and Images 13-th Biennial International Conference Biosignal '96* Proceedings Brno, June 1996, pp. 38-40.

Vlček, K. (1983): Multiplier - Accumulator with Directed Data Flow. *The 2-nd European Signal Processing Conf. EUSIPCO-83,* Erlangen, Germany (Sept. 12 - 16, 1983, North-Holland Amsterdam, 1983), pp. 833-836.

Baraniecki, A., Parikh, V. (1996): Efficient Implementation of Wavelet Transform and its Inverse. *Workshop on Design Methodologies for Signal Processing,* Zakopane, Poland, (Aug. 29 - 30, 1996), pp. 16-20.

TMS320C5x Digital Signal Processor. *Texas Instruments* (1994).

Kucera, R., Vlček, J., Vlček, K. (1998): DSP Implementation of Image Compression by Multiresolutional Analysis. Radioengineering, Vol. 7, No. 1, (April 1998), ISSN 1210-2512, pp. 7-9.

Cornelis, J., Munteanu, A., Salomie, A., Schelkens, P., Declerck, R., Christope, Y., Enescu, V.(1998): Medical Image Compression: Options for the Future, Proc. of the 14th Biennial Internat. Conf. Biosignal'98, ISBN 80-214-1169-4, pp. 1-12.

Vlček, K. (2000): Compression and Error Control Coding in Multimedia Communications. BEN Prague (2000), ISBN 80-86056-68-6 (In Czech)

IFAC

Publications

www.elsevier.com/locate/ifac

SPECIFICATION AND SYNTHESIS OF PETRI NET BASED REPROGRAMMABLE LOGIC CONTROLLER

Marian Adamski

University of Zielona Gora, Institute of Computer Engineering and Electronics
50 Podgorna Street, 65-246 Zielona Gora, POLAND
tel. (+48 68) 3282 219; fax (+48 68) 3244 733; M.Adamski@iie.uz.zgora.pl

Abstract: The goal of the paper is to present a novel approach to Application Specific Logic Controllers realisation, which is suitable especially for small, embedded system designs. A discrete model of Logic Controller is derived directly from Hierarchical Control Interpreted Petri Net or related Sequential Function Chart (SFC) and synthesised as a dedicated modular microsystem. The unified model is reflected in Programmable Logic. The desired behaviour of the designed logic controller can be validated by simulation in VHDL environment. The paper covers some effective techniques for computer-based synthesis of Reprogrammable Logic Controllers, from Petri net level to the logic design level. *Copyright © 2001 IFAC*

Keywords: Sequential Control, Programmable logic controllers, Petri-nets, Rule-based systems, Digital systems, Logic design, FPGA.

1. INTRODUCTION

A Petri net that is related with Sequential Function Chart SFC (IEC 1131-3 standard) is a suitable design specification for RTL-level synthesis or logic synthesis of digital circuits, which realise given binary control algorithms (Adamski, 1998). In the paper it is shown how to implement Reprogrammable Logic Controllers (RLC) in FPL (Field Programmable Logic).

The main goal of the proposed design methodology is to continuously preserve the direct, self-evident correspondence among a control interpreted Petri net and several possible, but behaviourally equivalent, hardware implementations. The final design decision strongly depends on selected FPL structure and required technological parameters of embedded system. The well-structured formal specification, which is represented in the human-readable language, has a direct impact on the validation,

formal verification and implementation of digital microsystems in FPL. The declarative, logic-based textual specification of Petri net, reflected clearly in digital circuit structure, can increase the efficiency of the concurrent (parallel) controller design. The symbolic specification of Petri net is considered in terms of simultaneously holding local states and local events in the related Concurrent State Machine model. Petri nets express the causality as well as the concurrency of discrete control algorithms.

In *the transition-based specification* all the intended local state changes are recognized and distinguished as Petri net transitions, with their input places (preconditions) and output places (postconditions). In *the place-based specification* the description is local state oriented, and conditions for token entering and holding for any Petri net place are given. Both considered forms of specification are related, and can be expressed in symbolic logic conditional assertions –called *Decision Rules.*

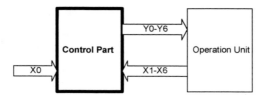

Fig. 1. Discrete control system

The paper covers some effective techniques for computer-based synthesis of Reprogrammable Logic Controllers, starting from modular Petri net level and finishing on logic design level. The logic controller model, as a part of discrete control system (Fig. 1) can be extended to Concurrent State Machine with Data Path, or on higher level of abstraction, to Concurrent Program State Machine with Data Flow Part. The symbols $x1, x2, ..., xn$ and $y1, y2, ..., ym$ represent the binary inputs and outputs of control part (sequencer). The operational part may contain digital subsystems, as well as electromechanical parts with appropriate sensors and actuators.

The paper presents the outline of formal methodology for Application Specific Logic Controllers (ASLC) design (Adamski, 1999). The previous work on that subject has been recently summarised in paper (Adamski and Węgrzyn, 2000).

2. INTERPRETED PETRI NET MODEL

Control Interpreted Petri Net (David and Alla, 1992) has been shown as a powerful tool to describe the intended performance of a parallel (concurrent) controller. In the considered model (Fig.2), Petri net places $\{p1, p2, ..., p8, p9\}$ stand for local states of the implemented logic controller.

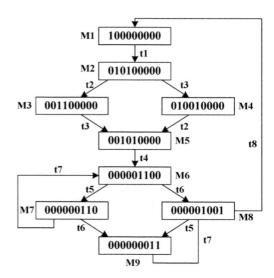

Fig. 3. Reachability graph

Petri net transitions $\{t1, t2, ..., t7, t8\}$ with associated Boolean expressions (guards) symbolize all possible local state changes and their external conditions. The distribution of Petri net tokens $\{M1, M2, ..., M9\}$ among places before the firing of any transition, can be regarded as identification of current global state of the modelled system (Fig. 3). In the considered interpreted Petri net model, particular combinations of Boolean external inputs, $\{x0, x1, ..., xn\}$ and eventually some internal conditions, have to be true, for the transition to be fired.

Petri net is mapped into the Boolean expressions (assertions) without explicit enumeration of all possible global states and all possible global state changes (Fig. 3, Fig. 4).

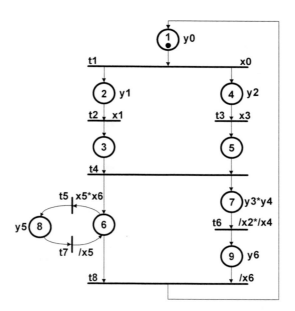

Fig. 2. Control Interpreted Petri Net

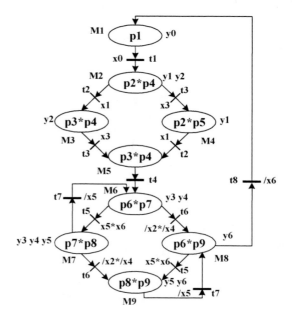

Fig. 4. Interpreted reachability graph

All the global states of logic controller can be eventually found by generating a reachability graph of the given Petri net. The reachability graph is interpreted (Fig. 4), by supplementing it with names of inputs {x} and outputs {y} of logic controller. Later it could be transformed into the Petri net description of an equivalent non-deterministic control automaton model. On the other hand, it is possible to look into the interpreted reachability graph as at transition system like version of logic controller behavioural specification. It also serves as an interpretation structure (Kripke model) of the related formal logic symbolic description. From the present global internal state M the modelled controller goes to the next internal global state $@M$, generating the desired combinational output signals {y} in the current state and registered {@y} in the next state.

The combinational Moore type output signals are linked with places (Fig. 2, Fig. 4). Some Mealy type combinational output signals might be related both with places and input signals (Bilinski, *at al.* ,1994).

The introducing of Mealy inputs usually makes the skeleton of Petri net simpler, but frequently the associated net interpretation becomes more complicated and difficult to understand.

The synchronous Petri nets (Kozłowski, *et al.*, 1995) are introduced to model binary systems, which are later implemented as digital circuits, synchronised by a global clock. Some specific output signals synchronously change their values together with the firing of a particular transition. That makes it possible to label the considered Petri net transition with dynamic (pulse-like) signals. In synchronous digital systems such subset of selected output signals will be usually stored in the special output register of the implemented state machine. On the other hand registered output variables can be treated as a partial state variables and use for Petri net place encoding (local state assignment).

The behavioural specification is presented only in terms of subsets of simultaneously holding local states in the current and the next global states, and related with them local state changes (local transitions). As an example the simplified version of logic controller behavioural model taken from paper (Adamski, 1998) is selected.

3. MODULAR STRUCTURED IMPLEMENTATION OF PETRI NET

Petri nets can provide a unified method for the design of discrete-event systems (Fig. 1) from an initial system description (Fig. 2) to possibly different hierarchical physical realizations. The hierarchically structured Petri net (Fig. 5) consists of subnets, which, except possibly the Base Net are well-formed blocks.

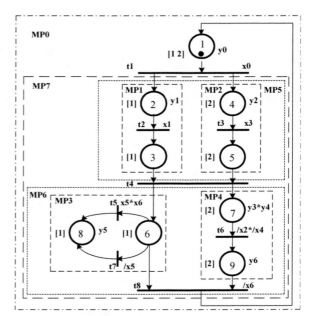

Fig. 5. Modular, hierarchical Petri net

The concurrency relation among subnets is depicted by means of colours, which are attached explicitly to the places, and implicitly to the transitions and arcs as well as to the tokens (Węgrzyn and Adamski, 1999). The colours (for example [1] and [2]), which are attached to places, may distinguish and validate the consistency of the intended sequences of local states (particular processes).

The set of subnets is partially ordered (Fig. 6). The coloured hierarchy relation tree graphically represents both hierarchy and concurrency relations among subnets. The Base Net *MP0* is on the root of the tree. It contains place *P1* and the double-macroplaces *MP1-MP7*, which stand for the hierarchically structured subnets at the appropriate level of hierarchy.

The structured local state assignment (place encoding) is used (Adamski, 1998) to map the places and macroplaces into logic conjunctions. Consequently, the proper state encoding transforms preconditions and postconditions of any transition into the consistent Boolean terms.

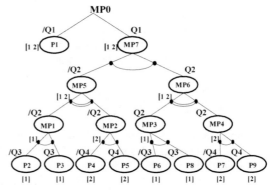

Fig. 6. Hierarchy tree

The Petri net is hierarchically encoded by means of state variables Qi, $i= 1,2,3,4$. The symbols Qi or $/Qi$, are distributed among particular paths in the hierarchy tree. The path directed from the root to the leave forms the unique encoding term for the selected macroplace or place in the considered level of abstraction.

4. FIELD PROGRAMMABLE LOGIC BASED IMPLEMENTATION

The Logic Controller RLC is implemented in Field Programmable Logic, for example as a FPGA based reprogrammable unit.

The use of FPL flexible structures instead of controllers with fixed architectures has several advantages. They create near-optimal controller implementations (in terms of performance), tailored to the exact requirements. Practical designs often fit in a single FPGA package, since there are no redundant elements. Particular additional functions can be taken from the library of operational blocks, which is effectively implemented and intensively tested. Controller outputs respond to inputs with high speed, in predictable and repeatable way, without glitches. The behavioural specification, which is mapped in the programmable logic can be formally verified or validated by means of advanced CAD tools.

The textual description of developed implementation in Field Programmable Logic is not procedural, as Petri net evidently is, but rather *declarative*.

The direct mapping of Petri net into FPL is based on correspondence between a transition, described as decision rule and clearly defined subset of CLBs (Węgrzyn, *at al.*, 1998).

5. CONTROL AUTOMATON AS AN INFERENCE SYSTEM

At the beginning of design process, the control-oriented specification of dedicated, reactive discrete-event systems is used. After analysis of some behavioural and structural properties of Petri net, a Petri net model is related with a knowledge-based, textual, descriptive form of representation, very easy for mapping into reprogrammable hardware.

In the proposed view, a control automaton (Concurrent State Machine CSM) is treated as a *dynamic inference system*, based on Gentzen logic. In reasoning system the knowledge base is structured as a group of decision rules and directly reflected in the hardware structure of Field Programmable Logic circuit. The symbolic logic assertions, *sequents-axioms*, describing the intended functionality of logic

controller may include some elements, taken from temporal logic, for example operator 'next' @ (Sagoo and Holding, 1991).

Complex sequents are formally transformed into the set of equivalent *sequent-clauses* (*simple sequents*), which are very similar to the *production rules*. The formally transformed decision rules are directly mapped into VHDL statements. The logic (Boolean) expressions, which are suitable for direct mapping into FPGA or CPLD, can be also derived as an input to other traditionally used tools.

6. PETRI NET SPECIFICATION IN SEQUENT LOGIC LANGUAGE

6.1 Transition-oriented declarative specification

The Logic Controller is considered as an abstract rule based reasoning system, implemented in re-configurable hardware. The *transition-oriented declarative specification* describes all possible internal events in concurrent state machine, when local states associated to considered transition change. The description is very closed to well-known production rules (Sagoo and Holding, 1991).

It should be noted that in following example names of transitions $T1, T2, ...,T8$ in this specification are not essential parts of logic decision rules. They serve only as labels, keeping readable correspondence among Petri net transitions and their symbolic descriptions.

```
T1: P1 * X0|-@P2 *@P4;
T2: P2 * X1 |-@P3;
T3: P4 * X3 |-@P5;
T4: P3 * P5 |-@P6 * @P7;
T5: P6 * X5*X6|-@P8;
T6: P7 * /X2*/X4|-@P9;
T7: P8 * /X5|-@P6;
T8: P6 *P9 * /X6|-@P1;          (1)
```

If the next value of the temporal variable, for example $@P1$, cannot be proved as *true* after the firing of any transition, it is considered that it takes the value $P1$, which holds in the current marking.

The static (level type) Moore type outputs depend directly only on related place markings:

```
P1 |- Y0;
P2 |- Y1;
P4 |- Y2;
P7 |- Y3*Y4;
```

```
P8  |- Y5;

P9  |- Y6;                    (2)
```

If the value of considered output signal is not proved in the current Petri net marking as *true*, it is *false* (Closed Word Assumption CWA).

6.2 Place oriented declarative specification

In some cases, like implementations with D flip-flops in FPGA, the declarative, *place oriented specification* is taken into account. It gives separately the conditions of token entering and holding for any place. For example, the sequents, which include explicit transition symbols *{T1, T2, ..., T8}*, after mapping the Petri net into VHDL statements in Martin Bolton's style, give economical implementations in FPGA structure:

Preconditions:

```
P1 * X0        |- T1;

P2 * X1        |- T2;

...

P6*P9*/X6      |- T8;
```

Next markings:

```
T8+P1*/T1      |- @P1;

T1+P2*/T2      |- @P2;


T8+P9 */T8     |- @P9;        (3)
```

In this kind of specification, if the next value of the temporal variable, for example *@P1*, cannot be proved in the current marking (global state) as *true*, it is considered that it takes the value *false*. The Moore outputs are defined as in the previous example.

7. PETRI NET AND LOGIC DESIGN

The direct mapping of a Petri net transition into Field Programmable Logic (FPL) is based on a self-evident correspondence between a place and a clearly defined bit-subset of a state register (collection of CLBs).

The simplest technique for Petri net place encoding is to use one-to-one mapping of places onto flip-flops in the style of a one-hot state assignment. In that case, a name of the place becomes also a name of the related flip-flop. The flip-flop is set into 1 if and only if the particular place holds the token. In general, places after encoding are recognised by conjunctions, which are formed from state variables, chosen from the set *{Q1, Q2, ... , Qk}*.

The local states, which are active simultaneously, have non-orthogonal codes. They are represented by the places, which hold the tokens concurrently and consequently belong to the same vertex from the implicitly or explicitly given reachability graph of Petri net.

The local states, which belong to the same sequential process, have orthogonal codes. They never belong to the same vertex of reachability graph.

The new, efficient structured method of place encoding is based on hierarchical decomposition of the net (Fig. 5). The state variables are distributed among the vertices of hierarchy tree, in the direction from the root to the leaves. The vertices having the same colour get orthogonal codes. The vertices with disjoint colour sets have to obtain non-orthogonal codes. The result of a heuristic hierarchical local state assignment [Q1,Q2,Q3,Q4] is as follows:

```
P1  = 0 - - -
                  P1 = /Q1
P2  = 1 0 0 *
                  P2= Q1*/Q2*/Q3
P3  = 1 0 1 *
                  P3= Q1*/Q2*Q3
P4  = 1 0 * 0
                  P4= Q1*/Q2*/Q4
P5  = 1 0 * 1
                  P5= Q1*/Q2*Q4
P6  = 1 1 0 *
                  P6= Q1*Q2*/Q3
P7  = 1 1 * 0
                  P7= Q1*Q2*/Q4
P8  = 1 1 1 *
                  P8= Q1*Q2*Q3
P9  = 1 1 * 1
            P9= Q1*Q2*Q4          (4)
```

The total code of the reachability graph vertex would be obtained by merging the codes of the simultaneously marked places. The code of the particular place or macroplace is represented by means of the vector composed from {0, 1, - , *} or it is given as a related Boolean term. The symbols 0, 1, - ('don't care') have the usual meanings, but the symbol * in vector denotes *'explicitly don't know'* (0 or 1, but not *'don't care'*).

For several practical reasons, during logic design of the controller it is recommended to manipulate with Boolean expressions (product terms), in which all the symbols of places are substituted by encoding conjunctions. For example, the code of place P2= [1 0 0 *] is described by Boolean term P2= Q1*/Q2*/Q3.

The transition-oriented specification of logic controller after substitution of place symbols by means of encoding terms would look as follows:

```
T1: /Q1 * X0|-@Q1*@/Q2*@/Q3*@/Q4;

T2: Q1*/Q2*/Q3* X1 |-@Q1*@/Q2*@Q3;

T3: Q1*/Q2*/Q4* X3 |-@Q1*@/Q2*@Q4;

T4:Q1*/Q2*Q3*Q4|-@Q1*@Q2*@/Q3*@/Q4;

T5: Q1*Q2*/Q3*X5*X6|-@Q1*@Q2*@Q3;

T6: Q1*Q2*/Q4*/X2*/X4|-@Q1*@Q2*@Q4;

T7: Q1*Q2*Q3*/X5|-@Q1*@Q2*@/Q3;

T8: Q1*Q2*/Q3*Q4*/X6|-@/Q1;        (5)
```

The simplified sequent specification, planned for implementations based on the state register with JK flip-flops, on the right sides does not contain signals, which conserve their values during the occurrences of transitions:

```
T1: /Q1 * X0|-@Q1*@/Q2*@/Q3*@/Q4;

T2: Q1*/Q2*/Q3* X1 |-@Q3;

T3: Q1*/Q2*/Q4* X3 |-@Q4;

T4: Q1*/Q2*Q3*Q4 |-@Q2*@/Q3 *@/Q4;

T5: Q1*Q2*/Q3*X5*X6|-@Q3;

T6: Q1*Q2*/Q4*/X2*/X4|-@Q4;

T7: Q1*Q2*Q3*/X5|-@/Q3;

T8: Q1*Q2*/Q3*Q4*/X6|-@/Q1;       (6)
```

For Field Programmable Logic with JK flip flops, symbols @Qi can be replaced by J_Qi and symbols @/Qi respectively by K_Qi. If it is necessary the separate expressions for each flip-flop could be deduced as follows:

```
X0 |- J_Q1;

Q2*/Q3*Q4*/X6 |- K_Q1;

Q1*Q3*Q4 |-J_Q2;

/Q1 * X0 |- K_Q2;

Q1*/Q2*X1+Q1*Q2*X5*X6|-J_Q3;

/Q1*X0+Q1*/Q2*Q4+Q1*Q2*/X5|- K_Q3;

Q1*/Q2* X3 + Q1*Q2*/X2*/X4|- J_Q4;

/Q1*X0+Q1*/Q2*Q3|- K_Q4;          (7)
```

The outputs after place encoding depends directly on state variables:

```
    /Q1|- Y0;

    Q1*/Q2*/Q3|- Y1;

    ...

    Q1*Q2*Q4|- Y6;                (8)
```

8. CONCLUSIONS

Formal logic language, which is complementary with Petri nets, is suitable in specifying system level designs of logic controllers, implemented in FPL. Simulating of Petri net model and its hardware implementation can be simplified by translating of rule-based description to VHDL. The simulation results, at circuit level and algorithmic level, can be compared immediately. The next design step concentrates on the automatic synthesis of Reprogrammable Logic Controllers from their VHDL descriptions. The paper presents the hierarchical Petri net approach for synthesis, in which the modular net is mapped into the Field Programmable Logic structure. The hierarchy levels are conserved and related with some particular local state variable subsets. A concise, understandable specification can be easily locally modified.

REFERENCES

Adamski, M. (1998). SFC, Petri Nets and Application Specific Logic Controllers. In: *Proc. of the IEEE Int. Conf. on Systems, Man, and Cybernetics*. IEEE, San Diego, USA, 11-14.10.1998, pp. 728-733.

Adamski, M. (1999). Application Specific Logic Controllers for Safety Critical Systems. In: *Proc. of the 1999 IFAC Triennial World Congress*, Beijing, China, Vol.Q, pp.519-524.

Adamski, M. and M. Wegrzyn, (2000). Interpreted Petri Net Approach for Design of Dedicated Reactive Systems. In: *Proceedings of the 6th IFAC International Workshop on Algorithms and Architectures for Real-Time Control, AARTC'2000*, Palma de Mallorca, Spain, 15 -17.04.2000, pp.179-184.

Bilinski, K., M. Adamski, J. M. Saul and E.L. Dagless (1994). Petri net based algorithms for parallel controller synthesis. *IEE Proceedings, Computers and Digital Techniques*, **141**(6), 405-412.

David, R. and H.Alla, (1992*). Petri Nets & Grafcet. Tools for modelling discrete event systems*. Prentice Hall, New York.

Kozlowski,T., E .L. Dagless, J. M. Saul, M. Adamski and J. Szajna, (1995), Parallel controller synthesis using Petri nets. *IEE Proceedings-E, Computers and Digital Techniques* **142** (4) pp. 263-271

Sagoo, J.S. D.J. Holding (1991). A comparison of temporal Petri net based techniques in the specification and design of hard real-time systems. *Microprocessing and Microprogramming*, **32**, No.1-5, 111-118.

Wegrzyn, M., M. Adamski and J.L. Monteiro (1998). The Application of Reconfigurable Logic to Controller Design. *Control Engineering Practice*, **6**, 1998,.879-887.

Wegrzyn, M. and M. Adamski (1999). Hierarchical Approach for Design of Application Specific Logic Controller. In: *Proc. of the IEEE Intern. Symposium on Industrial Electronics ISIE'99*, Bled, Slovenia, 12-16.07.1999, Vol.3, pp.1389-1394.

IFAC
Publications
www.elsevier.com/locate/ifac

A NEW COMPACT PROGRAMMABLE LOGIC CONTROLLER WITH INTEGRATED PROGRAMMING EQUIPMENT

Miroslaw CHMIEL
Edward HRYNKIEWICZ
Adam MILIK

Institute of Electronics
Silesian Technical University,
Akademicka 16; 44-100 Gliwice; Poland
chmiel@boss.iele.polsl.gliwice.pl; eh@boss.iele.polsl.gliwice.pl

Abstract: A paper presents design of compact logic controller with embedded programming function. There is presented design consideration and its influence on final construction. Analysis of required component for compact controller is carried out and optimal set of component based on our researches is presented. Compact logic controller with programming function faces new problem of efficient program presentation on small display but also possible fastest execution of it. *Copyright © 2001 IFAC*

Keywords: PLC, Compact PLC, PLC programming, PLC architecture, LD, Ladder Diagram

1. INTRODUCTION

In the middle of the eighties, apart from Modular PLCs, models with all internal circuitry, i.e. CPU, power supply, input and output circuits, integrated in a single case were introduced. They are mostly referenced as Compact PLCs. Of course the Compact PLCs were not suitable for controlling complex processes, but they were very cheap and efficient for small and medium automation applications. At present every major control equipment manufacturer offers, apart from big and fast Modular PLCs, at least one Compact PLC model. Several years ago, when those controllers appear on the market they combine features of controller with programming capabilities inside. Siemens has introduced new family of PLC C7 and simple and cheap "LOGO!". Minicontroller "LOGO!" achieved great success on market and starts new group of small and cheap logic controllers with embedded display and keyboard. Market accepted this new construction and competitors appear on market like EASY from Klöckner Moeller.

2. PROGRAMMING, METHODS AND COMPONENTS

Fundamental assumption made for compact logic controller construction was embedded programming interface that doesn't require use of additional programming equipment like personal computer. Programming is made possible by integration of small keyboard and LCD display into controller case. There are some technical issues that must be solved like architecture of electronic circuit, type and size of LCD display (textual or graphical), number of required keys in keyboard. We have assumed that programming should not influence in any way functionality and computing speed of controller. Designed automation device should be fast, functional, reliable and offer flexible programming possibilities.

As a modern construction can be considered controller based on FPGA devices configured by microprocessor programming system depending on entered program. In opposite to classical concept of serial-cyclic computation concurrent method of working is possible there . All output values can be determined almost immediate (Milik and. Hrynkiewicz, 1999). Unfortunately internal architecture of FPGA and configuration data format details are not available for customers.

For this reason we choose microprocessor architecture. As we know target platform further assumption was made concerning hardware architecture, graphical display and necessity of use additional microcontroller for user interface service. Additional microconotroler would be responsible for programming interface, program translation and communication with external equipment.

While programming language was considered, attention was paid on complication of programming and user friendly feature of the controller. One of the SIMENS' advertisement watchwords says that LOGO! controller is so simple even people without specialised technical education can program it. Programming tutorial uses ladder diagrams and shows how to translate it into logic gates network (Siemens AG). This example has convinced us that ladder diagram is the simplest and the most readable for the most people that would have programmed this device. When works over the controller idea was carried on an information about similar solution with embedded display and keyboard was not known. Few months later appears report in press about small compact controller from Klöckner Moeller called EASY (Klöckner Moeller, 1998). EASY was programmed with use of ladder diagram. Till that time not only was general architecture designed but also graphical symbols were ready and display area was assigned to particular function. We can say that EASY in any way did not influence our design.

We were fascinated with handheld programmer Modicon VPU192 designed for controller Modicon Micro that belongs to family Modicon 984. It is equipped with alphanumeric display containing four 20-characters lines. A network number is presented on the display as well as output number, part of ladder diagram and some other information (AEG, 1995). Diagram edition is possible but also some kind of debug view that shows "current flow" in switches network. Display area was divided into two parts. One of them is ladder diagram while second part contains switch parameters (address, state and so on). Rectangular shapes represent switches. When switch is on half of it is blacken. Idea of display arrangement and methods of schematic drawing in our controller is based on solution applied in Modicon VPU192.

Method of programming limits controller to basic logical operation. Basic logical operation, edge detection, pulses counting (counters), time interval measurement (timer) was specified as set of operation that should be implemented. Before this set of operation was established the authors reviewed of many programmable logic controller solutions. Those research shows that each of them offers different amount and type of components in ladder diagram language. Step5 language apart from switches normally open and closed offers two type of SR flip-flops, few types of timers and counters. Software

package Modsoft and Concept from Modicon doesn't have flip-flops and have only one timer type and two types of counters. In opinion of many users shorter list of components doesn't mean worse or significantly reduction of programming freedom. Selected for designed controller components of ladder diagram language that are functionally complete set are presented in table 1.

Table 1 Ladder Diagram language elements

ELEMENT	SYMBOL
horizontal connection	—
vertical connection		
direct contact	⊣ ├
inverted contact	⊣/├
coil	⊣( )├
counter	C Q CO R
timer	C Q TO R
positive transition-sensing contact	⊣↑├
negative transition-sensing contact	⊣↓├
branch out connection	⊤

3. PROGRAM EXECUTION

In previous paragraph we have specified ladder diagram components. At this point we start work over general diagram analysis algorithm. There are three methods of diagram analysis used in professional controllers. Siemens and GE Fanuc review schematic by rows that means that analysis go horizontally from left to right (figure1). If during analysis of ladder diagram row vertical connection was reached it means that the analysis of next row should begin (Legierski, et al., 1998).

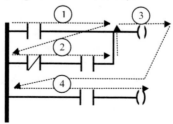

Fig.1. Row analysis

102

AEG Schneider Modicon software package interprets ladder diagram in different manner. Technical details are not known but we know that schematic is scanned column by column vertically from top of the column to it bottom. When all procedures connected with currently analysed component are finished, in opposite to horizontal analysis, component placed below is analysed. When bottom of the column is reached analysis starts from top of next column (figure 2).

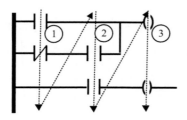

Fig.2. Column analysis

Column analysis enables signal distribiution into other rows gives freedom of vertical connection creation. Signal can be forked or joined in diagram without any limitation while in row analysis it is impossible. Presented method of a schematic representation has also disadvantages. It's possible that developed schematic can behave differently than it was expected. Example of such specific behaviour is given in figure 3. The coil 1 is defined at the bottom of schematic page, but it's state is calculated befor coil 2. It can lead to the situation when state of the coil 1 influence on the circuit defined above it. We will call this as an effect of column analysis.

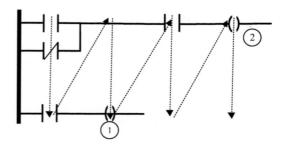

Fig.3. An example of column analysis

Third method of analysis is conversion of ladder diagram into Boolean equation that is written into controller memory. Controller executes program written into memory in appropriate format (Banasik, 1988; Kessler, 1991).

Presented methods of schematic analysis have advantages and disadvantages collected in table 2.

Table 2 Advantages and disadvantages of three types of schematic analysis

Analysis type	Advantages	Disadvantages
Row	• Siple requirements for CPU bit operation; • ON LINE debug mode is easy to implement; • No effect of column analysis.	• Complicated calculation of state in vertical connection; • Execution time depends on components placement on schematic sheet; • Procedures for signal distribution in network are complicated.
Column	• Great simplicity of the method; • Quick analysis can be compared with row analysis; • ON LINE debug mode is easy to implement; • Connection can be spread out and joined.	• Column effect; • Execution speed depends on component graphical function presentation (logical optimisation is required); • It demands flexible bit operation CPU.
Diagram to Boolean equation conversion	• Execution speed is independent from schematic graphical presentation (simplified Boolean equation are calculated); • When ideal translation methods are used execution speed can be compared to instruction list program execution.	• Extremely high complexity of translation methods; • Very demanding simplification procedures; • High caculation power is needed.

From all considered ideas we choose methods of column analysis. This method was evaluated as the simplest and allows achieving satisfying results. This method also gives user freedom in creation of network connection. User can create any type of forked connection and join any number of branches together. Signal and branch tracing (debug) for this method also can be realized relatively simple. State of traced branch is displayed with appropriate selection on display. Watching of "current flow" in circuit allow for easier and faster detecting of functional mistakes in automation design. Design effort was made in order to obtain high quality debug system for rapid software development and maintenance.

Each row of schematic diagram (rung) is assigned one bit (variable V.row number– see figure 4). Its value describes state of "current flow" in particular row during analysis stage (while column is being analysed). When this bit is set connection to power rail on the left side of schematic is available (any route). When bit is cleared connection to power rail is break.

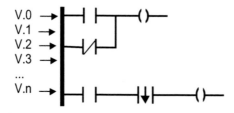

Fig.4. V bits assignment to row

At the beginning of analysis all V vector variables are set. Modification of V items is done by service functions of particular network components. Schematic scanning program iterates through all components in each column from its top to bottom. When component is recognised appropriate service function is called (figure 5). Service function carry out required operation (e.g. input state fetch) and it result are placed in V.row variable.

When end of the first column is reached the second column is being analysed. Column analysis is repeated till it reaches last column on the right. When entire schematic is analysed function assigned to serial-cyclic program execution are performed. Outputs are written to and input states are copied into process image memory.

Prototype version of schematic analysis fetch and identify all schematic cells. Computation time for schematic with only one component requires almost the same time that entire area of schematic filled with components. Of course this is great disadvantage and

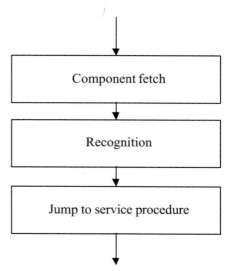

Fig.5. Analysis of column schematic

strongly impact computation speed. Final version of program was improved with electrical schematic mapping that force only analysis of cells that contains network component. Electrical schematic mapping speeds up analysis and works independently from scanning procedures that remain unchanged.

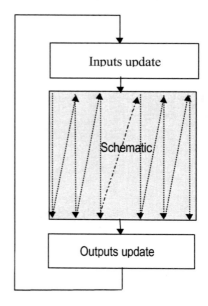

Fig.6. Control program execution method

4. DIAGRAM COMPONENT SERVICE FUNCTIONS

In this paragraph is given breath view of operation carried when specific component is reached during network analysis. Horizontal connection doesn't influence state of V.row variable. Horizontal connection service procedure is an empty call that does nothing. In ON LINE mode, state of V.row is fetched and appropriate information is displayed (figure 7). Empty cell is equal to brake in a network and V.row variable is unconditionally cleared. Wherever symbol of horizontal connection is placed variable V.0 remains unchanged. Empty cell brakes circuit connection and V.0 is cleared.

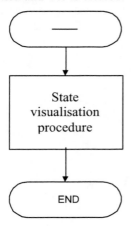

Fig.7. Horizontal connection service procedure

Vertical connection is equivalent logical OR operation of adjacent branches of circuit from its left side. Operation result is written to all connected together cells of V vector from right sight of the schematic. This operation is similar to logical or of all components of V vector that are joined together on the right side. Finally all connected together vector variables get operation result value (figure 8).

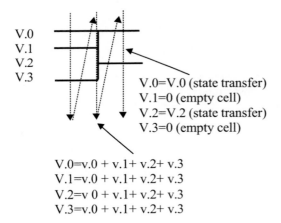

Fig.8. Vertical connection analysis

Switch normally open doesn't influence on V.row variable when switch is closed (high). This happens when assigned to switch variable is in high state (ON). In opposite case variable V.row is cleared (figure 9).

Normally closed switch differs from previous one in switch closing condition test. V.row variable remains unchanged when variable assigned to switch is in low state (OFF).

Switches sensitive to state changes (rising or falling edge detection) remains closed for time of one program scan cycle and they influence on set V.row

variable for on scan. In other situation V.row variable is cleared. Switch is closed when new state of variable differes that the previous old one (depends on direction).

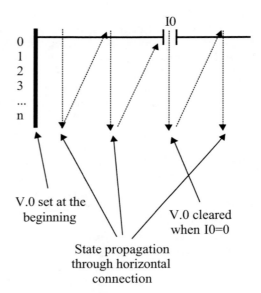

Fig.9. V.0 variable states during row analysis

Coil service procedure assigns state of V.row variable to coil variable (figure 10). This operation influences only on output image table in PLC memory. Assignment to output is done after completion of entire schematic analysis at the end of a program cycle.

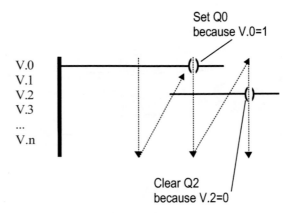

Fig.10. An example of coil setting and clearing

In case of encounter other components (counters, timers) execution is continued accordingly to rules of recognized component. For example variable V.row is set by counter when given pulses count is reached. In other cases it is cleared. Depending on timer type V.row variable is kept in particular state (high or low) from trigger event for given period of time.

5. Summary

Designed and build controller works correctly. We were able to meet all requirements made at the beginning of design process. Device can be programmed by means of embedded keyboard and display with use of ladder diagram (figure 11). Controller can save and restore written program. Controller is equipped in PC interface to program transfer to and from PC file.

Simplicity of keyboard and editing procedures drawing of larger schematic takes relatively long time. It was supprising for the authors that little, handheld device gives similar comfortof programming as the program editors for prsonal computers. Program execution speed can be compared with similar construction of compact controllers. Functionality of designed controller offers only basic function. Construction comprises the most important function for that type of controller. Grate advantage of designed controller is ON LINE debug system that traces branches value. Designed and build controller features 16 digital inputs, 8 relay outputs, 16 markers and 8 counters and timers. User program consists of up to 5 networks. Each network contains up to 8 rows and 10 columns – it gives maximum 400 element fields.

Time execution of elementary function (e.g. serial connection of two switches) is 60µs.

References

AEG (1995): *Modicon Micro / Modicon 984-120 Compact. Logic Controller Data Sheet, User Manual.*

Banasik K. (1988): *Ladder diagram based instruction generation for Pictronic PC system MS EE thesis.* Silesian University of Technology, Gliwice, Poland.

Kessler W. (1991): *Ladder diagram based instruction generation for Pictronic PC system continuation of prevoius work from 1988 MS EE thesis* (in polish). Silesian University of Technology, Gliwice, Poland.

Klöckner Moeller (1998): *EASY 412-DC-R, EASY 412-AC-R. User manual.*

Legierski T., Kasprzyk J., Hajda J., Wyrwal J.(1998): *PLC programming* (in polish). Wydawnictwo Pracowni Komputerowej Jacka Skalmierskiego, Gliwice, Poland.

Milik A, E. Hrynkiewicz (1999): "Reconfigurable Logic Controller". IWCIT. Ostrava, Czech Republik.

Siemens AG: *LOGO!. User manual. Third edition.* Siemens AG.

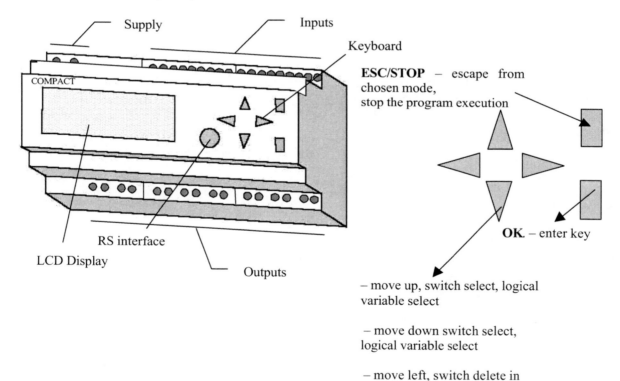

ESC/STOP – escape from chosen mode, stop the program execution

OK. – enter key

– move up, switch select, logical variable select

– move down switch select, logical variable select

– move left, switch delete in PLACE MODE

– move right, horizontal connection in PLACE MODE

Fig.11. The view of the conteoller an keys meaning

ON POSSIBILITIES OF THE B-BAC IMPLEMENTATION
ON PROGRAMMABLE LOGIC CONTROLLERS

Jacek Czeczot

Institute of Automatic Control, Technical University of Silesia
ul. Akademicka 16, 44-100 Gliwice, Poland, jczeczot@ia.polsl.gliwice.pl

Abstract: In this paper two important problems are considered. First, the principles of the B-BAC (Balance-Based Adaptive Control) methodology are presented. This methodology is dedicated to the adaptive control of a wide range of nonlinear processes. It is based on the simplified and generalised dynamical model of a process derived on the basis of mass and/or energy balance considerations. Due to its generality the form of the B-BAC ensures the possibility of its implementation on PLC devices. The second part of this paper deals with these possibilities and shows how to adjust the structure of the final control algorithm to a particular process that is to be regulated. The most important features of the control algorithm and of the estimation procedure are also discussed in term of the PLC-based implementation. *Copyright © 2001 IFAC*

Keywords: Model-based control, Control oriented models, Adaptive control, Least-squares estimation, Programmable logic controllers.

1. INTRODUCTION

In the field of the advanced automatic control the very important development has been taking place for last several years. Although the majority of control applications in the process industries still rely on the classical PID controller (Seborg, 1999), it must be said that the advanced control techniques and especially the model-based approaches have become more and more important area of the research activities (Henson and Seborg, 1997, Joshi et al., 1997, Seborg, 1999). These research activities mainly concentrates on the theoretical considerations and simulation experiments but it must be always kept in mind that the control algorithm is derived to be applied in industrial control loops. Of course, the stage of the theoretical and experimental researches is very important for every new control strategy that

is being developed but the practical application is always the main goal.

Nowadays there are two possibilities that should be considered for such a practical application. One of them is the implementation as a PC-based simulated controller: so-called virtual controller (Wolfe, 1993, Metzger, 2000). The other possibility, which is discussed in this paper, is the implementation of the new advanced control algorithm on a commercial PLC device. Since the characteristic feature of each model-based approach is that the final form of a control law depends on a model of a process that is to be controlled, the possibility of the PLC-based implementation must be discussed for each control law separately.

This paper deals with the possibility if the implementation of the B-BAC (Balance-Based Adaptive Control) algorithm on a PLC device. The B-BAC methodology is dedicated to the adaptive control of a wide class of technological processes. Its generality ensures that this methodology can be considered as an interesting alternative for the classical PID controller. The control performance of the B-BAC algorithm in the application to different technological processes was validated by computer simulation and can be found in (Czeczot, 1999, 2001, 2002). In (Czeczot, 2001a) the possibilities of the B-BAC implementation as the LabView-based virtual controller are discussed.

This paper is organized as follows. First the theoretical approach to the B-BAC methodology is given. Then the most important features of this methodology are discussed in terms of the PLC-based implementation. Concluding remarks complete the paper.

2. THEORETICAL APPROACH TO THE B-BAC METHODOLOGY

As it was said before, the B-BAC (Czeczot, 2001, 2002) is dedicated to control a wide range of technological processes for which it is possible to define the control goal in the following way: one of the parameters characterizing a process, defined here as $Y(t)$ and called the controlled variable, should be kept equal to its pre-defined set-point Y_{sp}. $Y(t)$ can be chosen as one of state variables (a component concentration or the temperature) or as a combination of two or more state variables. A number of isothermal or nonisothermal biochemical reactions and/or heat exchange phenomena with unknown kinetics can take place due to a process. A process itself takes place in a tank of time varying volume $V(t)$ [m^3].

The dynamical behavior of $Y(t)$ can be described by the following well known general ordinary differential equation written on the basis of the mass or the energy balance considerations:

$$\frac{dY(t)}{dt} = \frac{1}{V(t)} \underline{F}^T(t)\underline{Y}_F(t) - R_Y(t) \qquad (1)$$

The vector product $\underline{F}^T(t)\underline{Y}_F(t)$ represents mass or energy fluxes incoming to or outcoming from the reactor tank. The elements of the vector $\underline{F}(t)$ are the combination of the volumetric flow rates and, consequently, the vector $\underline{Y}_F(t)$ is the corresponding vector to $\underline{F}(t)$ and its elements are the combination of the inlet values of $Y(t)$ and of the value of $Y(t)$ itself. $R_Y(t)$ is a positive or negative time varying term with an unknown expression form. It represents

one global reaction including all reversible and/or irreversible reactions or heat exchange and/or production with unknown and nonlinear kinetics that influence the value of $Y(t)$. Let us note that in the case when $Y(t)$ is a state variable, the equation (1) is a generalized form of a state equation describing $Y(t)$ and taken directly from a mathematical model of a process. However, if $Y(t)$ is a combination of two or more state variables, a part of a mathematical model must be combined and rearranged to obtain the equation (1).

Once the equation (1) has been obtained, it can be a basis for the B-BAC under the following assumptions:
- the control variable must be chosen as one of the elements of the vectors $\underline{F}(t)$ or $\underline{Y}_F(t)$,
- the other elements of the vectors $\underline{F}(t)$ and $\underline{Y}_F(t)$ as well as the value of $Y(t)$ must be measurable on-line at least at discrete moments of time or they should be known by choice of the user.

If the above requirements are met, at this stage we can apply the same methodology as in the case of the 'model reference linearising control' (Isidori, 1989; Bastin and Dochain, 1990). For our control goal let us assume the following stable first-order closed loop dynamics:

$$\frac{dY(t)}{dt} = \lambda\left(Y_{sp} - Y(t)\right) \qquad (2)$$

where λ is the tuning parameter. After combining the equations (1) and (2) we can obtain the following equation:

$$\underline{F}^T(t)\underline{Y}_F(t) = \lambda V(t)\left(Y_{sp} - Y(t)\right) + V(t)R_Y(t) \qquad (3)$$

Once the control variable has been chosen, the above equation can be rearranged to obtain the control law describing its value. This control law has the form of the 'model reference linearising controller' and is very well known in bibliography.

In order to provide the adaptability to the control law resulting from the equation (3) there is a need to estimate the value of the nonmeasurable term $R_Y(t)$. This value can be estimated on-line at discrete moments of time by the recursive least-squares method with the forgetting factor α. The estimation procedure is also based on the discretised form of the simplified model (1) (Czeczot, 1997, 1998, 1998a). For this purpose let us consider the discrete form of the equation (1) with backward Euler discretisation and with the sampling time T_R [min]:

$$V^i\left(Y^i - Y^{i-1}\right) = T_R \underline{F}^{T,i}\underline{Y}_F^i - V^i T_R R_Y^i \qquad (4)$$

where i denotes the discretisation instant. Let us define the auxiliary variable y^i:

$$y^i = V^i \left(Y^i - Y^{i-1} \right) - T_R \underline{F}^{T,i} \underline{Y}^i_F \qquad (5)$$

Note that, in line with the assumptions made above for the controller design, y^i is only a function of measurable or known parameters so its value can be easily calculated at each discrete moment of time. The following equation can be derived by combining the equations (4) and (5):

$$y^i = -V^i T_R R^i_Y \qquad (6)$$

Since the equation (6) is linear with respect to the parameter R^i_Y that is the only one value, which is to be estimated, it is possible to apply the recursive least-squares method. It leads us to the equations describing the value of $\hat{R}^i_Y$ that is the discrete time estimate of the term $R_Y(t)$:

$$\hat{R}^i_Y = \hat{R}^{i-1}_Y - V^i T_R P^i \left(y^i + V^i T_R \hat{R}^{i-1}_Y \right) \qquad (7a)$$

$$P^i = \frac{P^{i-1}}{\alpha} \left(1 - \frac{V^{i2} T_R^2 P^{i-1}}{\alpha + V^{i2} T_R^2 P^{i-1}} \right) \qquad (7b)$$

where α is the forgetting factor.

When we replace $R_Y(t)$ by its discrete time estimate $\hat{R}^i_Y$ in the equation (3), it can be rewritten in the following discrete form:

$$\underline{F}^{T,i} \underline{Y}^i_F = \lambda V^i \left(Y_{sp} - Y^i \right) + V^i \hat{R}^i_Y \qquad (8)$$

The equation (8) is a basis for the B-BAC. There is only a need to implement it together with the estimation procedure to calculate the value of the control variable. Let us note that the suggested in this paper controller is general since it is derived on the basis of the generalized form of the balance equation (1). It is also adaptive since the unknown term R^i_y is replaced by its estimate $\hat{R}^i_Y$. The estimation procedure is very simple to carry out and also general (since it is also based on the general equation (1)) and thus there is even no need to assume any form of a nonlinear expression describing the term $R_Y(t)$. Moreover, there is no need to know a complete mathematical model of a process. Both the B-BAC and the necessary estimation procedure are derived on the basis of the general mass and/or heat balance considerations in the form of the equation (1).

3. IMPLEMENTATION OF THE B-BAC ON PLC DEVICES

The methodology, presented in the previous section, is a very general approach to the problem. Due to this fact the control law (8) itself has also very general form and therefore it is practically impossible to implement this control algorithm in this form on any PLC device. Let us shortly explain the reasons why.

First of all the elements of the vectors $\underline{F}(t)$ and $\underline{Y}_F(t)$ must be determined for each implementation. Especially it is important to define a control variable because its value is calculated as the controller output and therefore it must be "connected" to the analog output module.

It must be also said that the indirect form of the B-BAController law (8) cannot be used for calculation of the current value of a control variable. This form should be rearranged into a direct form, which allows a control variable to be calculated directly. Such a rearrangement cannot be made in a general way and it must be made for each implementation separately.

Due to the above facts let us show on the simple and practical example how to derive the final form of the B-BAControllers on the basis of the general B-BAC methodology, given in the previous section.

3.1 Practical example: continuous fermentation process

As the illustrative example let us consider the continuous fermentation process (for instance bacterial production of amino acids, e.g. lysine) that takes place in the biotechnological reactor with the tank of the constant volume V. The simplified diagram of this process is presented below.

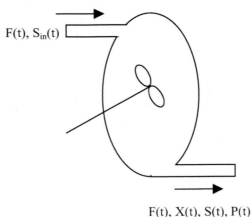

$F(t)$, $S_{in}(t)$

$F(t)$, $X(t)$, $S(t)$, $P(t)$

Fig. 1. Simplified diagram of the example biotechnological reactor

For our process F(t) represents the volumetric flow rate of the medium flowing through the bioreactor tank, V is the constant volume of the bioreactor tank, $S_{in}(t)$ is the inlet substrate concentration, X(t) is the outlet biomass concentration, S(t) is the outlet substrate concentration and P(t) is the outlet product concentration.

In practical applications of such systems usually the yield-productivity conflict occurs and thus there is a need to manage this problem by properly designed control loop. It was shown in (Bastin and Dochain, 1990; Van Impre and Bastin, 1995) that the effective control of the lysine production process consists in regulating the outlet substrate concentration S(t) at the properly chosen set-point S_{sp} during the bioreactor activity. Notice that this control goal is in line with the one defined for B-BAC in Section 2.

As it was said before, the B-BAC methodology is very general and thus it allows us to derive the model-based adaptive control algorithm without any knowledge about the complete mathematical model of a process itself. If one is interested in the control performance of the B-BAController in the application to this particular example, it can be found in (Czeczot, 1999).

In the case of our example the controlled variable is the outlet substrate concentration S(t) and, as it was stated above, the control goal is to keep this value equal to its set point S_{sp}. Let us note that the dynamics of the controlled variable can be easily described by the general dynamic state equation written in the form of the equation (1) on the basis of the mass balance considerations:

$$\frac{dS(t)}{dt} = \frac{1}{V}\underline{F}^{T}(t)\underline{Y}_{F}(t) - R_{Y}(t) \qquad (9)$$

Now we can define the elements of the vectors $\underline{F}(t)$ and $\underline{Y}_{F}(t)$ in the following way:

$$\underline{F}(t) = [F(t)] \qquad (10a)$$

$$\underline{Y}_{F}(t) = [S_{in}(t) - S(t)] \qquad (10b)$$

The value of the time varying parameter $R_{Y}(t)$ represents the nonlinearity resulting from the biotechnological reaction taking place and for this particular process it can be called the "substrate consumption rate". This value cannot be measured on-line and therefore it must be estimated by the estimation procedure described in Section 2.

For the considered process there are two possible control variables: the flow rate F(t) and the inlet substrate concentration $S_{in}(t)$. Therefore, even if it is possible to derive the control law in the general form

of the equation (8) with the vectors $\underline{F}(t)$ and $\underline{Y}_{F}(t)$ defined above, it must be said once again that this form of the B-BAControl law is useless for implementation on any PLC device. We need to rearrange it into the direct form and this direct form always depends on the choice of the control variable.

The general form of the B-BAController for our example is given by the equation (11). This form is based on the form of the equation (8) with the vectors $\underline{F}(t)$ and $\underline{Y}_{F}(t)$ defined by the equations (10a – 10b). The vector product $\underline{F}^{T}(t)\underline{Y}_{F}(t)$ has also been calculated, which results in the final form of the equation (11).

$$F^{i}\left(S_{in}^{i} - S^{i}\right) = \lambda V\left(S_{sp} - S^{i}\right) + V\hat{R}_{Y}^{i} \qquad (11)$$

Now we have to consider two possible choices for the control variable and finally we can rearrange the above equation into two different B-BAControllers for two different control variables:

- for the flow rate F(t)

$$F^{i} = \frac{\lambda V\left(S_{sp} - S^{i}\right) + V\hat{R}_{Y}^{i}}{S_{in}^{i} - S^{i}} \qquad (12a)$$

- for the inlet substrate concentration $S_{in}(t)$

$$S_{in}^{i} = \frac{\lambda V\left(S_{sp} - S^{i}\right) + V\hat{R}_{Y}^{i} + F^{i}S^{i}}{F^{i}} \qquad (12b)$$

Of course, for both B-BAControllers there is still a need to apply the estimation procedure for on-line estimation of the value of $R_{Y}(t)$ (namely, to obtain its estimate $\hat{R}_{Y}^{i}$).

3.2 On-line measurement data acquisition

For each PID controller there is a need to have the measurement of a controlled variable and the possibility of the adjustment of a control variable. For the B-BAC it is important to provide additional measurement data. As it was said before, all the elements of the vectors $\underline{F}(t)$ and $\underline{Y}_{F}(t)$ (excluding a control variable) must be measurable on-line or at least known by choice of the user. Thus, if the B-BAC is to be implemented on a particular PLC device, it must be noticed that it is very important to define the elements of the vectors $\underline{F}(t)$ and $\underline{Y}_{F}(t)$ and then to know exactly how many measurement signals are needed because it results in the number of the analog input modules that must be connected to our PLC device. The number of these signals depends on the particular process that is to be regulated.

In the case of our example the vectors $\underline{F}(t)$ and $\underline{Y}_F(t)$ are defined (10a – 10b) but we also need to define the control variable to know exactly which parameters must be measured:

- for the control variable $F(t)$ we need measurement data for the inlet substrate concentration $S_{in}(t)$ (disturbance) and for the outlet substrate concentration $S(t)$ (controlled variable) – B-BAController (12a)
- for the control variable $S_{in}(t)$ we need measurement data for the flow rate $F(t)$ (disturbance) and for the outlet substrate concentration $S(t)$ (controlled variable) – B-BAController (12b)

In both cases the volume of the reactor tank V is constant and can be known by choice of the user.

Let us note that the same choice of the measured parameters is necessary both for the controller and for the estimation procedure. Therefore, once the choice of the measured parameters for the B-BAController has been made, it is always suitable for the estimation procedure as well. However, the estimation procedure additionally needs the current value of a control variable but this value is the calculated output of the controller so there is only a need to store it in the memory of a PLC device without any necessity of measuring it additionally.

The analog input modules that must be used to provide measurement data must be chosen to be compatible with sensors. If it is necessary, some additional recalculations should be made for re-scaling some measurement signals. Moreover, some external signal conditioners may have to be used for the same purpose.

3.3 Implementation of the estimation procedure

The estimation procedure, described in details in Section 2, is based on the very well known recursive least-squares method. However, since in the B-BAC methodology there is only one time varying parameter to estimate (namely $R_Y(t)$ that represents all the nonlinearities in a process), the equations for this method have the scalar form instead of the common matrix form. Due to this fact it is possible to avoid all the difficulties resulting from multiparameter identification. Moreover, there is no need to apply any additional exciting signals and the influence of the measurement noise can be decreased by suitable adjustment of the forgetting factor α. All these features ensure very good convergence and estimation accuracy, which, in consequence, ensure very good adaptability of the B-BAC.

For the implementation of the estimation procedure it is important to note that the procedure itself is recursive so there is a need to store the current values of the parameter P^i and of the estimate $\hat{R}_Y^i$ in memory of a PLC device. These values must be calculated at each run of the programming loop according to the equations (7). The other difficulty that results from the recursive nature of the calculations is the problem of the initial values. The initial value P^0 should be chosen as $P^0 \gg 1$ and its value has no significant influence on the estimation accuracy in the case of the scalar form of the equations (7). As far as the initial value $\hat{R}_Y^0$ is concerned, it must be said that the choice of its value has very significant influence on the initial stage of the estimation run. Although the estimate $\hat{R}_Y^i$ always converge to its true value $R_Y(t)$, there is a transient at the initial stage of estimation due to choice of the initial value $\hat{R}_Y^0$. Therefore, in order to avoid the influence of this estimation inaccuracy on the control performance, there is a need to start the estimation procedure in open loop and then to close the control loop with B-BAC after the estimate convergence.

3.4 Switching between manual and automatic mode

Sometimes, while using PLC devices in control loop there is a need to ensure the possibility of switching between the manual control (open loop control) and the automatic control (closed loop control). For this purpose the bumpless switching (Hanus *et al.*, 1987; Trybus, 1992) is an important feature that must be discussed for the B-BAC.

The idea of bumpless switching consists in the necessity to ensure that the value of a control variable calculated from a control law is the same as the one adjusted manually by the user. Only in this case the switching between the manual and the automatic mode is always bumpless.

Let us note that in the case of the B-BAC the bumpless switching is always possible because the final form of the control law (8) is based on the general form of the simplified dynamic state equation (1) and therefore it is always valid, even if the value of a control variable is adjusted manually. Of course, the equation (8) is satisfied only if the value of the estimate $\hat{R}_Y^i$ is always up-to-date. In other words, even if the B-BAC is switched into the manual mode and the control application works in open loop, the estimation procedure must be in run all the time to ensure bumpless switching. It also allows us to avoid the difficulties with the initial stage of the estimation procedure due to choice of the initial value $\hat{R}_Y^0$, described in the previous subsection.

4. CONCLUSIONS

In this paper the most important features of the B-BAC algorithm are discussed in terms of its application on commercial PLC devices. To summarize, let us state that due to its generality and very good control performance the B-BAC algorithm can be considered as an interesting alternative for the classical PID controller in practical control applications for a wide class of technological processes. However, although the classical PID controller has always the same structure and once it has been implemented on a PLC, it can be used for each control application, the form of the B-BAC depends on the nature of a process that is to be controlled and therefore it must be derived for each control application separately and then implemented on a PLC device. The other difference between the classical PID controller and the B-BAC algorithm consists in the number of necessary measurement signals. The B-BAC needs additional measurement data from disturbing parameters but in return it provides the adaptability and feedforward action. The classical PID controller has three independent settings. In the case of the B-BAC algorithm there is a need to store in a PLC memory two settings: the tuning parameter λ and the forgetting factor α for the estimation procedure. These two settings together with the sampling time T_R can be adjusted separately to ensure the best possible control performance.

Finally, let us state that that in our opinion it is possible to implement the B-BAC algorithm on any commercial PLC device. This conclusion allows us to suggest the B-BAC to be in use in practical control applications in the cases when the application of the classical PID controller gives unsatisfying results and further improvement of the control performance is necessary.

Acknowledgements: This work was supported by Polish Committee of Scientific Research, (KBN), project no 8 T11A 001 16.

REFERENCES

Bastin G., Dochain D. (1990). *On-line estimation and adaptive control of bioreactors.* Elsevier Science Publishers B.V.

Czeczot J. (1997). On possibility of the application of the substrate consumption rate to the monitoring and control of the water purification processes. Ph.D. Thesis, Technical University of Silesia, Gliwice, Poland. (in polish).

Czeczot J. (1998). Substrate Consumption Rate Application to the Minimal-Cost Model Based Adaptive Control of the Activated Sludge Process. *Wat. Sci. Tech.* 37 (12), pp. 335-342.

Czeczot J. (1998a). Application of the recursive least-squares method to the estimation of the substrate consumption rate in the activated sludge process. In: *Proceedings of the 9th International Symposium on "System-Modelling-Control",* ed. P.S. Szczepaniak, Zakopane, Poland

Czeczot J. (1999). Substrate consumption rate application for adaptive control of continuous bioreactor – noisy case study. *Archive of Control Sciences,* **9**, No. ¾, pp. 33 – 52.

Czeczot J. (2001). Balance-based adaptive control of the heat exchange process. *7th IEEE International Conference on Methods and Models in Automation and Robotics, MMAR 2001,* Miedzyzdroje, Poland.

Czeczot J. (2001a). On possibilities of application of B-BAC methodology to LabView-based virtual controller. ESS 2001, Marseille, France (accepted).

Czeczot J. (2002). B-BAC: Generalised approach to the nonlinear adaptive control of industrial processes. IFAC World Congress, Barcelona 2002 (proposed for publication).

Hanus R., Kinneart M., Henrotte J.L. (1987). Conditioning technique, a general anti-windup and bumpless transfer method. *Automatica.* **23**, No. 6, pp. 729-739.

Henson M. A., Seborg D. E. (1997). *nonlinear process control.* Prentice Hall, Englewood Cliffs, NJ.

Isidori A. (1989). *Nonlinear control systems.* New York, Springer-Verlag.

Joshi N.V., Murugan P., Rhinehart R.R. (1997). Experimental comparison of control strategies. *Control Eng. Practice,* **5**, No. 7, pp. 885-896.

Metzger M. (2000). *Modelling, simulation and control of continuous processes.* Edition of Jacek Skalmierski, Gliwice.

Seborg D. E. (1999). A perspective on advanced strategies for process control. *ATP* **41**, No. 11, pp. 13-31.

Trybus L. (1992). *Multifunction controllers.* Warsaw, WNT (in polish).

Van Impre J.F., Bastin G. (1995). Optimal adaptive control of fed-batch fermentation processes. *Control Eng. Practice,* **3**, No. 7, pp. 939-954.

Wolfe R. (1993). Virtual instruments in VXI. *Proceedings of the IEEE Conference on Instrumentation,* pp. 183-190.

IFAC
Publications
www.elsevier.com/locate/ifac

A MATLAB-BASED PETRI NET SUPERVISORY CONTROLLER FOR DISCRETE EVENT SYSTEMS

Jana Flochová, Ronald Lipták, Rene K. Boel

*Department of Automatic Control Systems, Faculty of Electrical Engineering and Information Technology
Slovak Technical University, Ilkovičova 3, 812 19 Bratislava, Slovak Republic
email: flochova@kasr.elf.stuba.sk
Group Systems,Universiteit Gent, Technologiepark Zwijnaarde 9, 9000 Gent, Belgium,
email: rene.boel@rug.ac.be*

Abstract: This paper presents a methodology for modelling and implementation of discrete event control systems using a Petri net formalism. As an implementation platform a programmable Petri net-based controller is proposed which operates in an event-driven manner, delivers fast response to the incoming external stimuli and power efficiency. A tool for constructing a supervisory controller of discrete event systems has been designed and two methods have been included in the programs – a method based on P-invariants and a method based on reachability tree analysis. Programs in Matlab 5.3/6 have been written to solve off-line supervisory analysis and simulation of discrete event dynamic systems. A real time on line PLC control can be written and used to control a large discrete event system after the off-line analysis. *Copyright © 2001 IFAC*

Keywords Discrete-event dynamic systems, Petri-nets, Supervisory control, Reachability, Invariants, Matlab based PN-supervisory controller

1. INTRODUCTION

Discrete event systems (DES) describe the behaviour of a large plant as it evolves over time in accordance with the abrupt and asynchronous occurrence of events. The complexity of the plant models forces the modeller to use a countable (or even finite) state space. The evolution in time of the different modules of the plant is typically asynchronous and non-deterministic. Such systems are encountered in a variety of fields, for example manufacturing, robotics, computer, communication, networks and traffic. The supervisory control of a DES is based on feedback of the occurrence of events and must be implemented via a real time supervisory controller. The supervisory controllers must be designed so that the controller specification do not contradict the behavioural specifications of the plant model, i.e. the closed loop system is nonblocking, and such that the closed loop system is maximally permissive within the specifications (all events which do not contradict the specifications are allowed to happen).

The problem of supervisory control of discrete event dynamic system (DEDS) was presented by Ramadge and Wonham (1987) and since then has been studied extensively. Ramadge and Wonham, (1989), Ho, Cassandras (1990) described the dynamic behaviour of DEDS with a possibly infinite set of states and a finite set of events. The adopted level of abstraction assumes that events appear spontaneously, asynchronously and instantaneously and that always only one event occurs in some instant of time. For a given DEDS, it is of interest to synthesise a supervisor that prevents the occurrence of undesirable states of the DEDS and that guarantees that certain termination states are reached. Disabled events are certainly prevented from occurring and enabled events are not forced to occur. The mathematical framework of Ramadge and Wonham's approach is based on finite state machines and formal languages. A DEDS is regarded as a finite state machine formal language generator.

One method for simplifying the control synthesis algorithm, and for generating easily implementable control laws is to use a Petri net as a model for the DES plant. The assumption then is that only some of the transitions can be blocked by an external controller, while some other uncontrollable transitions can always be executed when they are state enabled. Holloway and Krogh (1990) specified a class of Petri nets called cyclic controlled marked graphs for which they obtained an easy control design algorithm. The states of the system are presented by the Petri net markings. The specification of the behavior of the controlled system is determined with help of the forbidden markings in a Petri net. The forbidden markings are described by means of so-called place conditions. More generally the set conditions or the class conditions can also be treated. The next-state supervisory control is completely solved in the paper of Holloway and Krogh.

Boel et al. (1995) treated the forbidden state problem for the class of DEDS's modelled by controlled state machines. State machines form a dual class to the marked graph class; they are capable of modelling choice but not process synchronizations. The authors characterized forbidden markings through general constraint sets obtained from unions and/or intersections of simpler constraint sets expressing that some places of the net cannot contain more than a certain number of tokens. The control laws require that one can observe the marking in certain influencing nets, containing all the places from where a token can uncontrollably reach a place involved in the specification of the forbidden markings. In Stremersch and Boel this approach has been extended to general Petri net models, and to more general forbidden sets, allowing the specification of a minimal number of tokens in some places. The method of Stremersch (2000) uses integer linear algebraic algorithms for describing the set of states (markings) reachable from the present state under certain control settings. It is shown that if the Petri net satisfies certain structural conditions then a simple linear algebraic algorithm exists for enumerating all the markings reachable from a given state, when the control law blocks certain transitions.

Yamalidou et al. (1996) described a method for effectively constructing a Petri net controller realizing the supervisory control. The controller is given by a Petri net attached to the Petri net model of the process. The method is based on the concept of Petri net place invariants. The supervisory control is specified using a single matrix multiplication without any state enumeration. The constraints are based on place invariants, expressed as linear inequalities. The methods become computationally more cumbersome, and the maximally permissive control law cannot sometimes not be found when one takes into account that some transitions are uncontrollable or unobservable.

Guia and DiCesare (1994) defined a class of Petri nets, called elementary composed state machine net (ECSM). To describe concurrent systems the model is extended by composing the state machine modules through concurrent composition, an operator that requires the merging of common transitions. The final model can model both choice and concurrent behaviour. The reachability problem for this class can be solved by a modification of the classical incidence matrix. The set of reachable markings is given by the integer solution of a set of linear inequalities. The approach of Stremersch for supervisory control design is closely related to this approach.

Screenivas (1997) presented a necessary and sufficient condition for the existence of a supervisory policy that enforces liveness in arbitrary controlled Petri nets. Mossig and Stäble (1995) solved the supervisory control problem for free choice Petri net models. Hrúz et al. (1996a, 1996b) described a class of interpreted conflict free Petri net models for real-time control of DEDS's and its possible use in the supervisory control theory.

A general problem for modelling a supervisory control problem using Petri nets can be stated as follows. Given the specifications for a closed loop plant, and a Petri net model of the open loop discrete event system design a feedback control law that disables as few controllable transitions as possible but that guarantees that the extra safety properties expressed by the specifications (such as boundedness, liveness and reversibility) are satisfied (Zhou et. al. 1992ab). In a manufacturing example these three properties imply the absence of overflows and deadlocks, and guarantee repeated execution of critical tasks and successful completion of production cycles.

This paper describes two methods for constructing such supervisory controllers. The paper demonstrates that it is possible to construct a practically useful program tool for synthesis of supervisory controllers. The controlled system and the supervisors are modelled as bounded Petri nets.

The first method synthesizes the supervisor that consists only of places and arcs and is computed based on the concept of P-invariants. The second method is based on the Reachability Tree Analysis Algorithm. A tool for reachability tree analysis was designed in Matlab 5.3 and implemented in the environment of the operating systems Windows98, 2000/NT. The novel aspect of this approach, that makes it potentially useful for control synthesis for large plants, is the fact that the reachability tree algorithm can be applied modularly in some practically useful cases. It is possible to develop the set of reachable states using the specifications of one component first, then for another component. Under certain conditions it will be possible to generate a maximally permissive supervisory controller using this modular approach. The presentation will include some examples of controllers designed using the proposed algorithm.

2. SUPERVISORY CONTROL BASED ON THEORY OF PETRI NETS

Petri nets are graphical and mathematical modelling tools for describing and studying discrete event systems such as manufacturing plants, information processing systems. A Petri net is represented by a bipartite oriented graph, where one type of nodes represents a system condition (so called places) and the other types of nodes represent the events (so called transitions). They are interconnected by oriented arcs. The number of tokens in a place marks the current system state. The occurrence of an event corresponds to the execution or firing of a transition; it changes the marking of places according to the weights of graph arcs - i.e. tokens are added or removed from places according to the weight and the direction of the related arcs.

Formal definition of a Petri net: (Murata 1989, Češka 1994, David and Alla 1994).

The ordinary Petri net PN is the quintuple

$$PN = (P, T, F, W, M_0) \qquad (1)$$

Where

(1) $P = \{p_1, p_2, ..., p_n\}$ is a finite non-empty set of places.
(2) $T = \{t_1, t_2, ..., t_n\}$ is a finite non-empty set of transitions.
(3) F is a relation given by union of two sets formed by the Cartesian products, the set of directed arcs connecting places and transitions.

- $W : F \rightarrow \{1,2,3...\}$ is the weight function of the arcs.
- $M_0: P \rightarrow \{0,1,2,3...\}$ is the initial marking.

Methods of analysis of Petri nets may be classified into the following four groups: the coverability (reachability) tree method, the matrix-equation approach, reduction or decomposition techniques and languages.

The controlled Petri nets is septuple:

$$S = (P, T, F, W, C, B, M_0) \qquad (2)$$

Where:
- P, T are the non-empty sets of state places and transitions.
- F the set of directed arcs connecting state places and transitions.
- C is the finite set of control places, most one per transition.
- $B \subset (C \times T)$ is the set of oriented arcs associated the control places with transitions.
- M_0 is the initial marking.

A control u: $C \rightarrow \{0,1\}$ assigns a binary token count to each control place. The possibility to fire a transition depends on the marking of Petri net (a state enabled transition) and on the value of $u(c)$. The transition is said to be control enabled under a control u if $u(c)=1$, the transition is said to be control disabled under a control u if $u(c)=0$.

3. PROGRAMS

The software consists of several modules. The first group of modules is used to edit the Petri nets models, validate these models, to compute the reachability graph and the incidence matrix of the Petri net model. The second group of modules implements the methods of finding positive P and T invariants and the analysis of boundedness, safeness and L1-liveness of the process Petri net. The third group of modules implements the supervisory control analysis methods described in the following sections, and the fourth the real-time control simulation.

The approach of the programs consists of the following steps:
1. The system to be controlled is modelled by a controlled Petri net (consisting of one or more subnets) and the reachability graph for this net is designed.
2. Analysis of the oriented paths in the reachability graph and determination of forbidden nodes (markings) and illegal edges (transitions).
3. Determination of values of control places marking.
4. Real time control of analysed discrete event dynamic system.

The reachability tree design and analysis and the supervisory control design are solved by the program modules SUPCON written in Matlab 5.3/6. The programs need as processor at least PC Pentium. The user interface is written in English or Slovak and accepts inputs from the graphical simulation tool

PESIM (Česka 1994, Urbášek, Česka 1998). The modules allow the solution of the following problems:

1. The complete Petri net analysis of the system (plant) to be controlled, reachability tree construction, incidence matrix design and correction, PN analysis (boundedness, L1-liveness, deadlocks, P/T-invariants).
2. The supervisory control problem formulation.
3. The supervisor design and testing.

A part of the program solutions tool is an on line help. The main menu is fully keyboard and/or mouse driven and consists of the following items:

- File and system - Incidence matrix, initial marking and controllability of transitions: design, display, print, modify and/or correct.
- P/T invariants – methods for calculating integer P/T-invariants.
- Restrictions – constraints of the process Petri net.
- Reachability tree construction.
- Control - two methods of supervisory control (algorithm based on reachability tree analysis and algorithm based on P-invariants).
- Simulation and Random real events generators.
- On line help.

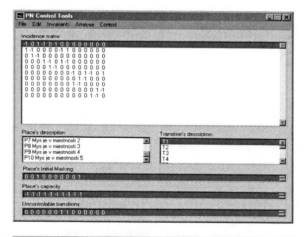

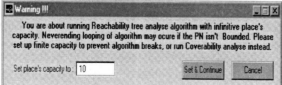

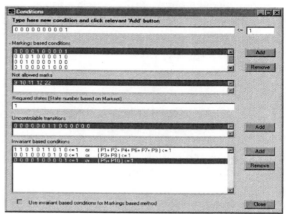

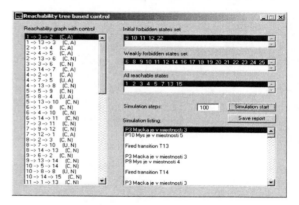

Fig. 3.1.a-e. The windows of the program SUPCON.

4. METHODS

Two methods of supervisory control have been solved by programs. The first method synthesizes the supervisor that consists only of places and arcs and is computed based on the concept of P invariants (Yamalidou et al. 1996), the second method is based on the reachability tree analysis algorithm (Flochová, Hrúz 1996). A part of the program solution tool is on line help.

4.1. The supervisory control design based on the place invariants of the Petri net

In this method (Yamalidou et al. 1996) the controller consists only of places and arcs, and is computed based on the concept of Petri net place invariants. The size of the controller is proportional to the number of constraints that must be satisfied. This method is computationally efficient, and can accommodate constraints written as Boolean logic formulas in the conjunctive normal form or algebraic inequalities that contain elements of the marking and/or the firing vectors.

The system to be controlled is modelled by a plant or a process Petri net with the incidence matrix D_p: The controller net is the Petri net with incidence matrix D_c made up of the process net's transitions and a separate set of places. The controlled systems (controlled Petri net) are the Petri nets with incidence matrix D made up of both the original process net and the controller. The control goal is to force the process to obey constraints of the form

$$\sum_{i=1}^{n} l_i m_i \leq \beta,$$

where m_i represents the marking of the place p_i, l_i and β are integer constants. The inequality constraints can be transformed into equality by introducing nonnegative slack variables m_c into them. Each constraint will have a slack variable associated with it, and each slack variable will be represented in the controlled net *by* the control place. The incidence matrix D is composed of matrices D_p and D_c. The controller D_c is defined by

$$D_c = -LD_p,$$

where L consists of the elements l_i. The initial markings of the controller can be written

$$m_{c0} = \beta - \sum_{i=1}^{n} l_i m_{i0}.$$

The method may be applied to the systems whose constraints are expressed as inequalities or logic expressions involving elements of the marking and/or the firing vectors and uses two different ways of transforming constraints to the desired form.

4.2. The supervisory control design based on reachability tree analysis

The supervisory control method based on reachability tree analysis (Hrúz 1996b, Flochová, Hrúz 1996, Flochová et al. 1997). The algorithm consists of the following steps:

Without imposing the return to home states or nodes.
Forbidden nodes in the reachability graph are labelled as *IA_nodes*.
Repeat until new forbidden nodes or not allowed edges are found
- If the edge of the reachability three is uncontrollable and its successor is *IA_node*, the preceding node is labelled as *IA_node*.
- If the edge of the RT is controllable and its successor is *IA_node* the edge is not allowed and labelled as *NA_edge*.

- If all the edges going out of a node are not allowed the node is labelled as *IA_node*

Imposing the guaranteed return to home states or nodes.
Begin in a home state. Analyse from which admissible nodes it is possible to reach home states using only allowed edges. The admissible nodes from which it is not possible to reach home nodes are converted into forbidden nodes.

4.3. The modification - on line supervisory control method
The on-line supervisory control algorithm is the modification of the algorithm in the previous paragraph for the case of unacceptable state-space explosion. The program controls a process without enumerating all reachable markings by constructing the set of all known weakly forbidden markings $R_F(PN, M_0) = R_F$. It is possible to express the set R_F as a list of states. The user draws the Petri nets of the system to be controlled, and supplies the program implementing the real-time controller with the following inputs data: incidence matrix D, initial marking M_0, the set of forbidden markings R_F, the number and indexes of uncontrollable events. The program analyses the–current system state, extends the set R_F, allows or forbids events with help of the fundamental-flow equation of the controlled system calculating all possible new markings and comparing them with the set R_F.

5. APPLICATIONS

The cat and mouse example introduced by Ramadge and Wonham (1987) involves a maze of five rooms where cat and mouse can circulate (Fig 5.1.).

The rooms are connected with doors through which the animals can pass. The problem is to control the doors so that the cat and mouse can never meet in the same room. Each animal should be able to return to its original room and should have maximum freedom of movement. The rooms are connected with doors through which the animals can pass. Each doorway in the maze is either exclusively for the mouse, or exclusively for the cat, and must be traversed in the direction indicated. The example specifies that all doors are controllable, except of the uncontrollable door for the cat between room 2 and 4.

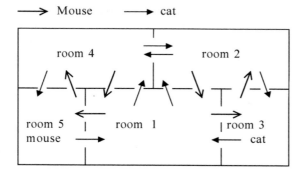

Fig.5.1. Cat and mouse maze - a supervisory control problem.

The Petri net model (Fig. 5.2.) of the above maze system consists of two subnets with one token in each subnet (one token in the subnet for mouse and one in the subnet for cat).

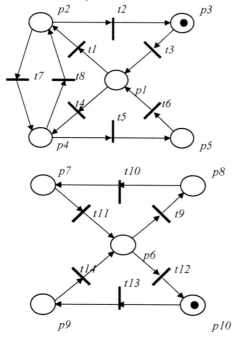

Fig.5.2. Petri net of the maze supervisory control example.

The initial marking is $M_0 = (0\ 0\ 1\ 0\ 0\ 0\ 0\ 0\ 0\ 1)^T$ and the incidence matrix D of the model for this system is:

$$D = \begin{array}{l} \text{-1 0 1-1 0 1 0 0 0 0 0 0 0} \\ \text{1-1 0 0 0 0-1 1 0 0 0 0 0} \\ \text{0 1-1 0 0 0 0 0 0 0 0 0 0} \\ \text{0 0 0 1-1 0 1-1 0 0 0 0 0} \\ \text{0 0 0 0 1-1 0 0 0 0 0 0 0} \\ \text{0 0 0 0 0 0 0-1 0 1-1 0 1} \\ \text{0 0 0 0 0 0 0 0 1-1 0 0 0} \\ \text{0 0 0 0 0 0 0 0 1-1 0 0 0 0} \\ \text{0 0 0 0 0 0 0 0 0 0 0 1-1} \\ \text{0 0 0 0 0 0 0 0 0 0 1 1-1 0} \end{array}$$

Both methods (the supervisory reachability three analysis, its on line modification and the P-invariants

method) are suitable for control and were successfully used to solve the cat and mouse maze problem. The computed nodes of the reachability graph (the set consists of all, i.e. 25, reachable markings) are:

```
0 0 1 0 0 0 0 0 0 1    1 0 0 0 0 1 0 0 0 0    1 0 0 0 0 0 1 0 0 0
1 0 0 0 0 0 0 0 0 1    0 0 1 0 0 0 0 1 0 0    0 1 0 0 0 0 1 0 0 0
0 0 1 0 0 0 0 0 1 0    0 1 0 0 0 1 0 0 0 0    0 0 0 0 1 0 0 1 0 0
0 1 0 0 0 0 0 0 0 1    0 0 0 0 1 0 0 0 1 0    0 0 0 1 0 0 1 0 0 0
0 0 0 1 0 0 0 0 0 1    0 0 0 1 0 1 0 0 0 0    0 0 0 0 1 0 1 0 0 0
1 0 0 0 0 0 0 0 1 0    1 0 0 0 0 0 0 1 0 0    0 0 1 0 0 1 0 0 0 0
0 0 1 0 0 0 1 0 0 0    0 1 0 0 0 0 0 0 1 0    0 1 0 0 0 0 0 1 0 0
0 0 0 0 1 0 0 0 0 1    0 0 0 0 1 1 0 0 0 0    0 0 0 1 0 0 0 0 1 0
0 0 0 1 0 0 0 1 0 0
```

In the initial state (0 0 1 0 0 0 0 0 0 1) the designed contol enables the firings ot transitions t_3 and t_{13}, then e.g. in the state (0 1 0 0 0 0 0 0 0 1) the control enables the firing of t_2, forbids the firing of t_{13} while t_7 is uncontrollable.

The real-time control simulation:
The firing of transition nr. : 13, Cat in room 3, Mouse in room 4
The firing of transition nr. : 14, Cat in room 3, Mouse in room 1
The firing of transition nr. : 12, Cat in room 3, Mouse in room 5
The firing of transition nr. : 13, Cat in room 3, Mouse in room 4
The firing of transition nr. : 14, Cat in room 3, Mouse in room 1
The firing of transition nr. : 12, Cat in room 3, Mouse in room 5
The firing of transition nr. : 3, Cat in room 1, Mouse in room 5
The firing of transition nr. : 4, Cat in room 4, Mouse in room 5
The firing of transition nr. : 8, Cat in room 2, Mouse in room 5
The firing of transition nr. : 2, Cat in room 3, Mouse in room 5
The firing of transition nr. : 3, Cat in room 1, Mouse in room 5
Etc.

The on-line control of flexible manufacturing cell
(Holloway and Krogh 1990) was solved with help of a set of algebraic equations, mathematical descriptions of forbidden states. The example illustrates the control of three workstations (two part receiving stations and one completed parts station) and five automated guided vehicles, that transport material between the stations and of four forbidden zones in which vehicles trajectories cross each other. The controller of this system is responsible for co-ordinating the departures of vehicles to prevent a collision of two vehicles in the same zone. In this system the set of reachable markings consists of 7 741 440 elements and the number of forbidden states overflows one million (1 345 563). This control problem was solved with help of the adjusted on-line control algorithm and with help of P-

invariant algorithm. The forbidden states were specified as the set of four equations and. the values of the control place markings were set according to the equations checking.

A robotic cell shown in fig. 5.3.and 5.4 Two robots R1 and R2 transfer two kinds of parts (A and B) in the cell. Robot R1 picks up a part A from the input conveyor C1 and transports it either through the rooms S2 and S3 to the machine M3 or through S4 and S5 to the machine M1. Then, after machining, R1 transports the part in the same way back to S1 and puts it on the output conveyor C2. The robot R2 transports a part B from input conveyor C3 through S7, S6, S4, and S3 to the machine M4 and back onto C4, or through S7, S6, S4 and S5 to the machine M2 and back onto S4. All transitions between rooms are controllable except those between the rooms S3 and S4. A forbidden situation arises when two vehicles are present in a room at the same time. Both methods described in the previous sections were successfully used to solve this problem. We received 80 nodes in the reachability graph. The windows programs solved fairly large robotic examples 10^{+6}-10^{+8} states.

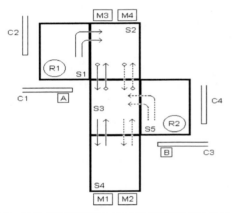

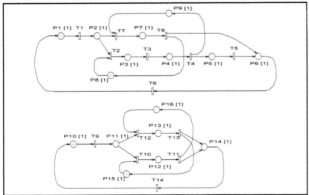

Fig. 5.3 A small robotic cell.
Fig.5.4. Petri nets of the robotic cell.

6. MODULAR APPROACH

The novel aspect of the reachability three analysis approach, that makes it potentially useful for control synthesis for large plants, is the fact that the reachability tree algorithm can sometimes be applied modularly (Hudák 1994). It is possible to develop the set of reachable states using the specifications of one component first, then for another component. Under certain conditions it will be possible to generate a maximally permissive supervisory controller using this modular approach.

7. CONCLUSIONS

This paper presents two program-oriented methods for synthesizing supervisory controllers for forbidden state problems for discrete event dynamic systems.

The first method is based on the reachability tree analysis and it guarantees that the closed loop plant never reaches a forbidden state and that it can always return to a home state. The method is enumerative by nature, but it can be applied efficiently in some cases as explained in the paper. Moreover it can be applied on-line and then it does not require enumeration of the full reachability tree.

The modification for applying the method on-line is based on matrix equations and can be used in control problems with state space explosion. The supervisory control synthesis method based on the reachability tree analysis can solve both the problem of the uncontrollable transitions and the problem of the required accessibility of the home places. Its disadvantage is a possible large cardinality of the reachability set, which complicates the use of this method. The fact, that the reachability tree algorithm can sometimes be implemented modularly, makes this method potentially useful for control synthesis for large plants.

The second method is based on P-invariants. This method is computationally efficient, and can be used for a very large Petri nets. The method doesn't solve directly neither the problem of the uncontrollable transitions nor the problem of required accessibility of home places. These drawbacks can be mitigated partially by transforming the systems specifications to include all uncontrollably reachable states. However it is difficult to obtain maximally permissive controllers via this method. The method may be applied to the systems whose constraints are expressed as inequalities or logic expressions involving elements of the marking and/or the firing vectors. There is also the possibility to transform the constraints from other forms into the form required by the method.

Both approaches provide particularly simple methods for constructing feedback controllers for untimed discrete event systems modelled by Petri nets. Several program modules were written in Matlab 5.3/6, that could be used to solve the OFFLINE supervisory control synthesis for discrete event dynamic systems, and to analyse and check the proposed controller. The outputs of the programs can be used to solve the real-time supervisory control of the discrete event dynamic systems with help of a quick PLC controller or a quick PN-dedicated controller.

8. REFERENCES

Balemi, S., Hoffman,G.J., Gyugyi, P., Wong-Toi, H. and Franklin, G.F. (1993): Supervisory control of a rapid thermal multiprocessor. *IEEE Trans. on Automatic Control,* **38**, pp. 1040-1059.

Boel, R.K., Ben-Naoum, L., Van Breusegem, V. (1995): On Forbidden State Problem for a Class of Controlled Petri Nets. *IEEE Trans. on Automatic Contro*l, **40**, no. 10, pp. 1717-1731.

Boel, R.K. (2001): Diagnosers and Adaptive supervisory control, SCODES, Paris 2001, p.1

Bulach, S., Brauchle, A., Pfleiderer, H.J. (2000): architecture of a Petri net based event-driven controller. In: *Discrete event systems. Analysis and Cotrol.* Kluwer Academic Publishers. Dortrecht, the Nederlands, pp. 383-390.

Češka, M. (1994): *Petriho sítě.* Brno: Akademické nakladatelství CERM, 1994. 94 s. ISBN 80-85867-35-4.

David, R., Alla, H. (1992): *Petri nets and Grafcet.* Cambridge: Prentice Hall international (UK) Ltd, ISBN 0-13-327537-X.

Flochová, J., Hrúz, B. (1996): Supervisory control for discrete event dynamic systems based on Petri nets. *Proceeding of International conference on Process control,* Horní Bečva, **2**, 80-83.

Flochová,J. Hrúz,B., Jirsák, P. (1997): Program solution of supervisory control based on Petri nets. In: *Proceedings the of 2nd IFAC Workshop on New Trends in Design of Control Systems,* Smolenice, 278-282.

Giua, A., DiCesare, F. (1994): Petri Net Structural Analysis for Supervisory Control. *IEEE Trans. On Robotics and Automation,* **10**, no. 2, s. 185-195.

Ho, Y.C., Cassandras, C. (1983): A new Approach to the Analysis of Discrete Event Dynamic Systems. *Automatica,.* **19**, s. 149-167.

Holloway L.E., Krogh, B.H. (1990): Synthesis of Feedback Control Logic for a Class of Controlled Petri Nets . *IEEE Trans. on Automatic Control,* **35**, pp. 514-523.

Holloway, L.E., Guan, X., Zhang, L. (1996): A Generalization of State Avoidance Policies for Controlled Petri Nets, *IEEE Trans. on Automatic Control,* vol. 41,. **6**, s. 804-815.

Hrúz, B. (1996): Design of process control based on Petri nets. In: *Preprints of the IFAC Workshop on Petri nets in industrial automation: modeling, performance, scheduling and control,* San Francisco, CA USA, 29.júna.1996, pp. 84-96.

Hrúz, B., Niemi, A.J., Virtanen, T.: (1996): Composition of conflict-free Petri net models for cotrol of flexilble manufacturing systems. *Proceeding of the 13 th World Congress,* San Francisco.

Hudák, Š. (1994): DE-compositional reachability Analysis. *Elektrotechnický časopis,* **45**, 11, pp. 424-431.

Mossig, K., Stäble, M. (1995): Steuerungssynthese mit kontrollierten Free-Choice Petri-Netzen zur Prozessbeschreibung, *Automatisierungstechni*k, **43**, 11, pp. 506-513.

Murata, T.(1998): Petri Nets: Properties, Analysis and Applications, In*: Proceedings of the IEEE,* **77**, no.4, pp. 541-580

Ramadge, P.J., Wonham, W.M.(1998): The Control of Discrete Event Systems. In: *Proceeding of the IEEE,* **77**, no.1, pp. 81-98.

Ramadge,P.J.G.-Wonham, W.M (1997): Supervisory control of a class of discrete event processes, *SIAM J. Control and optimatization.* **25**, pp. 206-230.

Screenivas, R.S. (1997): On the Existence of Supervisory Policies that Enforce Liveness in Discrete-Event Dynamic Systems Modeled by Controlled Petri Nets. *IEEE Transaction on Automatic Control.* **42**, no.7, s. 928-945.

Stremersh, G.: Linear and integer programmes in supervisory control of Petri nets. In.: *Discrete event systems. Analysis and control.* Dortrecht, 2000, Kluwer Akademic publishers group, 484 pages, ISBN 0-7923-7897-0.

Urbášek, M., Češka, M.(1998): Extension of the Pesim simulation tool. In: *Proceeding of the XXth International Workshop Advanced Simulation of Systems,* Krnov, Czech republic, pp. 81-86.

Yamalidou, K.-Moody, J. , Lemmon, M., Antsaklis, P. (1996): Feedback Control of Petri Nets Based on Place Invariants. *Automatica.* **32**, No. 1, pp.15-28.

Zhou, M.CH., DiCesare, F., Desrochers A. (1992): A Hybrid Methodology for Synthesis of Petri Net Models for Manufacturing Systems. *IEEE Trans. on Robotics and Automation,* **8**, no. 3, pp. 350-361.

Zhou, M.CH., DiCesare, F., Rudolph. D.L. (1992): Design and Implementation of a Petri Net Based Supervisor for a Flexible Manufacturing System. *Automatica,* vol. 28, **6**, pp. 1199-1208.

EVALUATION OF FAULT EFFECTS IN PROGRAMMABLE MICROCONTROLLERS

P. Gawkowski, J. Sosnowski

*Institute of Computer Science, Warsaw University of Technology,
ul. Nowowiejska 15/19, Warsaw 00-665, Poland
Email: gawkowsk[jss]@ii.pw.edu.pl*

Abstract: The paper discusses an experimental method of microcontroller dependability evaluation. It is based on software implemented fault injection adapted to real-time applications. The presented approach is illustrated with results for a developed microcontroller project. *Copyright © 2001 IFAC*

Keywords: error analysis, real-time systems, fault tolerance, safety, reliability.

1. INTRODUCTION

The increasing presence of computing equipment in numerous domains of daily life calls for a reduced rate of failure and pushes the dependability standards higher. This is observed in many real-time applications (microcontrollers, embedded systems - developed quite often as hardware/software co-design). Systems are exposed to various faults (permanent, transient, and intermittent) which may influence their behavior. This behavior (fault effects) depends upon system implementation, application features and fault types. Hence arises the problem of fault effect analysis in the designed system. In the literature various approaches to this problem has been presented. They base on fault insertion techniques e.g. Young *et al.*, 1993; Benso *et al.*, 1998; Carreira *et al.*, 1998; Aidemark *et al.*, 2001; Vakalis, 2001; and references. The published results mostly relate to calculation oriented applications. Real time applications especially those embedded into programmable controllers involve new issues related to time limitations, event driven programming and environment interactions. This was observed in checking fault robustness for some developed real time applications in our Institute.

Fault inserter FITS developed in our Institute (Sosnowski and Gawkowski, 1999) was used to trace fault effects. This system emulates faults by modifying instruction codes, memory cell and register contents. Real time and embedded applications require special organization of fault injection. An important issue was finding test scenarios covering the operation of the controller automata and then simulating system environment for the experiments. Another problem related to the specification of correct and wrong behavior of the analyzed controller. It is worth noting that in real-time controllers some output glitches caused by faults are not critical.

Section 2 of the paper describes main features of fault insertion testbed FITS, its adaptation to real time applications and the problem of setting representative experiment scenarios. These ideas are illustrated in section 3 which shows experimental results related to developed car immobilizer system.

2. SETTING TEST SCENARIOS FOR FAULT INSERTIONS

The best way to analyze fault effects in a system is to simulate faults and observe their influence on the system operation. This approach is used in validating system dependability. The main goal of fault insertion experiments is to determine the coverage of the error detection mechanisms and to identify dependability bottlenecks. There are mainly two fault injection techniques: software based and hardware based. Both approaches can also be combined in hybrid methods. The most universal is software implemented fault injection. This technique „mimics" the consequences of hardware or software faults. The basic idea is to disturb, in a controllable way, the hardware and/or software state of the system and in this way emulate faults by their effects in the executed program. This technique is used in our system FITS.

For each test (fault injection) FITS sets trap in appropriate triggering point within the analyzed program. During the test execution FITS takes over the control after the trap, performs fault insertion (e.g. by changing register state, instruction code or memory location) and traces the target application execution for a specified number of instructions. The set of faults to be inserted can be specified explicitly or generated in a pseudorandom way. It is also possible to specify fault type (single, multiple) and its duration (permanent, transient). The exit code and all generated events, exceptions and other result data are registered in result file and database. In general 4 classes of test results are distinguished: C - correct result, INC - incorrect result, SYS - fault detected by the system (FITS delivers the number and types of collected exceptions), TO - time-out. If the analyzed program generates user defined messages (e.g. signaling incorrect result), they are also monitored by FITS and specified in the final test report.

Adapting FITS to real time applications new problems appeared. They were not so important in classical calculation oriented applications. These problems relate to two issues:

- activation of the analyzed program in the environment of the fault injector,
- result classification in correlation to the application.

In calculation oriented applications it is sufficient to specify representative input data for the analyzed program. In real time applications environment interactions have to be taken into account, which are expressed in the form of events triggering various activities of the analyzed program (i.e. they influence activation of appropriate instruction sequences in the program). Moreover all this activity may be correlated with time clocks. Hence, supplementary environment simulators are needed. An important issue is to define representative scenarios of the environment activity so as to cover all typical operations as well as non-typical (e.g. related to external faults etc.). Fault insertions have to be performed for all these scenarios.

Another problem relates to the analysis of experiment results. In classical fault injection experiments results are qualified on the basis of program exit code as correct, incorrect etc. In real time system the result analysis is more complex. There are several reasons for this. Such systems deliver their services in a continuous way (e.g. for controlling some external process) moreover many output signals can be used and activated in appropriate time moments. In the consequence of faults (especially transient) erroneous states may appear. Depending upon the application, some erroneous states (e.g. temporary) are not critical e.g. covered by the inertia of the controlled object and subsequent control interactions. Moreover some fluctuations in value and timing of outputs is usually acceptable. Hence deviation of generated results from the ideal case does not necessary denote system failure (as it is usually assumed in classical fault insertion experiments). On the other hand, beyond output signals an important issue is the state of the analyzed application. In the calculation oriented applications practically final results are observed and can be captured by the fault inserter. In real time applications the problem is more complex because a fault may be latent for a long time with no impact on generated output signals within the observed time of the fault inserter. To overcome these problems special procedures to qualify experiment results are needed. Moreover fault effect analysis should cover not only output signals but also internal state of the application. This last issue can be resolved by observing appropriate internal variables.

Quite often a set of cooperating microcontrollers is used. In this case checking system susceptibility to faults involves experiments with faults inserted into a controller (with correct simulator of the cooperating modules) and experiments with faults inserted in the environment (simulators) and observed effect in the correct module. This can be extended also for external environment of the analyzed system. Another problem relates to event time scale. In many applications operation scenarios involve long time periods. Most time system is inactive so fault injection experiment could result in unacceptable duration. To resolve this problem it is reasonable to scale down system timing (including time-out) as to speed-up the experiments and assure real operation activities. This approach was used in 3 projects: car immobilizer system (CIS), multiprocessor network realizing some transmission protocol and object search algorithm. In the sequel only results for the first project will be illustrated.

3. EXPERIMENTS WITH CIS CONTROLLER

In this section the problem of evaluating fault effect in developed car immobilizer (CIS) is illustrated. Its implementation and operation are outlined in section 3.1. Representative test scenarios derived for functional model of CIS and performed fault insertion experiments are presented in section 3.2.

3.1 CIS model

CIS is embedded in Atmel's AT89C51 microcontroller comprising program code memory and other resources (RAM, internal timers, external ports and interrupts). One micro-controller's port is used as an output interface for car ignition system, fuel pomp, alarm siren, buzzer and LED diode. The other port is used to handle control keyboard (12 keys organized in rows and columns). After typing a secret keyword CIS goes to the state in which car appliances are turned on and the car can be driven. If a user puts a wrong keyword the car will be operational only for a short time (e.g. 1 minute) and after this period all car systems will be turned off and the alarm turned on. To go back to the normal state another special key sequence is needed.

CIS software (around 800 lines in C) is implemented basing on events. Typical events are pressing a key, time-out (software implemented timers), etc. The occurrence of events is checked periodically every 50ms. If no event is present CIS goes to the idle mode (reduced power consumption) for the rest of the 50ms period. When the system is armed and there is no event present for a specified period of time CIS goes to the power down state (power consumption reduced to less than 1mA). CIS wakes-up by pressing any key. The main automata implemented in software consists of 15 states. There are several supplementary simple automata responsible for software implemented timers, going into power down state, keyboard scanning etc.

Figure 1 shows simplified state graph of CIS. The initial state is marked out. In this state all car systems are switched off as well as CIS siren. User cannot start engine in this state. This is a normal situation when a car is in garage or stays in a parking place. Additional state *Unexpected state* was added to provide the capability of the automata to go back to the safe state. The output functions of CIS are specified in tab. 1. In the developed microcontroller all states of fig. 1 except *Unexpected state* are equivalent to some sub-graphs. Moreover, additional automata are provided for handling timeout, keyboard scanning, driving LED and buzzer (used to signal the user the current CIS state – in contrast to the siren which is used to signal the hijacking trial). After

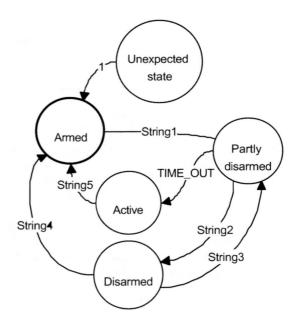

Fig. 1. Simplified state graph of CIS.

typing a string of numbers (denoted as *String1* on the graph) CIS enters *Partly disarmed* state in which siren is still switched off but all devices needed for driving a car are switched on. CIS can stay in this state maximum for a specified period of time. If *String2* will not appear before the specified timeout CIS will enter *Active* state in which all car devices are switched off and the siren is switched on. To leave the *Active* state the driver has to enter *String5*. In normal circumstances the driver will provide *String2* before the time out will occur and CIS will go to the *Disarmed* state. In this state all car devices are permanently switched on (no time out as in *Partly disarmed* state) and the siren is switched off. This is the normal situation while driving a car. The driver can change the state of CIS between *Disarmed* and *Partly disarmed* states at any time (by typing *String3*). This is helpful to avoid hijacking when we don't want to shut down the engine while opening the garage door. CIS can be turned back to the *Armed* state by typing *String4* sequence.

Table 1. Summary of CIS states and outputs

State	Car devices	Siren
Armed	Off	Off
Partly disarmed	On	Off
Disarmed	On	Off
Active	Off	On

Every 50ms event scanning procedure is activated in CIS. If there are any events to be served the system takes proper actions. Event scanning procedure is asynchronous to the event handling procedure so if new events occur during the event handling procedure, the procedure is reinvoked. CIS enters

IDLE mode (reduced power consumption) after finishing scanning procedure or event handling procedure. The whole system is driven by the two main events:
- *keyboard* –the user presses and releases single key on the keyboard
- *time-out* – the time out expiration.

Two other events in the system are used to serve buzzer and LED (they do not influence CIS state). When the system stays in the *Armed* state for specified period of time (T1) the system goes into the POWER DOWN mode (power consumption less then 1 mA). Since that, the only way to wake up CIS is pressing a special key on the keyboard (it resets CIS). The microcontroller used in this system holds the internal memory content during power down and reset.

For the most of the time CIS will stay in *Armed* state. This relates to the situation when the car stays in garage or at parking place. During driving the car CIS is in the *Disarmed* state. The *Partly disarmed* state is transitional and *Active* state relates to abnormal situations (e.g. theft of the car). High availability and safety of CIS is an important issue. Hence, CIS should provide fault detection capability and, if possible, fault tolerance. In case of unrecoverable error detection CIS assumes safe state (from the driver point of view) but still with the ability to prevent car hijacking.

3.2 CIS testing

When checking the safety and availability properties of the microcontroller applications it is very important not only to check the effects of disturbances in microcontroller modules during single iteration of the algorithm but also to find out if the system can operate correctly after recovery. In other words we are not only interested in finding undesired system outputs but also in the impact of faults on system operations in the future. This is very important since CIS stays in some states for a very long time (e.g. *Armed* state) and it should move to other states in specific situations. Some temporal disturbances in CIS operation can be accepted, i.e. short (less then 0.5 second) interruption of ignition or fuel pomp power won't interrupt car driving. On the other hand there might be situations that have to be identified due to their unsafe consequences. An example of this is unexpected (e.g. while overtaking another car) engine shutting off. Preparing fault insertion experiments appropriate CIS operation scenarios should be specified, which cover all its functional capabilities and environment interactions. The following scenarios starting in *Armed* state have been designed:

- S1 - **the system is in *Armed* state for one minute, then it goes to the *Partly disarmed* state for one minute – then it goes to the *Disarmed* state**,
- S2 - **the system goes to the *Disarmed* state and after a short period of time it goes back to the *Armed* state**,
- S3 - **the system goes to the *Partly disarmed* state, then after time-out occurrence it goes to the *Active* state**,
- S4 - **1 second delay, String1,** String2, String4, SCh // tests the *Armed* state and its leaving,
- S5 - String1, **time out, String5,** SCh // tests the *Partly disarmed* state and the first way of leaving this state,
- S6 - String1, **delay till 5 seconds before time out, String2,** String4, SCh // tests the *Partly disarmed* state and the second way of leaving it,
- S7 - String1, String2, **1 second delay, String4,** SCh // tests the *Disarmed* state and the first way of leaving it,
- S8 - String1, String2, **1 second delay, String3,** String2, String4, SCh // tests the *Disarmed* state and the second way of leaving it,
- S9 - String1, time out, **1 second delay, String5,** SCh // tests the *Active* state and the way of leaving it.

These scenarios assure checking state graph of CIS (visiting all states and transitions). Additional sequence denoted as "SCh" (Check Sequence) at the end of scenarios S4-S9 checks the whole system "vitality". This check scenario is as follows: after 5 seconds in *Armed* state (needed for initialization), String1 is typed, then after time out CIS goes to the *Active* state for 1 second, then String5 is typed followed by String1, String2, String3, String2 and String4. During experiments only the actions marked in bold were disturbed to achieve reasonably equal distribution of faults injected over instruction code.

To check the dependability properties of CIS fault injection system FITS was used. All output ports of the target microcontroller and some input ports (for observability and controllability reasons) were substituted with variables. So, the original source code was extended with some extra code responsible for reading a set of events from external file (a file with test scenarios) and modifying proper variables (e.g. port pins) to simulate the occurrence of desired event from the real world (e.g. pressing a key). This extra code was added at the beginning of the event scanning procedure. This code is transparent to the rest of CIS software and is responsible (at the run time) for simulation of the external microcontroller environment (interface - input/output) as well as for storing the states of CIS outputs and important internal variables in result file. Only data necessary to distinguish correct from incorrect CIS behavior is written into this file in each iteration (simulated 50 ms periods) of the event scanning procedure. This external result file is then analyzed (and compared with the golden run file) by FITS to identify correct, incorrect, etc. CIS behavior (see section 2). The supplementary code is not exposed to the fault injections (it is out of the fault injection region).

In the performed experiments faults were injected into instruction code, processor registers and data areas used by CIS (internal variables as well as simulated ports of the microcontroller). A special care has to be taken when performing experiments with fault injections into instruction code with FITS. Faults in this area are injected into the instruction just before its execution. This guarantees activation of a fault. Unfortunately, in case of controller code, some instructions are executed very often while other are executed rarely. For example, event scanning procedure is invoked every 50 ms in contrast to event handling procedures which are activated by events. In scenario S1, event scanning procedure is executed 2472 times while the event handling procedure is executed only 8 times and the procedure responsible for switching on car devices is executed only two times. It's worth noting that the executable code size (in microcontroller code) of the whole software is 1303 bytes while code executed every 50 ms is only 335 bytes long.

Checking CIS susceptibility to faults injected into registers and data area we can concentrate mainly on experiments covering most frequently used code (e.g. event scanning procedure). All these faults may generate latent effects (errors). To detect these situations test scenarios S4-S9 comprise SCh sequence - it checks overall CIS "vitality".

In case of faults in instruction code it is important to equally distribute faults within the whole code area and be sure to activate them. To do this, the original fault injection technology of the FITS (which corrupts instruction code with equal distribution but over dynamically executed code – not in a space manner) had to be changed. To achieve equal distribution of injected faults in the code area a special version of FITS was developed which lets the user to select (in time domain) code areas to corrupt. Performing a large number of fault injections (needed to evaluate the controller robustness) may lead to unacceptable experiment duration. This can be avoided by scaling down real life time test scenarios and by shortening the time of observing fault effect propagation. This last issue is resolved by checking if the controller is operative after fault disturbance (e.g. with test sequence *SCh*). Additionally it is reasonable to check the properties of CIS in situations with multiple faults especially when CIS has to operate without interruptions. It is very important in this case to find out if there is no unsafe output.

In this case to properly prepare fault injection experiments all acceptable situations must be identified and discriminated while examining the outputs of the tested system. In FITS user is able to define own procedure of examining the correctness of the executed scenario. This procedure is provided to the FITS as external dynamically linked library (DLL). The procedure has access to the result file from the golden run execution of the considered test scenario and to the result file created during fault injection execution. CIS behavior is assumed to be acceptable if the erroneous signal appear for no longer then 100 ms (two iterations of event scanning procedure). It's worth noting that this assumption is a strong one since some output combinations could be just unpleasant to the driver. The driver, as opposite to FITS, could eliminate this using proper actions to "manually" solve some problems (e.g. by going back to the *Partly disarmed* state and then once again to the *Disarmed* state). Unfortunately actions like this are human dependent and not easy to simulate.

Another problem relates to differences between golden run states (state graph) of CIS and the one obtained during experiments. Some deviations can be acceptable in correct behavior. For instance, changing the internal CIS state within the aggregated states (subgraphs) of the simplified graph does not disturb the behavior of the system in critical manner. These disturbances don't change CIS outputs and will be noticed by the user who can set correct CIS state. Similarly, starting or disturbing time out counters won't disturb CIS in critical manner.

In the presented experiments the correctness of the CIS behavior is made by checking the output states of the microcontroller and internal main automata state in every iteration of the event scanning procedure. Sample of results (in percents) is given in table 2 (notation as in section 2).

Table 2. Results for test scenarios S1-S3

Fault location	Result	S1	S2	S3
Data area	C	68.9	71.9	83.4
	INC	9.5	7.6	8.2
	SYS	21.5	20.5	8.4
	TO	0.2	0.0	0.0
Registers	C	71.9	68.4	72.8
	INC	0.0	0.3	0.4
	SYS	28.1	31.3	26.7
Code	C	45.7	46.1	41.4
	INC	6.8	6.0	12.8
	SYS	43.3	44.5	42.9
	TO	4.2	3.3	2.9

It's worth noting that CIS software is quite safe even without any special fault detection or tolerance mechanisms. Faults injected into registers result in less then 0.5% of incorrect outputs or corrupted internal state of CIS. In case of faults in data area or instruction code the results depend on test scenarios. This can be explained by the different utilization of disturbed resources in different scenarios. To improve the dependability properties of CIS software first of all the number of incorrect executions have to be decreased as well as the number of timed-out executions. Then, special care have to be taken to decrease the percentage of executions terminated by the operating system – on the target platform there are no similar mechanisms available so the situations that lead to system's on-line detectors activation have to be identified.

Inserting some control flow checking mechanisms into CIS software the availability of CIS decreased while no improvement was observed in systems safety (system safety even decreased slightly). For instance, including control flow checking at the procedure call level increased incorrect results to 7 % in case of faults in registers for scenario S1. Simultaneously the percentage of correct executions decreased to 51%. In case of faults in instruction code incorrect executions were observed in 11.6% while the correct ones only in 30.7%. However, the time-outs were decreased to 2.2%. In relatively robust systems inclusion of simple fault tolerance mechanisms (e.g. control flow checking) is not effective due to the code overhead and its susceptibility to faults. This effect disappears in techniques including more redundancy, e.g. Gawkowski and Sosnowski, 2001b; Pradhan, 1996.

4. CONCLUSION

In the paper it was shown that fault injection techniques can be used to evaluate dependability of real-time systems, however a special care is needed here. In particular system environment simulators and result qualification procedures has to be developed. These procedures are application dependent. An important issue is to develop representative test scenarios covering not only activities related to typical operational profile of the system but also to its rarely used functions (e.g. error detection and recovery). Handling experiments with many faults often needs scaling down time related events and time-outs. For the analyzed real-time projects relatively high robustness to transient faults was observed which resulted from the used algorithms and application specificity. Further research is related to checking the effectiveness of fault tolerance techniques and improving FITS system (Gawkowski and Sosnowski, 2001a).

Acknowledgment. This work was supported by Polish Scientific Committee grant no. 8T11C 02016.

REFERENCES

Aidemark, J., J. Vinter, P. Folkesson and J. Karlsson (2001). GOOFI: Generic Object-Oriented Fault Injection Tool. In: *Proc. Of the 2001 International Conference on Dependable Systems and Networks*, pp. 83-88. IEEE Computer Society, Los Alamitos.

Benso, A., P.L. Civera, M. Rebaudengo and M. Sonza Reorda (1998). An integrated HW and SW Fault Injection environment for real-time systems. In: *Proc. IEEE International Symposium on Defect and Fault Tolerance in VLSI Systems*, pp. 117-122.

Carreira, J., H. Madeira and J. G. Silva (1998). Xception: A Technique for the Experimental Evaluation of Dependability in Modern Computers. *IEEE Transactions On Software Engineering*, **24**, 125-136.

Gawkowski, P. and J. Sosnowski (2001a). Analyzing Fault Effects in Fault Insertion Experiments. In: *Proceedings of the 7th IEEE On-Line Testing Workshop IOLTW 2001*, pp. 21-24. IEEE Computer Society, Los Alamitos.

Gawkowski, P. and J. Sosnowski (2001b). Experimental Evaluation of Fault Handling Mechanisms. In: *Proceedings of the 20th International Conference SAFECOMP 2001* (Udo Voges (Ed.)), pp. 121-130. Springer, Germany.

Pradhan, D. K. (1996). *Fault-Tolerant Computer System Design*. Prentice Hall, Upper Saddle River, New Jersey.

Sosnowski, J. and P. Gawkowski (1999). Tracing fault effects in system environment. In: *Proceedings of the 25th Euromicro Conference,* pp. 481-486. IEEE Computer Society, Los Alamitos.

Vakalis, I. (2001). A comparison study behavior of equivalent algorithms in fault injection experiments in parallel superscalar architectures. In: *Proceedings of the 20th International Conference SAFECOMP 2001* (Udo Voges (Ed.)), pp.157-167. Springer, Germany.

Young, L. T., R. Iyer and K. K. Goswami (1993). A Hybrid Monitor Assisted Fault injection Experiment. In: *Proceedings of the DCCA-3*, pp. 163-174.

IFAC
Publications
www.elsevier.com/locate/ifac

DYNAMIC PERFORMANCES OF SMART SENSORS BASED ON SELF-ADAPTIVE FREQUENCY-TO-CODE CONVERSION METHOD

Nikolay V. Kirianaki [1], Sergey Y. Yurish [1,2], Nestor O. Shpak[1,2]

[1] *International Frequency Sensor Association (IFSA),*
[2] *Institute of Computer Technologies,*
Bandera str.12, Lviv, UA, 79013,
Tel.: + 380 322 97 16 74, fax: +380 322 97 16 41
E-mail: info@sensorsportal.com
Http://www.sensorsportal.com

Abstract: New modelling results of dynamic performances of programmable smart sensors based on the self-adaptive frequency-to-code conversion method of dependent count are described in the Paper. Modelling results of conversion time function for the method have completely confirmed theoretical and experimental researches of its dynamic performances. It has been shown that the conversion time is non-redundant and can be changed during measurements according to the required accuracy of conversion. Hence, such programmable sensors may be used in different real time applications, for example, ABS, as well as for measuring and data acquisition systems for various high-speed behaviour processes. *Copyright © 2001 IFAC*

Keywords: frequency conversion, frequency measurements, sensors, self-adapting algorithms, adaptation, quantization error, antilock braking systems, data acquisition, modelling, converters

1. INTRODUCTION

At creation of various frequency-time domain smart sensors and transducers with digital output, the correct choice of frequency-to-code conversion method, meets to requirements on speed and accuracy is an important task. In other words, it is desirable to have the non-redundant conversion time as well as self-adaptive opportunities in all frequency range. The last means an opportunity to program of smart sensors for exchange of accuracy to speed and on the contrary during conversion according to measuring conditions. One of example of such application is an automotive anti-lock braking system (ABS). The advanced ABS algorithm needs an automatic choosing of reference time interval

depending on the given quantization error of measurement. However, because of almost all known frequency-to-code conversion methods, like classical (direct counting method (Haward, 1979)) and advanced conversion methods (ratiometric, reciprocal (BurrBrown, 1994)), high resolution and high accuracy M/T method (Ohmae, et. al., 1982), method with constant elapsed time (CET method) (Bonert, 1989), single- and double-buffered (Prokin, 1991) and DMA methods (Prokin, 1993), have constant conversion time, they are unsuitable for the usage in self-adaptive smart sensors.

The non-redundant conversion time can be achieved at the usage of other classical method of measurement – the indirect counting method

(Kasatkin, 1966), but this method is effective only for low and infralow frequency ranges.

The most perspective method for application in self-adaptive smart sensors is the method of dependent count (Kirianaki and Berezyuk, 1980; Kirianaki *et al.,* 1998; Kirianaki *et al.,* 2001). It combines in themselves the advantages of classical methods as well as advanced methods ensuring the constant relative quantization error in a broad frequency range and high speed. It is suitable for a frequency conversion in a wide frequency range: from parts of Hz up to several MHz with the constant, beforehand given, quantization error and non-redundant conversion time. The method is suitable for single channel as well as for multichannel synchronous frequency conversions.

Therefore, the investigation of dynamic characteristics of smart sensors based on this conversion method, and, first of all, the conversion time is the urgent task.

2. NATURE AND PERFORMANCES OF THE METHOD OF DEPENDENT COUNT

With the aim to model of dynamic performances, the basic equations for the conversion time of method of dependent count should be developed.

The timetable of method is shown in Fig. 1. The method consists in the following. With arrival of the impulse of signals with the lower frequency f (this corresponds to the moment t_1 in Fig. 1), the counters start to calculate the impulses of both signals. The number of impulses N_i, stored in the up-down counter, which calculates the impulse with the frequency F, is compared with the number N_δ. This number is set up previously in the counter by the microcontroller. At some moment of time (this corresponds to the moment t_2 in Fig. 1), when the number of impulses, calculated by this counter, will be N_δ, e. g. $N_i = N_\delta$ with the arrival of next impulse (after the moment t_2) with the lower frequency f (moment t_3), the impulse count will be stopped. The number of impulses (of signal of lower frequency f) counted by one of the counters is n, and the number of impulses (of signal of higher frequency F), counted by the second counter, is $N = N_\delta + \Delta N$. The conversion time t_x always equals to an integer number of periods of a signal with the lower frequency f:

$$t_x = \tau \cdot n = \frac{n}{f} \qquad (1)$$

This interval can be given also like:

$$t_x = T \cdot N = \frac{N}{F} = \frac{N_\delta + \Delta N}{F} =$$
$$= \left(\frac{1}{\delta} + \Delta N \right) \cdot T \qquad (2)$$

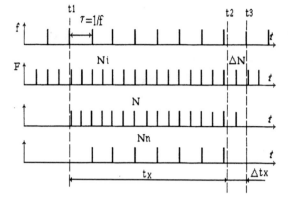

Fig. 1. Timetable of the method of dependent count.

From equations (1) and (2) follows, that f = F (n/N) or F = f (N/n). If the measured frequency f_x is lower frequency, e.g. $f_x = f$, and reference frequency f_0 is high ($f_0 = F$), that

$$f_x = f_0 \cdot \frac{n}{N}$$
$$\text{or} \qquad (3)$$
$$f_x = f_0 \cdot \frac{N}{n},$$

when $f_x = F$ and $f_0 = f$.

The embedded microcontroller calculates the unknown frequency from the formulas (3). The program for calculation is determined by the command, which to be prepared on the basis of earlier entered into the microcontroller information which from the frequencies is lower.

For the period τ or T, the conversion is carried out similarly. The microcontroller calculates it values from the following equations:

$$\tau = \frac{N}{f_0 \cdot n}$$
$$\text{or} \qquad (4)$$
$$T = \frac{n}{f_0 \cdot N}$$

The quantization error for considered conversions caused by that, the interval t_x (conversion time) is not equal to the interval, which is determined by the integer number N of periods of a signal with high frequency, e.g.

$$t_x \neq N \cdot T \neq N/F. \qquad (5)$$

At change of the lower frequency in known limits, the interval t_x will be changed (or, on the contrary, will be changed the N/F at change of higher frequency). It will result in change of the number of impulses N from $\Delta N = 0$ up to $\Delta N = \Delta N_{max}$ calculated by the counter. Here ΔN_{max} is the number of impulses in the interval $\Delta t_{xmax} = \tau$. Taking into account the fact, that the period of these pulses is equals to T, the following equation will be true:

$$\Delta N_{max} = \frac{\tau}{T} = \frac{F}{f}. \qquad (6)$$

The maximum quantization error arises in that case, when the number of impulses N counted by the counter is minimum and equal to the N_δ (one of frequencies is changed). Then

$$\delta_{max} = \frac{1}{N_{min}} = \frac{1}{N_\delta}. \qquad (7)$$

Hence, the maximum error is determined by the value N_δ only and practically does not depend on the measured frequency. The minimum value of the error will be at $N = N_{max}$. But as $N_{max} = N_\delta + \Delta N_{max}$, then

$$\delta_{min} = \frac{1}{N_{max}} =$$
$$= \frac{1}{N + \Delta N_{max}} \qquad (8)$$

Now let's consider the dynamic error and its main components. At frequency measurements, this error essentially depends on dynamic properties of the researched process and speed of a method of measurement. The decreasing of dynamic error is possible by increasing of speed and continuous correction of results of measurement.

The method of dependent count has the highest speed at measurement for all frequencies from the frequency range and consequently, allows reducing essentially the dynamic error The main components of the dynamic error are: the tracking error and approximating error.

The first component depends on the time t_x. The result of measurement of period duration or several periods of low frequencies composing the time t_x, corresponds to their average value on the time

interval equal to their duration. The inevitable availability of such average and also the change of researched value during the latency time if it is not eliminated, is the reason of rise so called the average error.

The second component - is the approximation error of continuously varying value of lattice function with appropriate approximation between points of its discrete values. The approximation error is determined by the digitisation interval and a kind of approximation. In one-channel and multichannel frequency-to-code converters the digitisation interval cannot be lower than the conversion time t_x.

The decreasing of dynamic error can be reached by minimisation the time t_x by the choice of algorithms, eliminating latency time of the next conversion cycle.

The dynamic error should not exceed the static one. Under this condition, the acceptable time t_x at the known frequency of change of the researched value and its maximum value at the selected t_x are determined. This algorithm can also be used at higher frequencies of the range D_f, if the time of measurement is large, even in case of the method of dependent count.

3. MODELLING RESULTS

For the conversion time modelling the following equations have been used:

$$t_x = \frac{N_\delta + \dfrac{f_0}{f_x}}{f_0}, \qquad (9)$$

in the case, when $f_x \leq f_0$ and

$$t_x = \frac{N_\delta + \dfrac{f_x}{f_0}}{f_x}, \qquad (10)$$

when $f_x > f_0$

Thus, with the aim to cover a lot of existing smart sensors, working in various parts of frequency range, the following size of variables changing were used: $f_x \in [0.1 \div 10\ 000\ 000]$ Hz; $f_0 \in [100\ 000 \div 1\ 000\ 000]$ Hz; $N_\delta \in [1\ 000 \div 10\ 000\ 000]$ (it is corresponds to the size of given quantization error changing from 0.1 up to 10^{-7} %; $\Delta N \in [0 \div \Delta N_{max}]$.

The modelling has been realized with the help of system of analytical calculations Maple 6.0. Modelling results of $t_x = \varphi(N_\delta, f_x, f_0)$ function are shown in Fig.2 – Fig.7. As the most expressed dependence of conversion time from f_x, f_0 at the fixed number N_δ is observed in the area of infralow frequencies, for the best visualization, diagrams of dependence $t_x = \varphi$ ($N\delta$, f_x, f_0) are constructed for this frequency range (the greatest conversion time). At increasing of converted frequencies from 1 Hz and higher, the conversion time is essentially decreasing (in some orders).

begins essentially to be visible approximately from $N_\delta = 25\ 000$. Plots of researched function at $N_\delta = 10^6$ and $N_\delta = 10^7$ are shown in Fig.4 and Fig.5 accordingly.

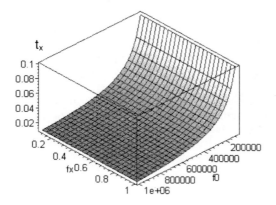

Fig. 2. Modelling result of $t_x = \varphi$ (N_δ, f_x, f_0) function at $N_\delta = 10\ 000$; $\Delta N=0$; $f_0=100\ 000 \div 1\ 000\ 000$ Hz; $f_x=0.1\div1$ Hz.

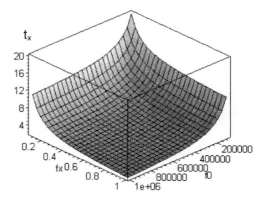

Fig. 4. Modelling result of $t_x = \varphi$ (N_δ, f_x, f_0) function at $N_\delta=10^6$; $f_0=100\ 000\div1\ 000\ 000$ Hz; $f_x=0.1\div1$ Hz.

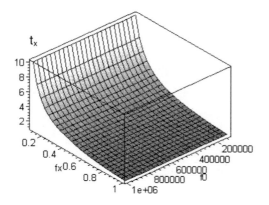

Fig. 3. Modelling result of $t_x = \varphi$ (N_δ, f_x, f_0) function at $N_\delta=10\ 000$; $f_0=100\ 000\div1\ 000\ 000$ Hz; $f_x=0.1\div1$ Hz.

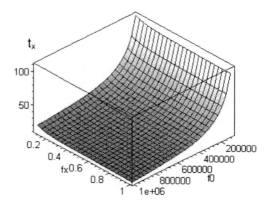

Fig. 5. Modelling result of $t_x = \varphi$ (N_δ, f_x, f_0) function at $N_\delta=10^7$; $f_0=100\ 000\div1\ 000\ 000$ Hz; $f_x=0.1\div1$ Hz.

As it was specified earlier, the number ΔN can be varied during the conversion from 0 (in the case of $N_\delta = N_i$ equality in the counter at the ending of current period of converted frequency f_x) up to $\Delta N_{max} = f_0/f_x$ (in the case of $N_\delta = N_i$ equality in the counter at the any moment of time). In the first case (Fig. 2), the conversion time is minimum possible for the appropriate infralow frequency. The hyperbolic dependence of conversion time from the reference frequency f_0 is well distinct in the plot. In the second case, the conversion time is increased in two orders.

For $N_\delta \in [1\ 000 \div 10\ 000]$ the conversion time t_x is sensibly constant and equal to $\cong 10$ sec in area of infralow frequencies. The dependence t_x from N_δ

Values of t_x at different number $N_\delta=1/\delta$ and reference frequency $f_0 \leq 200$ kHz for infralow frequency range are adduced in Table 1.

Table 1 Dependence t_x from N_δ at $f_0 \leq 200$ kHz

N_δ	δ, %	t_x, sec	
200 000	$5\cdot10^{-4}$	$\cong 12$	
400 000	$2.5\cdot10^{-4}$	$\cong 14$	
600 000	$1.7\cdot10^{-4}$	$\cong 16$	@
1 000 000	$1\cdot10^{-4}$	$\cong 20$	$f_0 \leq 200$ kHz
10 000 000	$1\cdot10^{-5}$	$\cong 100$	
100 000 000	$1\cdot10^{-6}$	$\cong 1000$	

Modelling results for the mode fx > f$_0$ are shown in Fig. 6.

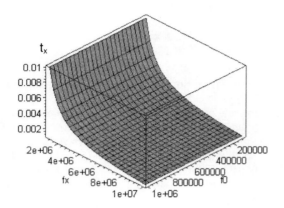

Fig. 6. Modelling result of $t_x = \varphi$ (N_δ, f_x, f_0) function at $N_\delta = 10^4$; $f_0 = 100\,000 \div 1\,000\,000$ Hz; $f_x = 1\,000\,001 \div 10\,000\,000$ Hz.

The plot of modelling results of function $t_x = \varphi$ (N_δ, f_x, f_0) in the wide frequency range from 1 Hz up to 10 MHz at relatively high accuracy of conversion $N_\delta = 10^6$ (the quantization error does not exceed 10^{-4} %) is shown in Fig. 7. This example illustrates one more essential advantage of the method of dependent count - an opportunity to convert frequencies exceeding the reference frequency $f_x \geq f_0$. Besides, the given example testifies about the necessity to use the high reference frequency only in low and infralow ranges. This opens a prospect for the adaptive control of reference frequency during the conversion that results in reduction of power consumption in smart sensors.

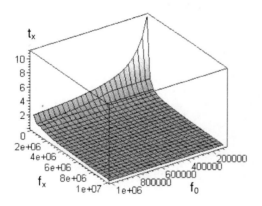

Fig. 7. Modelling result of $t_x = \varphi$ (N_δ, f_x, f_0) function at $N_\delta = 10^6$; $f_0 = 100\,000 \div 1\,000\,000$ Hz; $f_x = 1 \div 10\,000\,000$ Hz.

4. EXAMPLES

Let it is necessary to convert the frequency $f_x = 2 \cdot 10^4$ Hz at $f_0 = 10^6$ Hz and $N_\delta = 10^6$ ($\delta = 10^{-4}$ %). According to the method of dependent count the maximum conversion time (9) is

$$t_x = \frac{10^6 + \dfrac{10^6}{2 \cdot 10^4}}{10^6} \approx 1 \sec.$$

In turn, according to the standard counting method, the time of measurements necessary for the same accuracy is calculated according to the following formula:

$$t_x = \frac{1}{\delta \cdot f_x} = \frac{N_\delta}{f_x} = \frac{10^6}{2 \cdot 10^4} = 50 \sec.$$

Let's consider the conversion of same frequency, but with the help of indirect counting method. In spite of fact that it is a conversion method with non-redundant conversion time on the nature, in order to have the required quantization error for the given frequency range it will be necessary to convert much more than one periods of f_x:

$$N_T = \frac{f_x}{f_0 \cdot \delta} = \frac{f_x \cdot N_\delta}{f_0} = \frac{2 \cdot 10^4 \cdot 10^6}{10^6} = 2 \cdot 10^4.$$

In this case, the conversion time will be calculated according to the following formula:

$$t_x = N_T \cdot T_x = \frac{2 \cdot 10^4}{10^4} = 2 \sec.$$

The usage of other advanced conversion methods with the fixed conversion time, for example, ratiometric, reciprocal, SET, M/T, DMA, etc. methods also demands to increase the conversion time like to the standard counting method - up to 50 sec.

In other words, the time of measurement for the method of dependent count is non-redundant in all specified measuring range of frequencies. In the standard counting method and modern advanced, the time of measurement is redundant, except the nominal frequency. Moreover, for the method of dependent count the time of measurement can be varied during measurements depending on the assigned error.

5. CONCLUSIONS AND FUTURE RESEARCHES

Modelling results of conversion time function for the method of dependent count have completely confirmed theoretical and experimental researches of dynamic performances of the method. The received new data testify that the conversion time t_x essentially depends on the reference frequency only in the area of infralow frequencies. This opens a prospect for the adaptive control of reference frequency during the conversion that results in reduction of power consumption in smart sensors.

The conversion time is non-redundant and can be changed during measurements according to the required accuracy of conversion (quantization error).

The usage of method of dependent count in smart sensors results in an opportunity of self-adaptation of such sensors. This opens a prospect to use such programmable sensors in different real time applications, for example, ABS, as well as for measuring and data acquisition systems for various high-speed behaviour processes.

REFERENCES

Bonert, R. (1989). Design of a High Performance Digital Tachometer with Microcontroller. *IEEE Transactions on Instrumentation and Measurement*, Vol.38, No.6, December, 1104 - 1108.

BurrBrown Applications Handbook. (1994). USA, 409-412.

Haward, A.K. (1979). Counters and timers. *Electronic Engineering*, **51**, (630), 61-70.

Kasatkin, A.S. (1966). *Automatic Processing of Signals of Frequency Sensors.* Energiya, Moscow (in Russian).

Kirianaki, N.V. and B.M. Berezyuk (1980). Method of Measurement of Frequency and Period of Harmonic Signal and Device for its Realisation, *Patent* No. 788018, USSR (in Russian).

Kirianaki N.V., S. Y.Yurish and N.O. Shpak (1998). New Processing Methods for Microcontrollers Compatible Sensors with Frequency Output. In: *Proceedings of the 12th European Conference on Solid-State Transducers and the 9th UK Conference on Sensors and their Applications*, Southampton, UK, 13-16 September EUROSENSOR XII, (N. M. White, Ed.), Institute of Physics Publishing, Bristol and Philadelphia, Sensors Series, vol. 2, 883-886.

Kirianaki N.V., S.Y. Yurish and N.O. Shpak (2001). Methods of Dependent Count for Frequency Measurements. *Measurement*, Vol.29, Issue 1, January, 31-50.

Ohmae T., T. Matsuda, K. Kamiyama and M. Tachikawa (1982). A Microprocessor-Controlled High-Accuracy Wide-Range Speed Regulator for Motor Drives. *IEEE Transactions on Industrial Electronics*, Vol. IE-29, No.3, August, 207-211.

Prokin, M. (1991). Double Buffered Wide-Range Frequency Measurement Method for Digital Tachometers. *IEEE Transactions on Instrumentation and Measurement*, Vol.40, No.3, June, 606 - 610.

Prokin, M. (1993). DMA Transfer Method for Wide-Range Speed and Frequency Measurement. *IEEE Transactions on Instrumentation and Measurement*, Vol.42, No.4, August, 842 - 846.

IFAC

Publications
www.elsevier.com/locate/ifac

METHODOLOGY OF DESIGNING EMBEDDED CONTROL SYSTEMS

Jiří Kotzian, Vilém Srovnal

VSB - Technical University of Ostrava
Faculty of Electrical Engineering and Computer Science
Department of Measurement and Control
17. listopadu 15, CZ-708 33 Ostrava-Poruba, Czech Republic
jiri.kotzian@vsb.cz, vilem.srovnal@vsb.cz

Abstract: Progress on the area of embedded control systems is rapidly growing. This paper describes the methodology of their development. The output of modelling and design systems is the source code of used microprocessor. We use for the design the development environment Rhapsody by I-Logix. Applications with Motorola HC12 microprocessors are used the Rhapsody in MicroC by I-Logix integrating Metrowerks (Hiware) C-compiler and source debugger. *Copyright © 2001 IFAC*

Key words: Computer control, Model, Microprocessors, Fieldbus, Real time operating system.

1. PREFACE

Presently is standard to develop the programme system on the higher-level language. The assembler was replaced mostly with C-code to control applications. In the described development system C-code will be replaced with the automatic code generation based on the model. This option allows to focus on the main problem and not to deal with implementation details. This option also reducing time to market, reducing price and increasing reliability of designed control systems.

2. METHODOLOGY

The designing methodology consists of fifth stages. A model of a designed system with inputs and outputs is defined in the first stage. In the second stage a model is compiled by a using compiler and tested instead the final target. The third stage expands the second stage with models of an environment outside a modelled system. The fourth stage implements a model to the final target. The last stage tests an application in a real environment.

2.1 Model

In the first stage we have to define a structure and behaviour of the designed system (Fig. 1)

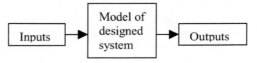

Fig. 1: Stage 1 – defining model

For the defining structure is using activity charts. Statecharts and flowcharts are defining behaviour of system.

2.2 Compiling and testing on PC

In the second stage is defined a configuration for PC (or any other platform supported by I-Logix Rhapsody) processor. After a generation of the application source code, it is compiled using Microsoft Visual Studio compiler (Fig. 2). For the testing our design we can use the Rhapsody graphical backup animation server, which animate statecharts or test drivers.

2.3 Testing with the model of environment

In the third stage a model of designed system is expanded with a model of environment. It means that inputs and outputs are replacing with simple buttons and indicators or more complicated models. For testing is possible use same tools as in the stage 2.

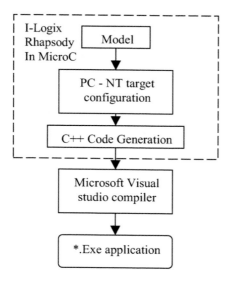

Fig. 2: Stage 2 – PC target

2.4 Implementation

The design on the final target is implement by a changed target configuration for a code generation. For the Motorola HC12 microprocessor is used the Hiware compiler. The Motorola SDI interface is used for the downloading generated source code and for system debugging. Process is shown on figure 3.

2.5 Testing in real world

In the final stage an application is tested in real environment. Testing tools from Rhapsody are increased with the In-system debugging, which is supported by Hiware debugger in cooperation with SDI interface. Using these futures is possible to observe and change any memory positions or registers, create breakpoints and trace an application.

3. DEVELOPPING SYSTEM FOR SELECTED MICROCONTROLLER

3.1 Rhapsody in Micro C (RiMC)

Rhapsody in MicroC is a graphical software design and implementation tool that supports the development of embedded real-time software for micro-controllers. The focus of the tool is to support the process of developing software pieces while targeting small micro-controllers. The support of design-level debugging and testing (both interactively) in batch mode and analysis of runs is implemented through various instrumentation's of the generated code. The output of the tool is a compact, readable ANSI-C code, with support to local extensions of the standard C, as well as automatically generated design documentation.

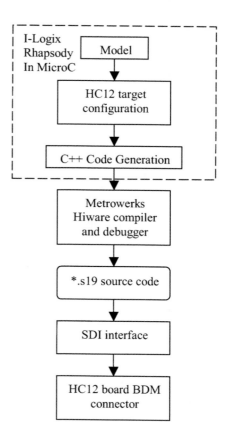

Fig. 3: Stage 4 – HC12 target

RiMC uses an operating system implementation (OSI) definition to describe the implementation of the software and hardware target environment for a given design. Any one OSI might support only a subset of the design concept referred to above. As a general rule, the tool tries to make use of any such design aspect/concept it encounters in the model. If the given OSI has no support for that design aspect/concept, an error message will be produced. Code is generated directly within RiMC based on a graphical model that represents the full functionality of the application being designed. There are four basic graphical tools used to define the application: State charts, Activity charts, Flow-charts, Truth tables.

Each graphical tool has an associated graphical design language that allows the designer to be very precise in defining the functional role of each graphical element. The graphical elements can be supplemented by linking in user supplied C and/or assembly language code. All of the graphical elements are stored in an internal database that contains associated data about each element. The Data dictionary tool is used to define and manage the various data elements as well as various other properties of both the textual and the graphical entities in the model. Properties can be applied to data or to tasks. Data properties are typically defined as exact type, although integer types can be byte-defined for as appropriate for specific machine architectures. Task properties are defined with a task priority in the model.

RiMC also includes a Check model utility that serves as a model checker (somewhat like a precompiler) to detect and warn of incomplete definitions as well as common design pitfalls to help reduce development time and increase the quality of the generated code.

RiMC is using different languages to make the model. The languages used in RiMC can be both graphical and/or non-graphical (i.e. textual).

Structuring Language - activity chart. The software structure is defined in the top-level activity chart. In this graphical view of the application, the architecture of the software being developed is determined. Tasks and Interrupt service routines (ISRs) are defined as well as the functional content of them. The bindings of signals to physical hardware ports and addresses is done using the flow lines to and from the various tasks and ISRs in the chart. The user, in this view defines the generated application architecture. Task and ISR code frames are generated, according to the specific properties of the TASK/ISR. A TASK/ISR code frame invokes the activities mapped underneath the TASK/ISR.

Decomposition Language - activity chart. This is a data-flow oriented graphical language. Functionality, in here referred to as activity behaviour is defined using the, well known, decomposition method. Each required functionality, i.e., activity is sub-divided into functions, i.e., sub-activities that might be further divided into even smaller sub-activities, until no further decomposition is needed. When no further decomposition is needed, we define the basic activities – those that implement certain functionality. The code generated for an activity is a function (or a C Preprocessor macro). For a non-basic activity, the function calls each of the activity's sub-activity functions. For a basic activity, the function contains the implementation code.

Statecharts. Statecharts are hierarchical state transition diagrams. That language is best in describing application modes and transitions between the modes, as well as application reaction to various events in each of the modes. This discrete behavioural language is very much powerful in describing such application modes and transitions between those modes. The implementation of statecharts in RiMC is compact and efficient. The application uses a state variable per each of the activities implemented by a statechart. States are encoded to reuse RAM bits. Several synthesis algorithms are used to reduce both the RAM and the ROM required implementing a statechart on a base of "pay for what is used". Therefore, it is

recommended to use that language whenever that information, i.e. the application state, is required.

Flowcharts. Here we refer to regular flowcharts. Iterative algorithms, if-then-else constructs, switch statements and direct calculations should normally be defined as a flowchart. That graphical language enables the user to graphically debug the algorithm. The code of a flowchart runs from beginning to end, without stopping. If the flowchart is ever run again, it starts from the beginning. The code generator tries to minimize the number of goto statements that are needed. This makes the code readable and structured.

Truth table. The functionality of activity might be directly defined using Truth table. Truth table is a table describing the inputs, the resulting outputs and the actions performed. Truth tables are recommended to use when the activity has much input to consider and few states/modes to be in. When the truth table is defined in a reduced form, it will be reflected in the generated code. This enables the user to build highly efficient implementations.

Reactive activities. When the functionality is best defined as pairs of triggers and actions, that language is the most suitable to define that behaviour. The syntax is exactly trigger and action: E/A thus directly expresses the required behaviour. This textual language allows most clear, compact and straightforward implementations, when the required functionality might be defined as a set of triggers and resulting actions.

Procedural activities. When the functionality is a pure calculation, defined as a sequence of "if then else", iterations and numerical calculations that language might be used. It is similar in its expressiveness to the flowchart graphical language, however it does not requires any graphics, thus might be faster to complete when the algorithm is already proved to be correct.

Time Model and Related Time Operators. RiMC has three model constructs that have a notion of time: Timeout and delay operators referring to Software counter(s), Schedule operator referring to a hardware timer and Periodic task referring to Timer.

Asynchronicity. For no synchronised actions is possible to use ISR and Task. ISR is a reactive component that models interrupt service routine, with associated data and functionality, defined as an activity sub-type. Task, is reactive component with associated interface, data and functionality defined as an activity sub-type. RiMC tasks run independent of each other. According to the environment, a task run might be interrupted and pre-empt by a higher priority task, or an interrupt.

Synchronisation. To synchronise actions is defined Semaphores and Signals. Semaphores are used to co-ordinate accesses to shared resources such as memory or hardware by asynchronous entities, modelled as condition sub type. Signal (Task event) is used to signal to a task on some occurrence like timer expiration, message arrival etc., modelled using event sub typed as task event. Used like regular events as trigger, to wait on the event, as action, to set (generate) the event.

Serial Communication / Messages: Messages are modelled using data item sub typed as message.

3.2 Codewarrior

Metrowerks CodeWarrior is integrated development environment (IDE) for embedded microprocessor. The software provides an intuitive graphical user interface (GUI) with project manager, source code editor and browser. CodeWarrior provide highly optimising C/C++ compiler, powerful macro assembler and SmartLinker that only link objects, which are really referenced. Burner is able to create Motorola S-records, Intel hex files or binary files. Codewarrior decoder is decoding object and absolute files. Using Libmaker is it possible generates libraries. Multipurpose debugger allows simulation and debugging and/or cross debugging of embedded and real time embedded applications. Debugger support multi-language debugging in assembly, C or C++. It is supported true time stimulation and simulation of a hardware design (such as a board, processor, or I/O chip).

3.3 Motorola SDI interface

Motorola's SDI is a serial in-circuit debugger that uses the background debug mode BDM on M68HC12, M68HC16, and M68300 micro-controllers, allowing quick verification and updating of embedded software applications. When used with compatible debug software, the SDI allows users to view and modify applications on the fly reducing development time and speeding time to market. BDM is a feature of Motorola's M68HC12, M68HC16, and M68300 MCU embedded breakpoint and trace hardware. A user connects the SDI in-line between the computer's serial port and the target system's BDM connector.

3.4 MC68HC912BC32

The MC68HC912BC32 micro-controller unit (MCU) is a 16-bit device, which is composed of standard on-chip peripherals. There are a 16-bit central processing unit (CPU12), 32-Kbyte flash EEPROM, 1-Kbyte RAM, 768-byte EEPROM, an asynchronous serial communications interface (SCI), and serial peripheral interface (SPI).

Other on-chip peripherals are an 8-channel timer, 16-bit pulse accumulator, a 10-bit analogue-to-digital converter (ADC), a four-channel pulse-width modulator (PWM) and CAN 2.0B compatible controller (MSCAN12). System resource mapping, clock generation, interrupt control and bus interfacing are managed by the Lite integration module (LIM). The MC68HC912BC32 has full 16-bit data paths throughout, however, the multiplexed external bus can operate in an 8-bit narrow mode so single 8-bit wide memory can be interfaced for lower cost systems. The MC68HC912BC32 support Single-Wire Background Debug™ Mode (BDM) and On-Chip Hardware Breakpoints.

4. REALIZATION RESOURCES OF EMBEDDED CONTROL SYSTEMS

4.1 Testing Lab

On our department is prepared testing lab for the testing of methodology and for students teaching. Aim of an application is to control level of liquid in two reservoirs using level sensors and electromagnetic valves. This lab is part of distributed control system (Kotzián,J. Kašík,V. and Srovnal,V. ,2000). The heartbeat of the control system is terminals and PC - CAN interface based of Motorola HC12 microprocessors. These applications are programmed and controlled using described metrology. For monitoring and changing desired values and parameters PC with real time operating system is used. Scheme of control system is on figure 4.

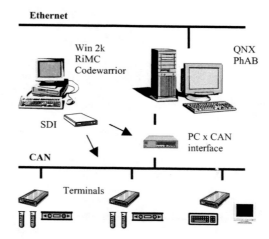

Fig. 4: Testing Lab.

136

4.2 PC- CAN interface

Standard or industrial PC usually has several standard interfaces that can be used to realize interconnection like parallel or serial ports. To ensure faster transaction we may use internal PCI card. We can choose from several of producers offering all above-mentioned types of interconnection. In case of some internal cards, in addition, are implemented protocols of application layer CANopen or DeviceNet with the help of intelligent CAN controllers. Most of producers, however, support only platforms DOS and Windows and only minimum of producers support QNX. These cards have one disadvantage – we have no model of their behaviour. In this reason we are using interface realized with HC12 with build in CAN controller. This interface can be modelled and controlled by described methodology.

4.3 CAN

The network CAN (Controller Area Network) is robust, very quick and cheap network for communication in real time for controlling systems of an automobile, industrial automation as well as in other applications where the network for communication between microprocessors is necessary. The CAN bus is given by international standard ISO 11898 which makes use of layer 1 and 2 of the ISO/OSI model. In the time being the CAN protocol is often implemented directly into products of foremost firms (e.g. MOTOROLA, INTEL, SIEMENS).

The CAN protocol is projected as *multimaster*. The transmitted message does not include the information about target station but only identification number of transmitting station. Each station has its own identification number. Further, each message includes the information concerning the importance (priority) of transmitted message. Thus, all equipment units receive the message, but on the basis of identification numbers it is possible to ensure the receiving only these messages which concern the given equipment (acceptance filtering). For control of the CAN bus there exist several standards defining the application layer of the CAN protocol (CANopen, DeviceNet ...). The layers 1 and 2 remain the same.

To the bus the 2^{29} equipment units can be theoretically connected. With the respect of the bus loading and for ensuring of static and dynamic relations on the bus the ISO 11898 standard sets the maximum number of equipment units as 30. The speed of transmission is set as 125 kBit/s up to 1 Mbit/s (according to the application layer) for which the maximum length of lines is set as 40 m.

4.4 QNX

The operating system QNX belongs to main representatives of operating systems for work in real time. It is based on operating system UNIX and has the architecture of microkernel.

The operating system QNX ensures the multitasking, by priorities controlled pre-emptive planning with quick switching of context what is suitable for the RT applications. There are 31 priority levels (16 operators' ones) and three methods of planning. The inter-processes communication is based on messages (reports). The commands of the UNIX, POSIX and QNX standard, which ensure the RT functions, are supported. For the creation of application software the translator of the C and C++ languages is for disposal. For the system visualisation it is possible to use the superstructure of operating system as X-Windows and especially Photon containing the powerful tool PhAB enabling the generation of application structure and users' interface.

5. CONCLUSION

The work describes the development of embedded control systems using design environment *Rhapsody* and integrated development environment *Metrowerks Codewarrior* for code generation. Both design tools are application orientated. Their version depends on used microprocessor and programming language. The development of control systems with these tools is much quicker then current methods.

Acknowledgement
The Grant Agency of Czech Republic supplied the results of the project 102/01/0803 with subvention.

REFERENCES

Kotzian,J and Srovnal,V.(1999): Distributed control system of technological processes using OS QNX and CAN bus. In: Proceeding International Workshop IWCIT'99. VSB TU Ostrava 1999, ISBN 80-7078-679-5, p.190-193

Kotzian,J. Kašík,V. and Srovnal,V. (2000): Heterogeneous Interconnection of the QNX Systens and PLC in Distributed Control Systems. In: Proceeding International IFAC Workshop PDS 2000. PERGAMON(Elsevier Science) 2000,Great Britain, ISBN 0-08-043620 X, p.251-254

Kočíš, T., Kotzian, J. and Srovnal V. (2000): Embedded Systems. In: Proceeding International Workshop IWCIT'01. VSB TU Ostrava 2001, ISBN 80-7078-907-7, p.204-210

IFAC
Publications
www.elsevier.com/locate/ifac

SOFTPLCS AND THEIR USING IN CONTROL APPLICATIONS

Jiri Koziorek, Lenka Landryova

Department of Measurement and Control
Faculty of Electrical Engineering and Computer Science
Department of Control Systems and Instrumentation
Faculty of Mechanical Engineering
VŠB – Technical University of Ostrava
Address: 17. Listopadu No. 15, 70833 Ostrava-Poruba, Czech Republic
jiri.koziorek@vsb.cz, lenka.landryova@vsb.cz

Abstract: This paper will bring up main information about PC-based control systems. The first part brings up general information about PC-based control systems and introduces products of the Siemens company. The second part of the paper reviews author's experiences with using SoftPLCs in different types of control problems and it presents the results of testing some SoftPLCs characteristics. The last part of the paper refers to communication capabilities of SoftPLCs and the possibilities of their connections to SCADA/HMI systems. *Copyright © 2001 IFAC*

Keywords: computer control, programmable controllers, process control, man/machine interface

1. PC-BASED CONTROL

Softlogic or SoftPLC recently became one of the new trends in a field of logic control. Their creation and development bear on the explosive expansion of IBM PC computers. Personal computers aren't used in an office only but they get to control applications in industry too, where they are usually used as a platform for process visualization (HMI/SCADA) (Mudroncik and Zolotova, 2000). The permanent increase of PC power and possibilities, together with other demands of the company, information systems, development of Internet, Intranet and industrial networks, make it possible to use personal computers in direct control. Direct control has been the domain of PLC systems so far, and the beginning of using personal computers in this field caused open discussions between protagonists of modern ways of PC-based control and sticklers for PLC's.

1.1 SoftPLC

What can we imagine under SoftPLC? SoftPLC is essentially a program running on a PC, which communicates with a set of distributed I/O devices by any standard industrial network. In other words, SoftPLC is every control system, which realizes PLC on the platform of a PC. We can find two basic modifications of such control systems in practice:

- SoftPLC is the program that is running usually under WinNT and it uses the microprocessor of a PC only for control activities. Software and HMI/SCADA producers often use this type.
- A control program doesn't use the main microprocessor of PC and runs under some RT-operation system on a stand-alone processor. The processor can be a part of a special card inserted in a PC or part of external peripheral device connected to a PC. The processor card is usually

called as SlotPLC. It should be very similar to a classical PLC. Companies producing PLC's often use this modification.

1.2 Software SoftPLC's

If we consider the operating system WinNT, that is frequently used in control applications, we can divide software SoftPLC to three categories (Bílek, 1998):

- Pure software SoftPLC – softlogic. It uses only services of standard WinNT. It isn't too advisable for real-time control applications because of the interruption system of WinNT that prefers system devices before the application level. Softlogic runs a control program as a high priority task under WinNT. It means that the control algorithm can be interrupted at any time by procedure calls, which is used to service NT system functions - hard drive access, mouse operations and so on. It can increase scan time over the acceptable limit (scan time is the usual sequence – reading inputs, processing programs, writing outputs). Softlogic does not provide the same level of deterministic control as a PLC. Generally, SoftPLC that are using for their activities only standard operating systems and have the above mentioned characteristics are not good for application with a requested scan time under 100ms.
- Soft PLC, that use the extended operation system WinNT, which is able to provide really deterministic behavior of control application in real-time. The enhancement is realized by software accelerators. They modify the Hardware Abstraction Layer (HAL – lowest level of operation system) so that the control application has a higher priority than system devices.
- SoftPLC doesn't use WinNT for control services but runs independently under any real-time operational system (QNX, OS-9 and so on) which is also a primary operation system of PC. WinNT can be used here as one of the tasks and runs simultaneously with SoftPLC.

1.3 Advantages and disadvantages of SoftPLC's

There is a number of arguments for using SoftPLC – the power of PC, big capacity of memory (RAM, HDD), availability, easy handling, good graphical interface, easy connection to network, number of peripheral devices.... It could look as if there isn't any problem. We will try to compare the main characteristics of both systems.

- The power of processors – PC processors are more powerful than PLC processors. But the power of the processor is not the main criteria for control application – much more important is the reaction time (after change, the inputs react adequately by setting the outputs). In the reaction time criteria wins PLC's that can assure its repeatability.
- Advantage of SoftPLC is the possibility of the creation of data archives. In PLC systems, we can solve this demand by periodical data transfer to a higher level but sometimes it is not possible.
- Availability and good price of PC – it is the only apparent advantage. We can hardly use the standard PC for industrial applications and special industrial PC's are considerably more expensive. Innovation cycles of PC's are very short and the question is how long can the producer guarantee the supply of spare parts.
- Reliability – SoftPLC's are rarely used in the control of important technology where there are hard demands on reliability and the safety of operation.

The trade-off can be in using a slotPLC, which eliminates the instability and dependency of SoftPLC on a PC.

1.4 View at the future

New strategic research from Frost & Sullivan (www.frost.com), World PC-Based Control Markets, provides numerous industrial forecasts. PC-based controls software market revenues increased 70 percent to $58.6 million in 1998. By 2003, the total market revenue is projected to reach $393.4 million in total revenue.

Currently, PLC's account for 60 to 70 percent of the market share, while PC-based controls account for a mere 15 to 20 percent. However, this study projects a market shift over the next three to five years with PC-based control systems outnumbering PLC's.

2. SIEMENS SOFTPLC SYSTEMS

Siemens Simatic PC-based control (Simatic WinAC) is a combined solution for control, visualizing, and communication on a common PC platform. Simatic WinAC is divided into five central components for solving a variety of automation tasks (Siemens, 2000):

- – WinAC Controlling (for solving control tasks)
- – WinAC Technology (for solving time-critical process-related tasks)
- – WinAC Visualizing (for interfacing to WinCC, ProTool/Pro and other HMI software)
- – WinAC Computing (data exchange for the further processing of process data using Windows programs)
- – WinAC Networks (for linking the PC to office and industrial networks).

WinAC Controlling contains two variants of SoftPLCs (www.siemens.com):

- – Software SoftPLC with WinLC (Windows Logic Controller) – WinAC Basis (softlogic, uses a standard operating system WinNT) or WinAC RTX (uses the real-time expansion for WinNT from VenturCom). Power characteristics are similar to PLC Siemens Simatic S7 300. It is intended for solving smaller control tasks in conjunction with projects where PC element predominates.
- – SlotPLC with CPU 416-2 DP ISA – control independent of WinNT, real-time behavior suitable for time critical applications, integrated Profibus-DP and MPI interfaces. It is intended for more demanding control tasks in conjunction with extensive PC tasks.

WinAC is connected to process via a communication interface for Profibus-DP. For programming of both modifications STEP 7 software is used as well as for Simatic S7. WinAC Visualizing enables visualizing the program to be linked to a PC. Access is optimized for WinCC visualizing system; other HMI software can be used via a OPC interface. WinAC Computing is the data interface between automation software and standard software. It uses WinNT communication mechanisms – Active X, COM and DCOM.

3. TESTING OF SOFTPLC SIEMENS WINAC BASIS

Authors have decided to test some main characteristics of SoftPLC. The testing was oriented to speed up the processing of logical and arithmetical (integer and floating-point) operations and compare

them to a similar PLC system. How much power from the PC influence to SoftPLC operation was also compared.

Hardware used:

PC1: Procesor Duron 800 MHz, 256 MB RAM, 20 GB HDD, 7200 ot./s

PC2: Procesor Pentium IBM Cyrix 233, 128 MB RAM, 15 GB HDD

SoftPLC: Siemens WinAC BASIS, CP 5613, ET 200M + module 4AI/2AO and module 16DI/16DO

Table 1 Measured and calculated values

Count of oper.	Sleep Time (ms)	Exec. Time (ms)	CPU Usage (%)		PLC Ex. Time (ms)*
			WinLC	PC	
1000x OR	5	1(3)**	0(40)	5(56)	0,98
	10	1(3)	0(32)	3(41)	
	20	1(3)	0(14)	3(17)	
5000x OR	5	2(13)	20(76)	25(81)	4,9
	10	2(13)	13(58)	17(62)	
	20	2(13)	6(40)	6(46)	
10000x OR	5	4(26)	41(88)	46(96)	9,8
	10	4(26)	23(77)	27(84)	
	20	4(26)	13(58)	14(64)	
1000x ADD (16b.)	5	1(3)	0(45)	6(61)	3,9
	10	1(3)	0(34)	5(44)	
	20	1(3)	0(14)	4(17)	
5000x ADD (16b.)	5	3(14)	29(80)	35(92)	19,4
	10	3(14)	17(59)	21(68)	
	20	3(14)	9(42)	10(48)	
10000x ADD (16b.)	5	5(27)	41(88)	44(97)	38,8
	10	5(27)	24(74)	28(83)	
	20	5(27)	13(57)	14(64)	
1000x ADD (32b.)	5	1(3)	0(46)	6(55)	< 54,2
	10	1(3)	0(31)	5(41)	
	20	1(3)	0(14)	4(17)	
5000x ADD (32b.)	5	3(14)	29(81)	35(89)	< 271
	10	3(14)	17(58)	23(68)	
	20	3(14)	9(42)	10(48)	
10000x MUL (32b.)	5	5(28)	44(89)	49(96)	< 542
	10	5(28)	28(75)	32(84)	
	20	5(28)	17(59)	19(65)	

*Execution times of the program in PLC Simatic S7 314 IFM was computed with values from Siemens manual (Siemens, 1998). Values for the floating-

point arithmetical multiplication depends on the value to be calculated and there are the worst possibilities in the Table.

** Values outside of the brackets are for PC1, values in the brackets are for PC2.

The testing program contains block OB1 for the cyclic execution and function FC1, FC2 and FC3 with testing types of operation. The FC's have a 100 of the same networks – for example FC1 has 100x logical OR. The FC's are called from OB1 – the number of calls is 10, 50 or 100.

PC CPU Usage rapidly increases when other applications run on the PC. When some common application (MS Word, MS Explorer) is running together with WinLC, PC CPU Usage may temporarily increase to 100%. Using other application, while WinLC is running has only a small influence on WinLC CPU using and execution time. The execution time of some single cycle should increase but the average execution time doesn't change.

3.1 Execution Time and CPU Usage

We can see from the Table that WinLC execution times for three types of operation almost don't vary. Arithmetic multiplication of two 32b. floating-point numbers has similar execution time and CPU usage as simple logic OR. This fact contrasts with the PLC where floating-point arithmetical operations have much longer execution times in comparison with logical operations.

We tested two PC configurations. The program execution of the PC1 was very fast and the processor still had good power reserve for other applications. The PC2, respectively, had quite a problem with processing higher counts of operations (5000, 10000) and the power reserve was quite small. Setting the sleep time to 50ms and more accomplished partial decreasing of CPU usage (for arithmetic addition, 10000 operations and 50ms sleeping time was CPU usage of WinLC 34% instead of 57% and CPU usage of PC 45% instead of 64%). However, it is much better to use a powerful computer for practical applications.

3.2 Communication capabilities of WinAC Basis

WinAC Basis has two general communication possibilities. Connection to distributed (remote) I/O is realized via Profibus. This connection is necessary for every control application. Other communication possibilities are used for access to process data by a higher system (such as man/machine interface). Access to process data is provided by WinAC Computing. The following possibilities can be used (Siemens, 1999):

– Using of standard ActiveX controls (OCX – OLE Control) – WinAC computing has S7SoftContainer for ActiveX controls, which can be used as a simple visualization tool.
– DCOM (Distributed Component Object Model) – for creating a distributed application over a network when WinLC can be an accessed control system from a remote PC.
– OPC server (OLE for Process Control), which allows any OPC client application to access data in the control device. This can be used, for example, for connecting to any SCADA system.

3.3 Price of the sample system

We will consider small control application with 30 DI, 30 DO, 3AI and 1AO. We will decide between a PLC Simatic S7314 IFM control system and a SoftPLC Simatic WinAC Basis.

For the PLC system we will need the power supply, CPU, 16DI/16DO module, 4AI/2AO module, din rail, connectors, programming adapter.

For the SoftPLC system we will need the power supply, WinAC Basis, Profibus interface, distributed peripheral ET 200M, 2x16DI/16DO, 4AI/2AO module, din rail, connectors, 20m Profibus cable, standard PC.

Both systems we designed with Step7 engineering software, which is not included in the price.

The prices, calculated with (Siemens, 2000), are:

– PLC system – 1802 Eur.
– Soft PLC system – 3064 Eur + standard PC (1000 Eur).

We can see that the SoftPLC solution is much more expensive. Since WinAC Basis is a softlogic system and we are considering standard PC, we can expect a lower level of reliability and non-deterministic operation. On the other hand, the WinAC Basis system contains a visualization tool and it is possible to easily connect to other Windows applications.

The PLC system has a lower price, guarantees a high level of reliability and deterministic operation. In the case that process visualization is demanded, additional costs would be necessary.

CONCLUSION

Powerful processors of a PC and the increasing reliability make it possible to use PC-based in control applications. We showed that SoftPLC program execution could be much faster than that of a PLC, especially when the program contains a number of floating-point operations. On the other hand, the price of the PLC system is often much lower for standard applications than the price of a SoftPLC system especially when we don't need process visualization. SoftPLC could be a good solution for the complex control of non-critical processes where we need controlling, visualization and communication via a net and with other applications.

This work has been supported by grant GACR 102/01/0803.

REFERENCES

Siemens (1999): *Simatic Windows Automation Center, WinAC Basis, Overview.* C79000-G7076-C219-03, Edition 3.

Siemens (1998): *Instruction List for S7 300,* 6ES7 398- 8AA03- 8BN0, Edition 2.

Siemens (2000): *Automation & Drives Catalog 12/2000.*

Bílek, K. (1998): Are softPLCs serious alternative to classical PLC ? *Automatizace,* **10/98,** p. 716-717.

Mudroncik, D. – Zolotova, I.: *Priemyselne programovatelne regulatory – konfiguracia, vizualizacia, kvalita softveru* (In Slovak). Elfa s.r.o., Kosice, 2000, pp. 169, ISBN 80-88964-45-8

www.siemens.com: Siemens company web site.

www.frost.com: Frost & Sullivan web site - international marketing consulting and training company.

IFAC

Publications
www.elsevier.com/locate/ifac

HYBRID SCADA SYSTEM WITH EMBEDDED PROGRAMMABLE CONTROLLERS

Piotr Laszczyk

Institute of Automatic Control,
Technical University of Silesia
ul. Akademicka 16, 44-100 Gliwice, Poland
tel. 48 (32) 2371473   Fax: 48 (32) 2372127
E-mail: laszczyk@ia.polsl.gliwice.pl

Abstract: In the Institute of Automatic Control of Technical University of Silesia exists an experimental laboratory installation of heat exchanger network. The installation is fully equipped with measurement sensors and actuators. Using computer with AD/DA cards and professional software it is possible to design SCADA system for this process. Research, which is made on the installation, includes testing standard and non-standard control methods. Available LabWindows/CVI software allows creating SCADA system with embedded programmable controllers. That task includes many aspects, such as: cascade control, switching between different controllers, real-time order of work as well as process visualisation, reacting on operator's action and storing the data. The developed SCADA system copes with all required tasks. *Copyright © 2001 IFAC*

Keywords: temperature control, programmable controllers, control system design, control algorithms, PI controllers, control oriented models

1. INTRODUCTION

Any research which is connected with many laboratory experiments requires good quality SCADA system that allows to program more complicated computations. Using standard PLC and ladder logic or statement list it is very difficult to implement more advanced control algorithms. Moreover that kind of programming is not easy for making changes, which is very frequent in that case.

There exists an experimental pilot plant, which is installed at the Institute of Automatic Control at the Technical University of Silesia. The installation represents main aspects of heat distribution plants. Instrumentation of the installation allows testing any control structures and algorithms as well as SCADA systems. In that case, there were chosen a

solution wit PC computer and A/D and D/A cards and LabWindows/CVI software.

2. SYSTEM DESCRIPTION

Constructed at the Institute of Automation of Technical University of Silesia experimental installation simulates on order of work of the real thermal center. Installation consists of two water circuits. In primary circuit water is served from electric heater to spiral-tube heat exchanger HE1 which transfers part of heat energy to secondary circuit containing plate type heat exchanger HE2 or double-pipe heat exchanger HE3. It is possible to switch using manual valves between HE2 and HE3 changing in this way receiver dynamics. Outlet water of heat exchanger HE2 or HE3 is served to mixer. It allows us building control structures containing six units. Automation system could consist of one level control unit and five temperatures control units. It is possible to built cascade control unit with master controller for temperature and slave controller for water flow rate.

Laboratory installation has 24 measurement points:

- 9 temperature measure points,
- 1 level measure point,
- 4 flow rate counting points,
- 10 water pressure-meter points.

Installation shows aspects of a thermal centre control. Controlling thermal centre is an important component in arrangement of state energy economy. In real thermal centre a thermal-electric power station is the source of heat. It serves steam or water under pressure (primary circuit) to heat exchanger which transfers part of heat energy to secondary circuit containing heat receivers (city quarters, public buildings etc.). Detailed description of installation could be found in another papers (Łaszczyk and Pasek 1995). Scheme of the installation is presented in further chapter on main panel of SCADA system (fig. 6)

3. CONTROL METHODS DESCRIPTION

One of the commonly task of heat exchanger control is to stabilise temperature of preheated fluid. In considered process there were considered control structure to stabilise temperature T_{sp}. To achieve that controller should influence on the quantity of heat energy transferred from first to second cycle. There are two possible structures of controllers. In the first case one controller is using T_{ec} as manipulated variable by influencing on the second controller which acts on heater power P_h. Though there are two controllers working in cascade as in figure 1.

In the second considered structure PFC_2 controller acts directly on flow F_c using valve V_1. Because temperature T_{ec} is strongly dependent on flow Fc there should be applied another controller to stabilise that temperature. In that case there are two independent controllers as shown in figure 2. PFC control of heat exchanger by acting on the flow, however for steam/fluid heat exchanger, was implemented previously by Ernst and Hecker (1996).

It should be mentioned that there exists also third method of control for heat exchanger. In that method controller acts on both flow and temperature at the same time. Such solution leads to an enthalpic control and is considered in order to optimise the quantity of energy transferred from heater to the heat exchanger. That method will be implemented and evaluated in a future work.

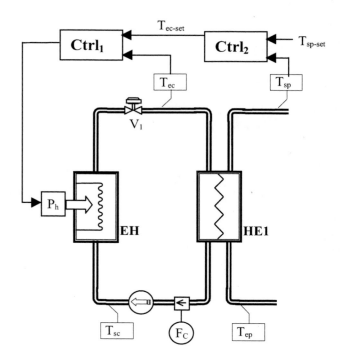

Figure 1. First considered control structure.

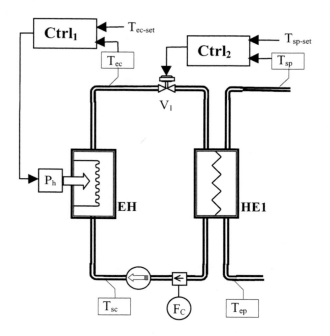

Figure 2. Second considered control structure.

In the paper the first solution was implemented and tested. There are many loops and correlation between process variables in the described process. For example increasing power of electric heater causes increase of temperature T_{ec} and farther T_{sc} that additionally gains the increase of temperature T_{ec}. That makes the process difficult to control unless that correlation is included in control algorithms.

There were implemented three different control types for stabilising output temperature of electric water heater. Since there is no other possibility for power control than switch on/off the full power of the water heater the simplest way is two-position control with histeresis. Experiments showed that this kind of control gives not satisfactory results. At the average flow F_c=0.5 [m^3/h] there temperature varies periodically with amplitude of 8 [°C]. Because of that there were also implemented another version of two position controller with inverted histeresis. In this case the amplitude of temperature changes varies with amplitude of 5 [°C].

Both types of two position control gives not stable work. There were considered and implemented another way of stabilising the output temperature. Good stabilisation of temperature on output of electric water heater is only possible when influencing electric power continuously. The idea is to manipulate on electric kettle switching it on and off with pulse modulated signal. The time period of that signal should be chosen in that case significantly shorter than time constant of the controlled system. It is obvious that time constant of the electric kettle assuming ideal mixing inside is described by:

$$\tau = \frac{V}{F_C} \qquad (1)$$

Where:
V – volume of the kettle
F_C – medium flow
For maximum flow the calculated time constant was $\approx$ 200 [s]. The period for pulse modulated signal was chosen 10 [s]. That secured despite on/off manipulation on the power, stable temperature on the output.

Basing on that type of continuous manipulation on electric power there were programmed two another controllers PI and PFC. The first of them is standard PI algorithm. Specific description of PI control is widely described in many publications. Discrete form of the control PI equation is following:

$$\varepsilon(n) = T_{EC-set}(n) - T_{EC}(n) \qquad (2)$$
$$y_{R1}(n) = y_{R1}(n-1) + \varepsilon(n) - \varepsilon(n-1) \qquad (3)$$
$$y_{R2}(n) = y_{R2}(n-1) + \frac{K_P \varepsilon(n)}{T_I} \qquad (4)$$
$$P_H(n) = y_{R1}(n) + y_{R2}(n) \qquad (5)$$

Where: K_P – proportional gain, T_I – integration time.

The second one control methodology PFC bases on a simplified process model (Richalet, 1993). The model has the structure presented on figure 3 and is described by following equations.

$$\frac{dT_{ec}}{dt} = k_1 F_C^{\gamma}(T_f - T_{EC}) + k_2 P_H \qquad (6)$$

$$\frac{dT_f}{dt} = k_3(T_{sc} - T_f)F_1 \qquad (7)$$

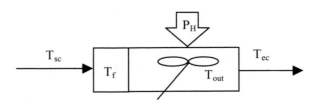

Fig. 3. Scheme of the internal model of the kettle

For control transfer function form was used and implemented as an internal model.
Transfer function form for the above model is following:

$$T_f = \frac{1}{s\tau_i + 1} T_{SC} \qquad (8)$$

$$T_{EC} = \frac{1}{s\tau_2 + 1} T_f + \frac{G}{s\tau_2 + 1} P_H \qquad (9)$$

Where: $\tau_1 = \dfrac{1}{k_3 F_C}$, $\tau_2 = \dfrac{1}{k_1 F_C^{\gamma}}$, $G = \dfrac{k_2}{k_1 F_C^{\gamma}}$

This form of model was implemented into controller as an internal model. More precise description could be found in Laszczyk (2001).

There were implemented only two types of controller as an upper one – controller 2. One is the standard PI described by the same equations (2, 3, 4, 5) as in the first controller. The second one is PFC control algorithm based on simplified model of heat exchanger. Obtained model of heat exchanger was implemented as internal model of the controller. Thus the controller is based on physical model and is self-adaptive to changes of process variables. Control algorithm for this unit was presented by Laszczyk and Richalet (1999).

4. SCADA SYSTEM STRUCTURE AND PROPERTIES

Presented in the paper, SCADA system is built using LabWindows/CVI programming tool. That software has many predefine graphical objects which makes building graphical user interface (GUI) quite easy. There are many objects, which are very useful for building SCADA systems, such as knobs, buttons, dynamic bars, sliders, graphical trends etc. Each object has defined number of events, which are serviced by callback functions. Those callback functions are programmed in ANSI C language.

It is possible to rum few subprocesses using timer object. Each timer could run with different pre-set time. In considered case there were used three independent process loops initiated by timers. First one is the main process loop. That loop is servicing reading from inputs of A/D cards, visualisation and writing output values to D/A cards. Another loop is running control algorithm. Thus process is rather slow period of that lop is set to 10 [s]. Very important in that case is that First loop should run more frequent than the second one. Another loop is running "writing data to file" process and trending process. That kind of classification of subprocesses might be questionable but LabWindows allows reconfiguring it very easily.

There is another set of functions in the program, which are called only once, or under some conditions. All initialisation of objects and setting start values of variables should be done in function *main()* which is called only once in the first moment of starting the system. Programming callback function for servicing operator's actions is very friendly because LabWindows generates on demand function with header and default content after creation of an object. Such feature makes programming more easy e.g. connecting value shown by knob or slider with some variable in C code takes almost no time.

It is possible to create multiple panels and bring them to front by special instruction placed e.g. inside callback function of any pushbutton. In that way there were created panels with controllers and trends.

On the trend panel it is possible to choose which trend lines are to be presented on the chart. This is to avoid some kind of confusion when too many variables are presented on the same chart.

It is obvious that the more sophisticated demands to GUI the more programming efforts and time it costs. Although most programming time has been consumed on implementing controllers than on creating GUI. It required many testing experiments and a lot of changes because that system was created for research purposes.

One of the problems that appeared during programming was switching between controllers. That task could not be done in the simple way because there were implemented also non-standard controllers. For PFC controller it is not proper to switch it completely off while another controller is active and running. PFC controller should be fed up all the time with process data to secure internal

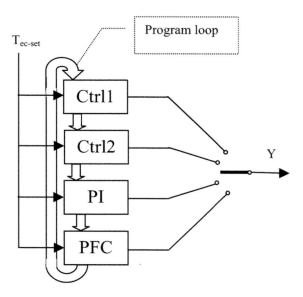

Fig. 4. Scheme of controllers switching.

model follows up the process. In other case switching to that controller could cause unexpected non-stable work in the very first moment. One solution is to run each control algorithm in the control loop but output of only one of them is send to the process (figure 4)

The big disadvantage of that solution is that it is not time optimised. The more alternative controllers much more time is consumed on running the control loop. It might be problematic and cause failing the set timer period for control loop. Another solution is to look inside each controller to choose which equations need to be calculated each loop run and which only if this or that controller is an active one. That kind of solution was implemented in the described SCADA system. While all the programming code is rather complicated user interface is rather simple and choosing an active controller is done by marking a bullet on the controller panel and setting the tunings and set value for controlled temperature (figure 5)

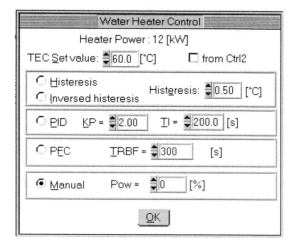

Fig. 5. Panel of the first controller

148

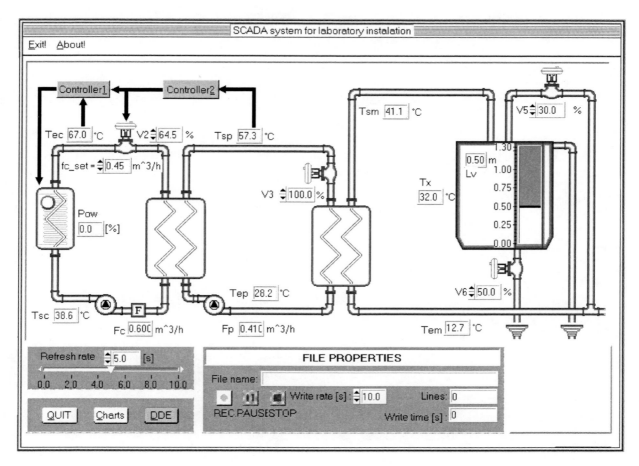

Fig. 6. Front panel of SCADA system

5. CONCLUDING REMARKS

There was described SCADA system for laboratory experimental installation. Front panel of the system is presented in the figure 6. That kind of system could be called hybrid because virtual controllers are implemented inside it. Usually other devices such as PLCs or single controllers perform control task.

Because the system was designed for research and didactic purposes there were implemented different control algorithms inside one program. That caused the problem of switching between controllers in smoothest way possible avoiding rapid changes of manipulated variable. The paper describes problem solution.

ACKNOWLEDGEMENTS

This research was supported by the National Committee for Scientific Research (KBN) under Grant No. 8-T11A-001-16

REFERENCES

Ernst E.-J., Hecker O., (1996), "Predictive Control of a Heat Exchanger", Manuscript, August, 1996

Laszczyk P. and Pasek K. (1995), "Modelling and Simulation of The Pilot Heat Distribution Plant", *System Analysis, Modelling, Simulation,* vol.18-19: 245-248

Laszczyk P. (2001), Simulative Comparison Of PI And PFC Control For Electric Heater, *MMAR Conference Procidings*, Miedzyzdroje, Poland

Laszczyk P. Richalet J., (1999), Application of Predictive Control to a Heat Exchanger, *Dynamic Control and Management Systems in Manufacturing Processes - Techniques for Supervisory Management Systems* (preprints, ed. D. Matko and G. Mušič), Bled, Slovenia

Richalet J., (1993), *Pratique de la commande predictive*, Hermes, Paris

IFAC

Publications

www.elsevier.com/locate/ifac

UNIQUE PROGRAMMING TECHNIQUES
DEDICATED FOR SOFTWARE CONTROLLED PROCESSES

Malczewski Klaudiusz, Nowara Andrzej, Oliwa Wojciech, Mikno Zygmunt

MSC Polska Sp. z o.o. Gliwice, Poland, Silesian Technical University Gliwice, Poland,
Welding Institute Gliwice, Poland

Abstract: This paper presents several programming methods of control applications and covers rarely published problems of software controlled systems working in severe conditions like dynamic restart hidden for a continuity of process. Such aspects like program and data structure, synchronisation of processes, interrupts, time monitoring, restores of control conditions will be taken into consideration. Usually the control application has to manage different processes and data structures to synchronise data exchange between them and other applications. Also the important aspects are restart of control program, user interface and diagnostic. The proposed methods were successfully used in difference control programs applied in weld controllers and PLCs. *Copyright © 2001 IFAC*

Keywords: control application, program, data, structure, programming, processes, synchronization

1. INTRODUCTION

In most cases control of the technological processes, automation lines and complex machines requires to apply specialised programmable controllers. Actually the usage of different programmable controllers designed to accomplish control in industrial systems become more and more popular. In general the main control tasks of the process are implemented in the form of control program located in the Programmable Logic Controller(s) (PLC) memory. The controller acts not only as device for realisation of control algorithm, but implements several additional operations. These operations like communication, diagnostic, visualisation, functionality checks and interfacing with other controllers become an integrated part of application.

As far as programmable controllers are concerned the behaviour of the application and its functionality are described by the control program. The functionality can be extended significantly if the special features of the processors are exposed and used, if the logical structure of the program is proposed and the data collected in the structures are protected.

All these aspects referred to control program supplemented with the internal consistency checks of data and program run provide an interesting method of programming that will be useful to control systems working in severe conditions like dynamic restart hidden for a continuity of process.

2. AUTOMATION SYSTEMS, CONTROLLERS AND CONTROL PROGRAMS

Conventional approach to automation systems distinguishes three levels in industrial control structure: management (visualisation and supervisory systems), control (PLCs, controllers) and field devices (sensors, actuators). The authors' attention is focused on the programmable devices and control programs in particular.

Usually PLC is the "main" programmable controller in automation system on the control level. However there are several specialised devices in the system that act as a standalone machines and are controlled by the dedicated controllers. This approach is intended to control complex sub-processes or sub-programs separately and to keep some kind of flexibility of the machine to adopt its features in different applications. The another positive aspect is that it simplifies control system. The controllers are equipped with interfaces for communication to other control units.

The controllers classified in the control level of the system hierarchy are the most important devices because they implement the control algorithm, they have access to the sensors and actuators and report to supervisory systems. They hardware configuration can be fixed or established by the user during specification phase of the project. Controller manufacturer offers wide range of different types of modules to be applied. The I/O modules enable to interface the controllers with the automation environment.

The control program written in form of instructions in the memory and realised by the CPU describes the functionality of the machine and furthermore automation system. The designer experience and knowledge means skills to write instructions to fulfil the functionality requirements and to capture and to solve those aspects of the process and the device, which determine quality-reliability of the application. The designer has to know not only the application algorithm but also the functions of the hardware, the system procedures and services as well for effective usage of the possibilities built in the hardware. The application program provides services to the users (alarms, events notifying) or automation environment (interfaces). Interpreting control program as a set of instruction creating logical structure and applying structuralized data, some important advantages referenced to reliability of the application and its functionality can be achieved. It is important in electro-magnetic filed environments like weld processes, thyristor control circuits (Mikno, 1999).

3. SOFTWARE ASPECTS AND REQUIREMENTS

The basic requirement of control program loaded into controller memory is to ensure the proper functionality of the designed machine or process. The further requirements are referenced to the effective maintenance and diagnostic. However in this paper the authors concentrate on the functionality and programming methods to provide the continuity of process and to reduce random influence of environment conditions. For this reason the following aspects can be taken into account and defined as requirements:

- transparency of the program structure
- fault recognition
- analysis of status of the machine regarding program status and "real" machine status
- definition of "certain" states of the machine to recover the run mode or to ensure safe restart
- handling and synchronisation of interoperating processes to protect against access to no valid data

Transparency and logical structure of the program is to be kept in each step of project absolutely. Especially it is very useful during tests and commissioning. The program structure affects on such aspects like process understanding, analysis of results of operations, signal state check, debugging possibilities. The structure requirement is important in case when two or more designers work together with the software. The precise definition and distribution of tasks and functions means better cooperation, less number of problems during merging of program modules and less designer effort in general. In same cases the mentioned modules dedicated to the special functions become reusable ones.

Fault recognition has two important aspects. One is visualisation of alarms and faults i.e. applying HMI panel, other is reaction on these faults to be done by the controller. The decision to be taken in this situation is whether the process is to be continued or is to be stopped. The answer to such defined questions can be moved to the user. It is common for PLC functionality that its operating system informs about occurrences of pre-defined alarm conditions by setting the dedicated signal in the memory or calling dedicated procedure (Siemens, 1996). In this way the user determines the reaction of the application writing proper actions or programming suitable procedures.

It can happen that "real" state of the machine or process differs from one seen by the control programs. The consequences of this mismatching can be unacceptable leading directly to faults in a controlled process. Typically the problem occurs in application where the behaviour of controlled object is implemented as sequence and states and transitions are programmed. If the conditions forcing the state change are improperly defined and programmed than the real state does not match to program one. The other reason of such mismatching is set and reset operations what means memorizing signals, states. If these operations are defined based on incorrect premises than above-mentioned problem occurs.

Sometimes such problem takes place during manipulating with the different operation modes i.e. automatic, manual or repair. The manual operations can cause the state change and the program does not notice this. This is one of the most typical cases of software faults. Of course the root cause is the wrong state/transition definition or missing so-called synchronisation procedures (Nowara, Chmiel, 1999). The problem of mismatching can be covered by programming the machine state conditions based on "real" signals from sensors and actuators and implementation of synchronisation procedure to match machine and program states.

The next requirement is referenced to the possibilities of recovery of status of the machine, it means to restart the machine. One of the most important fragments of the control application is initialisation procedure. This procedure declares and initialises variables, sets the default parameters i.e. communication rate. Any time the procedure is called the machine is restarted with ease and safe. However this procedure recovers only one specified machine states and to perform it the physical state of the machine is to be guaranteed. It has following disadvantages: the info about actual status is lost, the machine cannot continue the cycle, reaching the physical init position can be unexpected long. The solution can be obtained by introducing several reference points that are interpreted in the same way. More advanced proposal is to built the mechanism to recover each machine or process state.

Usually the control application is designed in such way that the CPU executes the main program loop in serial-cyclic way. The main tasks realized by the procedures are complimented with so-called background tasks (processes). As an example the fault identification and communication procedures can be given. The synchronisation mechanism is to be implemented to protect data access from different program processes (main loop and fault ones) and to execute predefined action. The well-known mechanisms are dual-port memory and semaphores.

4. PROGRAM STRUCTURES OVERVIEW

The base for the creating of different program structures is controller and its design and functionality. The conventional concept of program execution is cyclic processing.
The concept of processing using interrupts is applied in programmable controllers as well. It plays the important role in time determined control applications or to serve rare and not regularly occurring events.

There are 3 basic well-known structures applied in control programs:

- linear
- hierarchical (modular)
- processes

Usually the control program applies heterogeneous structures and it is connected to specifications of the applications, hardware features and program designer experience. In general the program structure is a hierarchical structure with event processing. Event procedure is to be written in a dedicated program module. The jump to the module can be controlled by designer or user or can be caused by occurring or disappearing event. Such mechanism seems to be applicable to categorise different tasks.

Another aspect is that today's programs are located remotely, the communications possibilities are not optional but standard one, so requirements are stronger. There is a trend observed to increase productivity of software engineering recently. Among many ideas intending to solve the problem, the approach of object-oriented technology, the concepts of the software libraries or standardized software structures were and are developed. Each proposal, which follows this trend "at least a little bit", is to be treated as positive one.

5. PROGRAMMING METHODS APPLIED IN AUTOMATION APPLICATION

The software aspect and requirements, program structures were presented above. In this chapter some program ideas are presented and they correspond to item discussed previously. All these applications were successful implemented in practice.

The first example is program applied in PLC and relates to the time functions programming. Such application can be treated as an untypical one, because it is necessary to know and consider PLC behaviour in context of system procedures and to ensure minimum delay time after event occurrence. Realisation of so-called untypical applications requires untypical approach of usage of hardware. The concrete example of such application is thyristor control one. The cycle time of controller varies depending on CPU processing capabilities and program task in general. Delay of firing impulses due to long cycle time (milliseconds) causes wrong (dangerous?!) functionality of the control circuit. This occurs in case of standard linear program structure including input/output refresh done once per cycle. The same result can be observed when system procedures are called and than the user program realization is "suspended". System procedures manages input/output image refresh, communication, autotest, cycle time monitoring.

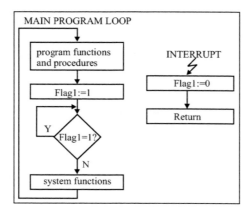

Fig. 1. Synchronisation method applied in time-determined programs.

The request to fire the thyristor occurs asynchronous to program scan. This implies to apply interrupt mechanism and program synchronisation method for interaction of both processes (program scanned cyclically and interrupt). Additionally the delay time required to call interrupt program and impossibility of interrupt of system procedures are to be taken into account.

The solution proposed is to enter additional time loop in program scanned cyclically. In this loop the program waits only for event to start signal processing for thyristor control. The signal processing can be initialised via external signal polled in the loop continuously or via interrupt signal what is more suitable solution. The next action after interrupt execution is call of system procedures.

The described idea of programming includes itself the mechanism of process synchronisation. Further functionality check connected to the cycle time monitoring is standard function of PLC. The program status recovery (to the certain level) can be

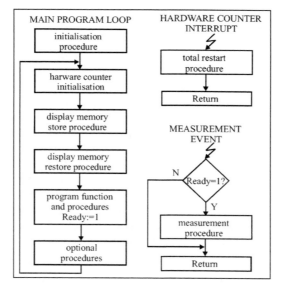

Fig. 2. Program method referring to built-in hardware functionality.

ensured using auxiliary signals from memory area named retentive flags. Another aspects of time functions programming and potential problems were discussed in (Pucher, Kania, 1999).

The next example corresponds to microprocessor application. The program designer has to take care about correctness of control functions, process synchronisation and status recovery. The program structure is typical; there are main program loop and other synchronous or asynchronous control processes. The cycle time monitoring methods uses hardware counter. The program instructions initialise hardware counter. The value can be constant or written to the variable. This creates the possibilities to control varied cycle time in different operating modes.

The example is dedicated to measurement unit. One of the important assumption of the project was to ensure fast status recovery mechanism or in worst case to reinitialise application program. Additionally the program exit setting are to be the activated during initialisation procedure. The fast recovery option is implemented based on time monitoring mechanism. If the time exceeds the declared one the program starts again executing the whole re-initialisation procedure. For this reason possibility of varied time definition obtains to be very useful to speed up this function.

To restore the exit settings the image of display is stored in memory area once per cycle at its beginning. Such proposal seems to be optimal because the probability of improper functionality at this program stage is relatively low. The memory contains information about display localisation in reference to main program to synchronise processes of displaying and parameters read/writes to data structures. The additional check bits are memorized as well to check consistency of data after reloading.

The main intention of the proposed solution is to ensure the fast recovery of the measurement unit on parameterisation (display) level and to achieve "ready" status. Further functions like data and results transmission to/from i.e. PC computer are programmed as a separate, independent processes and their start is allowed after interlocking the main program task.

The last example relates to PLC applications again, but the idea can be built in microcontroller application. The potential limitations concern memory size. This programming method seems to be more industrial, but it is only one of interpretations.

The concept improves SFC (Sequence Function Chart). Introducing some extended functions the concept can be interpreted even as a software package adopted in automation systems to control several machines (processes) treated as independent or interoperating together. Each machine process is a sequence. The programming method represents the behavior of machine as a sequence of steps and

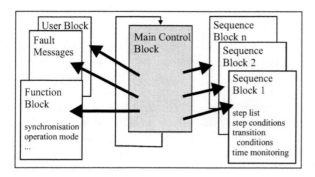

Fig. 3. Program method based on step/transitions.

transitions. The dedicated program module manages the sequences. The step represents a section of program and corresponds to the actual status of machine. Actions connected to output signal are based on activated step number. Transition comprises a set of conditions, which defines when the previous step becomes inactive, and the next or other is activated. It describes when the state of the process is to be changed. Information like step number, step status is memorized and accessible by other program processes. Such programming approach offers a lot of exact information needed to detect the potential cause of problems and the machine-sequence can be diagnosed quickly. To debug the cause of the problem only a concrete piece of control condition (step or transitions) is to be analysed. Step conditions are unique defined. For this reason the synchronisation procedure can be easily implemented.

The important advantage of described technique is separation of sequences and suitable software processes. Synchronisation and interoperation are done by proper definition of the step conditions. It is relatively easy to introduce time monitoring not only for whole cycle but also for each step separately. The same is valid for minimum step time definition. Besides sequences other program modules to perform specific functions like faults, visualisation can be adopted.

There is no need to apply interrupt mechanism for concept based on steps and transitions programming except special requirements or restrictions. However this method is very memory consuming and requires applying fast processors.

6. SOFTWARE CONCEPT TO CONTROL MACHINE STATUS AND TO ENSURE CONTINIUITY OF THE PROCESS

Based on the experiences collected by the authors to programs different application in industrial environment the proposal the concept and its implementation of control machine status to ensure continuity of the process was developed. The concept relates to the techniques discussed previously and

some aspects are similar in concept of Object-Oriented Technology.

The fundamental concept of presented programming style is breaking typical software program loops and treating stack stored information about process as temporary only.

In traditional approach to programming many looping techniques on many levels of procedure nesting are used. In this case all information about procedure return addresses are stored in stack area. Without copying that area, in case of system break it is not possible to recover process state. After detection of system malfunction, the only possibility is to reinitiate system and build the structure of stack stored data again, without any information on break status. In industrial control systems it is not acceptable.

In presented approach, all loops are broken i.e. no loop with ending condition depending on external event is allowed. As a consequence each program sequence flow defined as execution of ALL allowed on certain stage of process procedures, ends at end of main program loop (in case of C programming program ends on end of main procedure!). For this reason stack data on end of each program cycle are just flushed. Data on stack can be treated as temporary data. After reaching end of cycle, system can be restarted, most I/Os can be reinitiated and next cycle can start again. In this approach, system can even reset itself after each cycle without affecting virtual control continuity.

Concept of reaching end of main loop on each programming cycle was used because of assumption that CPU crash can occur at any moment of program execution. All data about process status are stored in data structures of static type. Structures are opened in well defined moment on the beginning of program sequence, updated within the sequence and closed on end of the sequence. Additionally on end of the sequence copies of closed data structures are stored and equipped with CRC or other check data. All

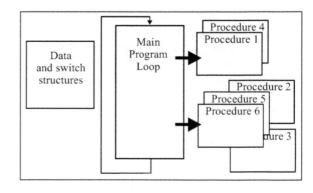

Fig. 4. Program structure referring to the broken loops concept.

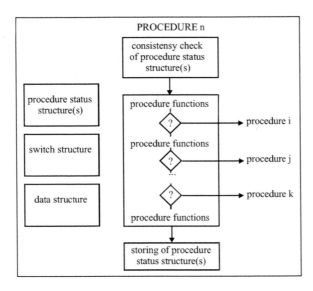

Fig. 5. Procedure structure.

described above measures enable to form general rule: on the beginning of each program cycle only closed and valid (based on CRC) data structures are used to control a process. In case of break within a program sequence actually processed data structure is not closed properly, so after re-initialisation of a system this data structure is not taken into consideration. Properly closed copy of data structure generated on the end of last correctly ended program sequence flow is accessed leading to re-processing broken cycle again.

Described programming style ensures absolute transparency to CPU brakes, but changes rules of programming totally. Because of rule, that each procedure cannot contain loops depending on process condition, internal structure of a procedure can contain only switch structures controlled by procedure descriptor stored as data of static type. These switch structures (program area) and data structures (data area) are equivalent to traditional loops with process dependent conditions. Scope of data structures is local to the procedure. Structure of procedures is hierarchical like in case of traditional approach.

According to the concept the synchronisation mechanism is built-in. Each process runs separately and uses local or global data. The standard processes like fault messaging can be embedded to the procedure directly.

The advantages of the proposed method are: logical and transparent structure, fault recognition and service can be programmed for each process separately, process status is easily recognised and diagnosed, processes start always (excluding hardware defect), the methods supports synchronisation and interoperation. The concept covers all requirements defined in the paper.

7. SUMMARY

The paper touches the rare published problem of control process applications in severe industrial environment. In most cases the programmable controllers control the industrial applications. Programmable means control algorithm is written as a program. For this reason authors focused mainly on the software aspects. Since the software is dedicated to the hardware it was necessary to consider in this paper hardware features, capabilities as well. Software designer aware of the hardware capabilities is able to create optimal program better suited to some defined aspects-criteria.

The proposed software requirements (structure, data, synchronisation, recovery) can be interpreted as some kind of criteria for judging software performance. Another important aspect is reusability of the software, implemented mechanisms or procedures. This leads to creating different concepts of control programs. The usefulness of such requirements was proved in real live applications that examples are described in the paper.

The mentioned aspects are considered in originally developed by authors concept of software controlled systems working in severe conditions like dynamic restart hidden for a continuity of process.

The practical verification of the concept took place in project of weld controller developed in Institute of Welding, Gliwice.

REFERENCES

Nowara A. and Chmiel M. (1999). *Program Structure and Diagnostic Aspects in Applications applied in PLCs*, International Conference Programmable Devices and Systems PDS'99, Ostrava, Czech Republic.

Mikno Z. (1999). Wrażliwość mikroprocesorowych układów sterowania procesu zgrzewania na zakłócenia elektromagnetyczne (in eng. Sensitivity of microprocessor control systems for resistance welding to electromagnetic interference. In: *Biuletyn Instytutu Spawalnictwa* 3/99. Welding Institute, Gliwice. p.39-42.

Pucher K., Kania D. (1999). Realizacja programu z uzależnieniami czasowymi (in eng. Time functions programming). In: *Maszyny, Technologie, Materiały.* p.18-22. Wydawnictwo SIGMA-NOT, Warszawa.

Siemens (1996). Simatic System Software S7-300 and *S7-400 Program Design*. Programming Manual.

IFAC
Publications
www.elsevier.com/locate/ifac

EASY PROGRAMMABLE MAPI CONTROLLER
BASED ON SIMPLIFIED PROCESS MODEL

Mieczyslaw Metzger

Institute of Automatic Control
Silesian University of Technology
ul. Akademicka 16, 44-100 Gliwice, Poland
E-mail: mmetzger@terminator.ia.polsl.gliwice.pl

Abstract: This paper presents an easy programmable version of the MAPI controller. The MAPI (model augmented and making astatic PI) controller which is a compromise between the standard (most frequently used in the industry) PI algorithm and advanced GMC linearising algorithm, includes standard proportional and integral actions (as PI algorithm) and additional non-linear feedback based on the process model. Basic mathematical derivations show the astaticising behaviour of the proposed non-linear feedback. Model-based controllers are usually dedicated for a particular class of processes. For the simplified to polynomial functions internal model, the proposed general-purpose version of the MAPI controller can be programmable by setting only a number of parameters from the operating panel. *Copyright © 2001 IFAC*

Keywords: Programmable controllers, astatic control, model-based control, PI controllers, model approximation, polynomial models.

1. INTRODUCTION

The PID feedback control remains the most popular control strategy in engineering practice. The major advantage of the PID controller deals with its applicability for a large spectrum of the process dynamics. The PID actions are well understood and accepted over several decades by process operators and control engineers due to the simplicity and intuitive interpretation of the algorithm and its setting parameters. Any kind of mathematical model of the process to be controlled is not required to successfully used PID algorithm.

Although the PID controller is still the most popular algorithm, not all industrial processes could be controlled by these controllers. For multivariable and especially for nonlinear processes some knowledge of the process behaviour is necessary to design more adequate particular-process-dedicated control algorithm. Especially chosen phenomenological information can be implemented in the control algorithm in the form of a mathematical model of the process to be controlled.

Several non-linear process model-based control algorithms have been proposed during last two decades for non-linear processes. The representative examples include (but are not limited to) the following: process model-based control (PMBC) (Riggs and Rhinehart, 1990; Rhinehart, 1994); linearising control (Isidori, 1989; Henson and Seborg, 1990); generic model control (GMC) (Lee and Sulivan, 1988); predictive functional control

(PFC) (Richalet et al., 1979; Richalet, 1993). One can find large overview of non-linear control strategies for example in Bequette review paper (1991) as well as in the book edited by Henson and Seborg (1997).

Model-based control can base on arbitrarily chosen mathematical models with estimated parameters but case that is more interesting deals with mathematical models that base on fundamental physical laws.

In majority of cases the control algorithms are based on this kind of mathematical model and the control responses are tested on the base of the same mathematical model by means of simulations in the batch mode (see for example Henson and Seborg, 1997; Viel et al., 1997). In this case, the control system allows obtain excellent control responses even in the working point regime when the open-loop response is unstable.

Because of the fact that the process model-based control is sensible for the changes of the process parameters, several sophisticated adaptive algorithms based on non-linear observers (see for example Dochain et al., 1992; Gibon-Fargeot et al, 1994; Henson and Seborg, 1997) as well as based on estimation of the whole reaction rate (Czeczot, 1998; 2000) can be developed to solve this problem. Nevertheless, in all of these investigations the same model is applied for control algorithm synthesis as well as for testing of control behaviour.

Application of more realistic mathematical models (including additional physical phenomena) realised as real-time simulators (Metzger, 1999, 2000) would be more challenging for testing of process model-based controllers. One of the interesting phenomena, that can be considered, is the imperfect mixing. Some suggestions how to include this phenomenon have been presented in (Luyben, 1968; Seborg et al., 1989; Luyben, 1990). The flexible real-time simulator of the exothermic CSTR with imperfect mixing is dedicated for testing of advanced controllers (Metzger, 2001 a, b).

Model augmented PI algorithm (MAPI algorithm) was proposed (Metzger, 2001 a, b) as a compromise between the standard PI algorithm and advanced linearising (for example the GMC) algorithm. The MAPI algorithm includes standard proportional and integral actions (with well-known setting parameters) as well as an additional astaticising element based on the process model.

For the CSTR the MAPI controller gives very promising control responses even for changes of process behaviour (Metzger, 2001 b). Nevertheless, the implementation of the mathematical model of the

CSTR requires possibility of a programmation of complicated nonlinear functions.

In this paper the MAPI controller with implementation of simplified internal models is proposed. The nonlinear functions should be approximate by polynomial functions. It allows to design easy programmable MAPI controller, in which the user should chose and set (for particular process to be controlled) only the proportional gain, reset time and coefficients of polynomials in astaticising element.

2. MODEL AUGMENTED PI ALGORITHM

The mathematical model of the non-linear single input-single output process under consideration can be described in a general form.

$$\frac{dx}{dt} = f(x) + g(x)u \qquad (1)$$

$$y = h(x) \qquad (2)$$

where x is an n-dimensional vector of state variables, u is a scalar control (manipulated) input variable and y is a scalar controlled (output) variable. The rate of the output variable can be derived in the following form.

$$\frac{dy}{dt} = L_f h(x) + L_g h(x)u \qquad (3)$$

where the Lie derivatives of scalar function h(x) with respect to vector functions f(x) and g(x) are defined as follows.

$$L_f h(x) = \frac{\partial h(x)}{\partial x} f(x)$$
$$L_g h(x) = \frac{\partial h(x)}{\partial x} g(x) \qquad (4)$$

The control algorithm, which is most frequently used in the industry (about 95% of industrial real-world applications) is the standard (linear) PI algorithm having the following form.

$$u_{PI} = k_P(y_{sp} - y) + \frac{k_P}{T_i} \int_0^t (y_{sp} - y)d\tau + u_0 \qquad (5)$$

The PI controller can be used with success for the non-linear processes as well. To improve the control responses for particular processes, several advanced algorithms mentioned above can be proposed. The

PMBC algorithm (Riggs and Rhinehart 1990) and similar GMC or linearising algorithms (Lee and Sullivan 1988; Henson and Seborg 1990) include the proportional and integral actions (as the standard PI algorithm) but with nonlinearly changed setting parameters based on the process model.

$$u_{GMC} = \frac{k_1}{L_g h(x)}(y_{sp} - y) + \frac{k_2}{L_g h(x)} \int_0^t (y_{sp} - y) d\tau$$
$$- \frac{L_f h(x)}{L_g h(x)}$$

$$(6)$$

The algorithm (6) is based on the state feedback linearising transformation (see for example Henson and Seborg book, 1997).

$$u = \frac{v - L_f h(x)}{L_g h(x)} \qquad (7)$$

where for new control (reference) variable v, the process is both linear and astatic. That means dy/dt = v. Applying for v the proportional and integral actions depending on the process error (e = y_{sp} − y) the algorithm (6) is obtained. The integral action is not necessary (the transformed process is astatic for exact model - hence the offset do not occurs), but it reduces offset for model mismatches.

Although the linearising GMC controller includes the proportional and integral actions, the setting parameters k_1 and k_2 do not have the well accepted technical interpretation such as proportional gain and reset time because the proportional and integral actions are nonlinear – see equation (6). Hence, this fact augmented with requirement of design of process model causes a little acceptance this family of control algorithms by practitioners.

In the papers (Metzger 2001 a, b) a model-augmented PI algorithm (MAPI algorithm), which is a compromise between the standard PI algorithm and the non-linear GMC algorithm, has been proposed. The proposed MAPI algorithm includes standard (linear) proportional and integral actions (with standard k_P and T_i settings well interpreted by practitioners) and an additional state feedback non-linear element, which is based on the process model.

$$u_{MAPI} = k_P(y_{sp} - y) + \frac{k_P}{T_i} \int_0^t (y_{sp} - y) d\tau$$
$$- \frac{L_f h(x)}{L_g h(x)}$$

$$(8)$$

The MAPI algorithm is based on another state feedback control transformation, making the control plant astatic only. This new astaticising transformation has the following form.

$$u = v - \frac{L_f h(x)}{L_g h(x)} \qquad (9)$$

where for new control variable v, the process is astatic. That means $dy/dt = L_g h(x)v$. Applying for v standard PI algorithm depending on the process error (e = y_{sp} − y), the algorithm (8) is obtained. This algorithm has the standard proportional and integral actions with settings well interpreted by engineers. The astaticising feedback improves control properties of the MAPI controller making the process astatic. Hence, for exact internal model, the integral action in the controller is not necessary but it reduces the offset for model mismatches.

The MAPI algorithm can be extended to MAPID controller according to general properties of the PID control.

3. SIMPLIFIED INTERNAL MODEL

The following description of the MAPI controller includes the switching key k_{MAPI}, which allows switching between standard PI algorithm and MAPI algorithm.

$$u_{MAPI} = k_P(y_{sp} - y) + \frac{k_P}{T_i} \int_0^t (y_{sp} - y) d\tau$$
$$- k_{MAPI} \frac{L_f h(x)}{L_g h(x)}$$

$$(10)$$

$$k_{MAPI} = \begin{cases} 0 & \text{for PI controller} \\ 1 & \text{for MAPI controller} \end{cases} \qquad (11)$$

In the majority of publications cited above, the process model is based on the phenomenological description with application of nonlinear functions such as exponent or power functions, and can include in general case other than controlled state variables as well. Such kind of models requires programming in higher order programming languages. For example, the MAPI controller for a CSTR presented in (Metzger, 2001 b) requires the reprogrammation in the G language (LabVIEW environment) and recompilation when changes of the form of nonlinear reaction rate should be considered. Although the application of phenomenological models with Arrhenius-type or Monod-type nonlinerities gives good results of control, it demands a big quantity of work.

That is why, easier programmable (only from the operator panel of the controller) will be very attractive. Such kind of controller must base on the more simplified process models.

For a class of processes in which one of the time constants dominates in process dynamics, the mathematical model can be approximate by the single phenomenological balance of the quantity, which is considered as control variable. That is, for y = x, when y and u are scalars, the process can be approximate by the equation.

$$\frac{dy}{dt} = f(y) + g(y)u \qquad (12)$$

A large class of nonlinearities can be approximated by three-piece polynomial functions for process variable $y \in [0,100]\%$.

$$f(y) =$$
$$\begin{cases} a_{f3}y^3 + a_{f2}y^2 + a_{f1}y + a_{f0}, \text{ for } y \in [0, y_a) \\ b_{f3}y^3 + b_{f2}y^2 + b_{f1}y + b_{f0}, \text{ for } y \in [y_a, y_b) \\ c_{f3}y^3 + c_{f2}y^2 + c_{f1}y + c_{f0}, \text{ for } y \in [y_b, 100] \end{cases}$$
$$(13)$$

$$g(y) =$$
$$\begin{cases} a_{g3}y^3 + a_{g2}y^2 + a_{g1}y + a_{g0}, \text{ for } y \in [0, y_a) \\ b_{g3}y^3 + b_{g2}y^2 + b_{g1}y + b_{g0}, \text{ for } y \in [y_a, y_b) \\ c_{g3}y^3 + c_{g2}y^2 + c_{g1}y + c_{g0}, \text{ for } y \in [y_b, 100] \end{cases}$$
$$(14)$$

The control algorithm for polynomial approximation has the following form.

$$u_{MAPI} = k_P(y_{sp} - y) + \frac{k_P}{T_i}\int_0^t (y_{sp} - y)d\tau$$
$$- k_{MAPI}\frac{f(y)}{g(y)} \qquad (15)$$

This algorithm can be implemented in the PI controller with standard proportional gain and reset time settings and with additional easy programmable nonlinear astaticising element by setting of the polynomial parameters. Although the programming of the controller is easy, the user must find the process model approximation (this is the most important problem for all model-based control algorithms).

4. PC-BASED IMPLEMENTATION

The PC-based MAPI controller for general purpose was realised as the real-time application in the LabVIEW environment. The I/O connection can be realised on the basis of appropriate DAQ-boards, with an application of the FieldPoint controller or using simple TCP/IP transmission by Intranet. The operating panel of the controller is shown in Fig. 1.

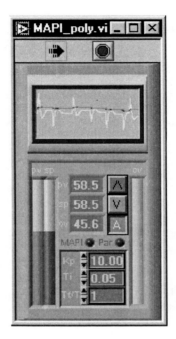

Fig.1. Operating panel of the general purpose MAPI controller with polynomial approximation based process model.

The pushbutton "MAPI" (see Fig. 1) allows to switch-off the model-based non-linear element in the algorithm (15) leaving standard linear PI actions.

The pushbutton „Par" (see Fig. 1) activates additional panel, in which the user must sets the coefficients of the piecewise polynomials (13) and (14).

5. CONCLUDING REMARKS

The model-based algorithms when implemented as real-world applications are always a compromise between the complexity of the model (necessary for good control quality) and the simplicity (dealing with possibilities of controller programmation). The presented general-purpose version of the MAPI controller is based on the fixed polynomial approximation of the internal model for a class of processes. Although the programming of the controller is easy, the user must find the process model approximation.

Some extensions of this controller can go to including feedforward actions for measurable variables or including in model polynomials additional dependence of control variable - in certain cases it can improve control results – that demonstrate the results obtained for PFC and GMC controllers (Janik, 2001; Pazurek, 2001).

ACKNOWLEDGEMENTS

This research was supported by the National Committee for Scientific Research (KBN) under Grant No. 8-T11A-001-16.

REFERENCES

Bequette, B.W. (1991). Nonlinear control of chemical processes: a review. *Ind. Eng. Chem. Res.* **30**, pp. 1391-1413.

Czeczot, J. (1998). Model-based control of the substrate concentration in an aeration basin using the wastewater level. *Water Science and Technology*, vol. **37**, No. 12, pp. 335-342.

Czeczot, J. (1999). Substrate consumption rate application to the adaptive control of the continuous bioreactor – noisy case study. *Archives of Control Sciences*, Vol. **8** (XLIV), No 3-4, 1999, pp. 33-52.

Dochain, D., M. Perrier and B.E. Ydstie. (1992). Asymptotic observers for stirred tank reactors, *Chem. Engng. Sci.*, **47**, 4167-4177.

Gibon-Fargeot, A.M., H. Hammouri and F. Celle. (1994) Nonlinear observers for chemical reactors. *Chem. Engng. Sci.*, **49**, 2287-2300.

Henson, M.A. and D.A. Seborg. (1990). Input-output linearization of general nonlinear process. *AICHE J.* **36**, 1753-1757.

Henson M.A. and D.E. Seborg (ed.). (1997). *Nonlinear Process Control*. Upper Saddle River, Prentice Hall PTR.

Isidori, A. (1989). *Nonlinear Control Systems*. Springer Verlag, New York.

Janik, J. (2001). Design and evaluation of the PC-based programmable PFC controller in the LabVIEW environment. MS Thesis, Silesian University of Technology, Gliwice.

Lee P.L. and G.R. Sullivan. (1988). Generic model control (GMC). *Comput. Chem. Engng.*, **12**, 573-580.

Luyben, W.L. (1968). Effect of imperfect mixing on autorefrigerated reactor stability. *AICHE J.* **14**, 880-885.

Luyben, W.L. (1990). *Process modeling, simulation and control for chemical engineers*. McGRAW-HILL.

Metzger, M. (2000) *Modelling, simulation and control of continuous processes*. Edition of Jacek Skalmierski, Gliwice.

Metzger, M. (2001). Modelling of imperfect mixing for real-time simulation and process model-based control of CSTR. Proceedings of the 7th IEEE Conference on MMAR, Międzyzdroje, vol. **1**, pp. 421-426.

Metzger, M. (2001). Comparative simulation studies of model-augmented PI control of CSTR. Proceedings of the SCS European Simulation Symposium 2001, Marseille, October 2001, SCS Publications, (in print).

Pazurek, D. (2001). Design and evaluation of the PC-based programmable GMC controller in the LabVIEW environment. MS Thesis, Silesian University of Technology, Gliwice.

Rhinehart, R.R. (1994). Model-based control. In: *Instrument Engineers Handbook* (B.G. Liptak – eds), 66-69.

Richalet, J., A. Rault, J.L. Testud and J. Papon. (1979). Model predictive heuristic control: applications to industrial processes. *Automatica*, **14**, 413-418.

Richalet, J. 1993. Industrial applications of model based predictive control. *Automatica* **29**, No 5, 1251-1274.

Riggs, J.B. and R.R. Rhinehart. (1990). Comparison between two nonlinear process-model based controllers. *Comput. chem. Engng.* **14**, 1075-1081.

Seborg, D.E., T.F. Edgar, and D.A. Mellichamp. (1989). *Process Dynamics and Control*. Wiley Series in Chemical Engineering.

Viel, F., F. Jadot and G. Bastin. (1997). Global stabilization of exothermic chemical reactors under input constraint. *Automatica,* **33**, pp. 1437-1448.

www.elsevier.com/locate/ifac

RECONFIGURABLE LOGIC CONTROLLER
ARCHITECTURE, PROGRAMMING, IMPLEMENTATION

Adam MILIK
Edward HRYNKIEWICZ

Institute of Electronics
Silesian University of Technology,
Akademicka 16; 44-100 Gliwice; Poland
milik@boss.iele.polsl.gliwice.pl; eh@boss.iele.polsl.gliwice.pl

Abstract: The paper presents state of the work carried over implementation and programming of control algorithms in FPGA circuits. We present new developed hardware architecture of Logic Controller. It is capable to observe and control up to 512 binary objects. Presented architecture brings together high computation speed and rational logic resources usage. Programming tools based on ladder diagram are presented too. We also present controller synthesis process from designed ladder diagram. *Copyright © 2001 IFAC*

Keywords: PLC, Logic Controller, Programming, Logic Synthesis

1. INTRODUCTION

FPGA devices give opportunity to design hardware equipment that meets user specific requirements and constraints using the same or similar semiconductor devices (Brown, *et al.*, 1992). In comparison to a standard microprocessor construction or a custom hardware solution based on sequential instruction execution, industrial logic controller designed with use of FPGAs has extremely high computing capability. Its main feature based on spread out computing method instead of centrally controlled process. There are as many as possible concurrent computing threads that are executed concurrently by hardware. Main and the greatest disadvantage of a hardware solution is difficult and time-consuming design and implementation process. (Kumar, *et al.*, 1992; Sasao 1993). Those reason made hardware solution less popular in those areas where logic controllers are fast enough to solve the problem.

In this paper we present new approach to hardware design. Our design starts from high level programming ladder diagram (LD), which will be implemented in hardware structure of the reconfigurable logic controller. We are going to

present rules for hardware implementation from control algorithm that is a key issue. There are many opposite request and expectation that should come together in final solution. The basic opposites are program speed execution and available hardware resources. Problem should be solved as fast as possible but also with as little as possible amount of hardware. There are also technical constraints like number of pins available in a chip package in comparison to number of observed and controlled signals (objects). In paragraph 2 we present a hardware solution, which were developed, to meet presented above requirements and constraints.

For an automation designer we have to deliver programming tools that translate abstract design into hardware structures. We are developing ladder diagram editor with set of tools for design processing. One of them allows generating synthesizable Verilog HDL (Palnitkar, 1996). This code can be used for implementation (Verilog) or for simulation purposes. We work also on synthesis improvement from the ladder diagram. For this purposes package for BDD diagram generation was developed (Akers 1978; Bryant 1986; Minato 1993, Sasao 1993).

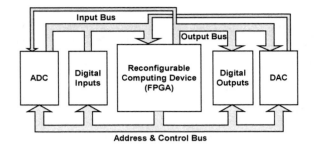

Fig.1. Reconfigurable controller general block diagram

2. RECONFIGURABLE LOGIC CONTROLLER HARDWARE STRUCTURE

The hardware construction of the controller is designed to implement in symmetrical FPGA families. Design work was carried out on Xilinx XC4000E (Xilinx, 1998) family as a representative family of modern FPGA device architecture. As a configuration storage is used static RAM that enables unlimited number of reprogramming cycles. An array of device is filled with a Configurable Logic Cells (CLB). Each of them contains Look Up tables that realize combinatorial functions and edge triggered D flip-flops. CLBs are also equipped with additional hardware that simplifies realization of an arithmetic addition and subtraction.

General block diagram with input and output modules is presented in figure 1. Architecture of controller allows connecting up to 512 digital inputs and outputs. There are 16 groups of 32 bits in input and output area. This is an open architecture that allows to connect binary control equipment, digital to analog and analog to digital converters or other equipment that sends data in byte, word or even in 32 bit length packages. There is also possibility to connect more then one computing unit together.

Computation unit block diagram is given in figure 2. Control program is executed as sequence of action. Computation starts from data fetch and dispatch. Internal control unit is addressing all address location in range from 0 to 15. From input area data are read and only necessary in the algorithm are stored in input registers. Signal ILE[15:0] enables group of input registers that should store data from the input bus. Content of results registers are written to output area in the same transfer cycle. Appropriate group of output registers is addressed by signal OE[15:0] that opens three state buffer and places data on the output bus. All read and write operations are completed in one burst transfer that covers all addressing area. When new data are fetched to the controller calculation cycle starts. Currently for computation purposes is assigned one clock cycle that can be extended to several clock cycles when needed. When computations are finished new results are stored by activating RLE signal in the result registers. Cycle starts a new after result calculation.

2.1 Timers and Counters Implementation

Timers are basic components for implementing time dependencies and they are derived from counters. Both devices count signal changes. Timers are clocked with a signal that period (frequency) is known while counters not.

FPGA devices are equipped with a dedicated block RAM memories or easy convertible look up tables (LUTs) into an edge triggered RAMs. Typically such memory contains from 16 to 32 words usually one bit length. In design we will consider use of the memory with 16-word storage capacity. From designer point of view it can be presented as an addressable bank of edge triggered D type registers. Similar structure designed with use of general-purpose D flip-flops consumes about 16 times more logic resources. Base on Xilinx architecture (Xilinx, 1998) one 16-bit memory can be implemented into one LUT while at least two exists in one CLB. Also two D flip-flop registers are available in one CLB cell.

Figure 3 presents a block diagram of counting unit. In the design was used idea of a resource sharing. Common components for all counters are shared between them that reduce occupied area in the FPGA

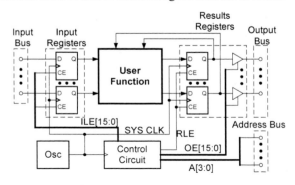

Fig.2. Computation unit block diagram

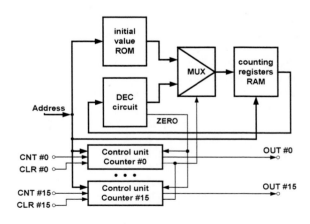

Fig.3. Counter unit block diagram

Table 1. Logic resources requirements comparison for different counting units architectures

Number of used counters	Logic resources in CLB		Efficiency in CLB
	RAM	Classic	
1	13,5	5	-8.5
2	15	10	-5
4	18	20	2
8	24	40	16
16	36	80	44

device. Main component of counter is the RAM storage, which holds count value of up to 16 counters. Content of selected counter cell can be written from an arithmetic increment-decrement circuit or a ROM cell that stores initial counting value. The counter register block and the initial value block are addressed from the same system address bus. All counters are down counting. Counting down generates carry from the oldest bit (LSB) when a counting value reach 0. This modification eliminates additional comparison circuitry while we base on features of a subtraction circuit. The control circuit is independent for each counting unit. Based on input signals it determines control for other blocks. This circuit also holds the counting result flag that shows counting progress. Important is fact that this flag can be continuously accessed by other surrounding devices especially during update cycle when result registers are written to. Modification of control unit allows changing features and obtaining different counter types. Such features like input signal response, and on counting value reach event response can be modified. In the basic solution control unit covers area of 1.5 CLB for each used counter. What is important the control unit takes the same area independently from counter size in bits. Counting unit cluster complexity with RAM and ROM is 1.5 CLB per bit (architecture can handle up to 16 independent cells inside). When we assume 8 bits long counter it gives counting unit complexity with 12 CLB.

The last thing but not the least important is a sequential access selector circuit to counter cells. The logic controller has implemented data exchange unit that sequentially addresses all devices in input and output areas in each computation cycle. External data area has capacity of 16 words. The same addressing circuit can address counters unit forcing its update. There is no need to design an additional finite state machine while already existing can be used. Execution time aspect is also important. There is full concurrency in data fetch and dispatch but also in counting unit update. This three level of independence is impossible to implement in typical microprocessor unit that can carry out only one operation in one instruction cycle (for CISC CPU it can take several clock cycles). We carry out some

experiments in order to determine speed of such solution. We carry out it on Xilinx SPARTAN family that is descendant of XC4000E. Maximal clock frequency for counting unit was 47MHz (XCS05-3). Achieved results allow to estimate that there is possibility to count signals that frequencies can reach up to 2 – 2.5 MHz without special hardware solution. Logic complexity of traditional counter implementation with proposed solution is compared in table 1.

2.2 Implementation of Mathematical Operation

Modern FPGA devices are equipped with helpful extensions that simplify implementation of a mathematical operation. Usually architectures offer easy design of adder and subtractor. Large FPGA devices even offer combinatorial multipliers that are capable to multiply two 18 bits signed integer (e.g. Virtex II family). In most cases multiplication unit must be implemented as sequential add and shift operation for unsigned or specific algorithm for signed numbers.

Typical mathematical operation consists of some steps that must be completed in order to obtain a result. Circuits for simple mathematical operation can be represented as a box with input signals and result of operation as output. This allows designing a complex computation with use of unified blocks. Results from a block can be obtained after few nanoseconds for function like addition or subtraction or can take several clock cycles for multiplication or complicated calculation as for PID module. In general a result calculation may takes different amount of time. For proper operation of an arithmetical block network designed from unified blocks synchronization circuitry that controls data circulation is required. Shorter branches of a circuit must wait for longer in order to complete all calculation. An operation should be hold till next operation is ready to take its results. Diagram of the basic block for arithmetic operation is presented in figure 4. Not only is there function block but also additional circuitry. As number of register is relatively high in FPGA devices result of operation is stored in output register. This register enables building pipeline processing. Results of previous calculation remains unchanged while next calculation

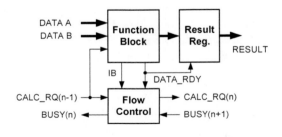

Fig.4. Simple arithmetic operation with flow control unit schematic diagram

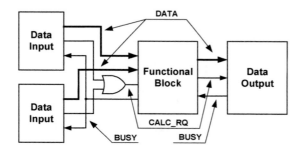

Fig.5. Arithmetic operation network diagram

is performed. There is also an additional circuitry that keeps calculation in order and controls a data flow. Signal CALC_RQ(n-1) informs about new data arrival on inputs and requests calculation update. While block is performing operation its BUSY(n) flag is raised that informs all predecessors cells to hold new result delivery until operation is completed. While input BUSY(n+1) flag is raised and calculation of current block are in progress they should be completed, new result written to data register and CALC_RQ flag should be raised till BUSY flag go inactive. One clock after this event CALC_RQ flag can change its state.

The schematic diagram of an arithmetic operation network is presented in figure 5. There are two interface modules that deliver data from external sensors or other sources. On delivered data is carried out two-argument operation. Arrival of at least one of the arguments forces calculation update by logic OR operation on both CALC_RQ lines of data input blocks. While calculation is performed both input blocks are informed by BUSY signal to suspend its activity. Similar rules are applied to output data block. Here CALC_RQ signal request dispatch of calculation result to connected device. In case a device is not ready to receive data BUSY flag is set.

3. PROGRAMMING WITH USE OF LADDER DIAGRAM

For programming purposes was written special ladder diagram editor which take into consideration specific construction of the controller. From user point of view it is similar to standards used in industry solution (Legierski et al. 1998; Michel 1992). User can built its own design from following components:

- Switches normally open or closed
- Counters with programmed pulse counts
- Timers with programmed time basis and interval
- Coils
- Mathematical component for integer operation
- Interfacing modules for AD and DA converters.

Described components allow building automation circuits that comprise binary control and

mathematical operation. Timing dependencies also can be used in a circuit design. Connection between components can be done as scalar or vectored signal (useful for word operation). Example of the schematic diagram is presented in figure 6.

Drawn by designer schematic is analyzed against design rule check. While this check is performed designer is informed about mistakes and possible errors in design like:

- Hanging wires
- Source less connection
- Missing of complementary coil in SR coil system
- Multiple use of the same execution component

Based on presented schematic editor we present tools for simulation and implementation, model generation with use of hardware description language (HDL).

3.1 Model generation

The simulation model generator is first step, which allow obtaining hardware equivalence of designed control algorithm. As a target HDL was used Verilog due to its simplicity and high readability for an electronic designer. The Verilog description is generated on register transfer level (RTL) and gate level. Generated model is fully synthesisable for FPGA devices. It gives opportunity to synthesize and implement a design with use of external tools. Basic and main module of controller at the beginning is equipped with bus architecture ports that define input, output and address buses. There are also additional ports that are hidden and they are vital for controller behavior like clock or reset signals. Inside the controller module is instantiated main control unit responsible for data exchange and control signals generation. To the controller framework are added components that user has drawn in its design. Diagram is drawn on worksheet that contains 10 columns and 32 rows. Design can be drown on multiple worksheets that are gathered in the design

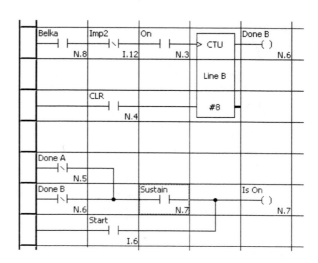

Fig.6. Schematic in ladder diagram editor

166

manager. Each cell can contain only one component (switch, coil, connection). There are also components that cover more then one cell like counters timers, mathematical operations. Generally one cell can contain only one signal input and output.

Diagram analysis starts from the upper left cell and goes to the last cell located in bottom on the right. Cells are examined sequentially cell by cell from top to bottom of column. In case of found component in cell following actions are taken:

- Switch – logical AND operation is performed on signal that drives input of the switch from neighboring cell and variable that controls analyzed switch. When signal is logic constant value simplification is made and new value is calculated. When signal belong to input area input data register is created.
- Net and junction – performed is logical OR operation with all signals that are driving the same net. Value is propagated to all driven cell inputs.
- Internal coil – Is used to named signal propagation inside the design. No additional hardware is generated.
- Output coil – is connected to the output signal. Named node is created. Output register with a three state buffer is also created (output cell). Input of a cell is driven from selected node. Output is connected to appropriate output bus line. Control signals are connected to the register and three state buffer.
- Counter – a counter control unit is created. Control signals and output signal are connected to the network of controller. The counting unit is check for presence and available space. In case there are not the counting unit or all free spaces are used new counting unit is created. Counter parameter are written to the counting unit
- Timer – a timer control unit is created. Control signals, time basis signal and output signal are connected to the network of controller. The counting unit is check for presence and available space. Set up of the timer parameters.

Generation method is illustrated on simple example with a counter that is presented in figure 7. Example presents switch-designed flip-flop with counter

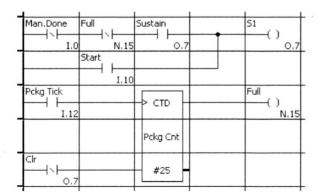

Fig.7. Example LAD diagram for code generation

```
//Internal signals declaration
wire I_00,I_10,I_12;
wire O_07;
wire N_15;
wire A_04,A_06,A_07;
//Input registers declaration
IN_REG IFD_(CLK,.ILE[0],I[0],I_00);
IN_REG IFD_010(CLK,.ILE[0],I[10],I_10);
IN_REG IFD_012(CLK,.ILE[0],I[12],I_12);
//Output registers declaration
OUT_REG
OFD_007(CLK,RLE,OE[0],A_06,O[7],O_07);
//Counter instance
CNT #(0,25)C00(CLK,I_12,O_07,A_07);
or  (A_06,I_10,A_04);
and (A_04,O_07,~N_15,~I_00);
```

Fig.8. Generated Verilog HDL for example presented in figure 7

dependencies. Verilog HDL code that is a hardware model of designed control circuit is given in figure 8. Based on presented rules of generation and architecture of reconfigurable logic controller input signals are stored in input registers. Signal ILE[0] latches appropriate input bus signals into input registers. All signals are in range from 0 to 31 and belong to slot 0 of controller. Computation result is connected to the output register. The output register instance has two outputs. One O_07 that works as feedback to the circuit and O[7] that is connected to the output bus through three state buffer. Three state buffer is controlled by the signal OE[0]. Data is transferred from the output register to output area slot 0. Counter instance is parameterized (#(0,25)). Address in counting unit is set to 0 and initial count value to 25. Follow the component instances is generated sequence of logic function that represent switches and wires drawn on the schematic.

3.2 Synthesis and implementation aid tools

Based on experience that we have got during implementation of simulation model compiler, a BDD generation tool for combinatorial function was implemented. The BDD is generated as result of logic operation (Akers 1978, Bryant 1986, Minato 1995). In the first step are generated a trivial BDD for each variable. A trivial BDD consists of single node that pints to a terminal 0 for 0 edge and points to a terminal 1 for 1 edge. When all variables are created analysis of schematic dependencies starts. Instead of using an intermediate variable tree root pointer is used. Following rules are used during BDD generation:

- Switch – logical AND for a BDD from previous cell and a variable that controls current switch
- Connection – logical OR of all BDDs that are joined together
- Coil – root of BDD is assigned to that variable.

Results for Boolean function BDD generation from example in figure 7 are presented in figure 9. BDD

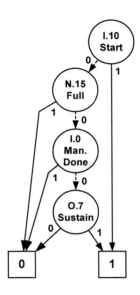

Fig.9. BDD diagram for example from figure 7

example covers only flip-flop implementation. One of the biggest problem in a BDD is variables ordering. It is difficult to answer if obtained result is already optimal one. One of the greatest application for BDD in synthesis is Boolean function decomposition for LUT architecture. In LUT architecture main limitation is set on number of inputs but complication of logical expression doesn't matter. Based on available BDD programming packages we currently work on suitable decomposition method for reconfigurable logic controller specific hardware.

4. SUMMARY

The paper presents the hardware solution that allow implementing control algorithm in the FPGA architecture. Presented solution make possible to control large number of digital inputs and outputs. Word or byte operations are also possible in presented architecture. Efficient data exchange system with concurrent computation assures possibly the fastest result calculation. Extremely important feature of the reconfigurable logic controller is constant response time that is independent from a control algorithm complexity due to parallel concurrent execution. Presented programming tools package enables creation of control programs for controller with use of standard ladder diagram. Drawn diagram is converted into hardware description language that can be synthesized. Works over the reconfigurable logic controller are still on going.

REFERENCES

Akers S.B. "Binary decision diagram", IEEE Transactions on Computers, Vol.C-27, No.6, str.509-516, June 1978

Brown S.D., R.J. Francis, J. Rose, Z.G.Vranesic "Field-Programmable Gate Arrays" Kluwer Academic Publisher 1992

Bryant R.E. "Graph Based Algorithms For Boolean Function Manipulation", IEEE Transactions on Computers, Vol.C-35, No.8, str.677-691, August 1986

Hrynkiewicz E. A. Milik "Reconfigurable Logic Controller", Signals and Electronic Systems, ICSES'2000, Ustroń 2000

Kumar S., J.H.Aylor, B.W.Johnson, and W.A.Wulf "The Codesign of Embedded Systems" Kluwer Academic Publisher 1992

Legierski T., J.Kasprzyk, J.Wyrwał, J.Hajda „PLC Programming" (in polish) Wydawnictwo Pracowni Komputerowej Jacka Skalmierskiego Gliwice 1998

Michel G. "Programmable Logic Controllers – Architecture and Applications", John Willey & Sons, 1992

Minato Shin-ichi, "Binary Decision Diagrams and Applications For VLSI CAD" Kluwer Academic Publisher 1995

Palnitkar S. "Verilog HDL" SunSoft Press 1996

Sasao T. "Logic Synthesis and Optimization" Kluwer Academic Publisher 1993

Xilinx. "The Programmable Logic Data Book", San Jose 1998

IFAC

Publications
www.elsevier.com/locate/ifac

FUZZY COPROCESOR IN PLC SYSTEM

Nowara Andrzej, Rakowski Piotr

Abstract: In this paper the project and implementation of fuzzy system is presented. The system consists of controller unit and software with parameterisation and simulation options. Additionally the fuzzy controller can be applied as fuzzy module in hardware configuration of PLC. It is possible to parameterise the fuzzy logic controller as MIMO 4 inputs/2 outputs unit or 2 independent MISO units with 2 inputs/1 output. The advantages of proposed solution are: independent inference time regarding configuration, free definition of the membership function, total number of fuzzy rules is 4096. The software tools runs in graphical environment. *Copyright © 2001 IFAC*

Keywords: fuzzy logic, fuzzy control, controllers, regulator, software tools

1. INTRODUCTION

The increasing interest of designers, researchers and users to develop regulation circuits and complex control systems in the industrial field application can be observed recently. The first group of solutions is based on wide and successfully used PID controllers, the second applies fuzzy logic ones with fuzzy inference mechanism. Conventional linear model–based controllers can be designed according to some optimal criteria, the optimality and stability can be proved. It is worth to mention that design of closed-loop control can not be proceed without some implicit or explicit knowledge about the process to be controlled. Knowledge means a concept to capture those aspects of the process, which represent its behaviour. That is one of the reasons why fuzzy control is suitable and acceptable for many applications and ensures satisfied control. Fuzzy logic control is a knowledge-based control, which utilises human experience by describing the control strategies with linguistic rules.

Industrial development, automation in particular implies usage of programmable controller (PLC) to control whole, complex technological processes. The programmable controllers are designed to accomplish control in industrial systems and become more and more popular. The PLC are equipped with software and hardware fuzzy modules and it enables implementation of fuzzy algorithms in such controllers.

In this paper the project and implementation of fuzzy system is presented. The system consists of controller unit and software with parameterisation and test options. Additionally the fuzzy controller can be applied as fuzzy module in hardware configuration of PLC developed in Institute of Electronics, Technical University, Gliwice.

2. FUZZY LOGIC NOMENCLATURE

The fuzzy set theory enables to describe complex or nonlinear events or processes using a higher level of abstraction originating from our knowledge and experience. Fuzzy logic provides an alternative way of thinking in comparison to Boolean logic. Controllers based on the fuzzy set theory have been reported to be successfully used for a number of complex processes (Ross, 1995).

The authors do not intend to present a formal and full introduction to fuzzy logic, which can be find in many publications (Yager and Filev, 1995), but it is useful to give conceptual description of nomenclature and notions that are of importance to understanding the application of fuzzy logic an controllers.

The rule base of Fuzzy Logic Controller (FLC) is written as a set of rules (k - rule's number):

$$R_k \; AND \; y \; is \; B \; THEN \; z \; is \; C \qquad (1)$$

The IF clause (x is A AND y is B) named antecedent is a condition in the application domain, the THEN clause (z is C) named consequent is control action given to the process under control.

In the fuzzy logic we define the linguistic variables (x, y, z), which can be considered as the process signals. These signal are characterised by the linguistic values (terms) with the ordered fuzzy sets. The fuzzy set can be described by the membership function (MF) μ_A in the universe of discourse X: A={x, $\mu_A(x)$ | x⊂X}. For example the set A of the terms can be as follow: A={small, medium, high}.

Generally the validity of the statement *IF x is A* is continuously graded from 0 to 1. Just as Boolean logic has operators for combining logic variables, so does fuzzy logic. In both system (classical and fuzzy) variables have different definitions but they use the same operators AND, OR, NOT (i.e. *Temperature is high AND Humidity is medium*.) The definitions most commonly used operators are: μ_{Ap}=min (μ_{Ai}, μ_{Aj}), μ_{Ap}=max(μ_{Ai}, μ_{Aj}) where the membership values for the sets A_i, A_j, A_p are equal to μ_{Ak}, μ_{Ai}, μ_{Aj}.

The rule base is a kind of mapping of input variables to output variables and in this way the characteristic of the controlled machine or process is defined The fuzzy approach does not similarize plant model, but tries to incorporate the nonlinearly by a limited number of IF...THEN... rules, which approximate the characteristic of the process.

The rule set in connection with inference procedure gives possibility to determine the output value z for the given input value x. The most popular inference methods in the hardware implementations are MAX-MIN and MAX-PROD. In both methods the so-called fire grade of the rule is calculated as a result of values of membership functions for the linguistic variables in the antecedent part. Next the fire grade value is used to determine the fuzzy set for the corresponding output term. In case of MAX-MIN method the fuzzy set is reduced to the above-mentioned value. Finally the analogue action is performed for the other rules in the rule base with non-zero fire grade values. The result is

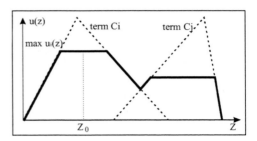

Fig. 1. Graphical interpretation of MoM method.

a fuzzy set, which is to be transformed to a crisp, executable value. The representative crisp output value is calculating applying the defuzzyfication mechanism. The Centre Of Gravity (COG) method is most representative one. The crisp value is calculating by division the sum of multiplication discrete output values with respective membership function values and the sum of these membership function values. The Mean of Maximum (MoM) method generates a control action, which represents the mean value of all local control actions l whose membership functions reach the maximum w (2).

$$z_0 = \sum_{j=1}^{l} \frac{w_j}{l} \qquad (2)$$

Fig.2 shows the basic structure of fuzzy systems. The fuzzyfier transforms the measured input signal into linguistic variables. The defuzzyfier acts in opposite way it means generates crisp output values from obtained in inference procedure membership function. The rule base can be processed simultaneously by a set of fuzzy input variables. This means that full parallel processing can be achieved. The synchronisation action is met in defuzzyfier unit, where all rules participate in crisp value generation. So the next important positive aspect of implementation of fuzzy logic is possibility of high speed processing.

Usually it requires special organisation of the rule base but is done once during the parameterisation and compilation of fuzzy controller. The scale(normalisation) units presented on fig. 2 are optional.

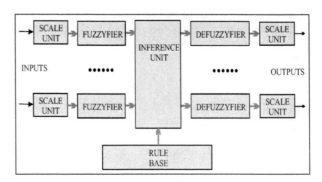

Fig.2. Basic structure of the fuzzy controller.

3. FUZZY LOGIC REGULATION

In general software, hardware and software-hardware implementation of fuzzy control is distinguished. The practical realisation requires special (dedicated) software and/or hardware for the parameterisation and data pre-processing additionally.

The interesting classification of the fuzzy logic hardware is presented in (Eichfeld *et al.*, 1996) and distinguishes the implementation based on the standard or special hardware.

Standard hardware is considered as hardware that is not especially designed for fuzzy system implementation but it is possible to program it to act as a fuzzy system. For example the fuzzy software installed on a PC computer can be used to run or simulate fuzzy logic algorithms. The processors, microprocessors or microcontrollers can execute fuzzy logic algorithms in the form of programs or procedures prepared in C or assembler languages.

The fuzzy system implementation based on special circuits applies the circuits and devices especially designed for the fuzzy algorithms realisation. For example the fuzzy logic processor (FLC) is a device, which is designed to execute fuzzy logic processing. All typical elements fuzzyfier, inference unit, defuzzyfier are included inside.

It was already mentioned that the production lines, technological processes are controlled by the PLC. Nowara and Chmiel (1998) discussed the software implementation of fuzzy control in PLCs. PLC controllers especially regarding their construction, programming and functioning allow to program applications for fuzzy control realisation without additional equipment. That is the simplest method of fuzzy control it means the software one. The world-known PLC (Omron, Siemens) suppliers propose and deliver dedicated fuzzy modules that perform operations specific for this kind of control. This can be interpreted as a sign of spreading fuzzy control in these controllers. The users get the powerful hardware and software and thus area of applications of PLC is growing significantly.

The other story is configuration of fuzzy unit in the regulation circuit. Basically the regulation loop comprises the process and the regulator operating in the feedback-loop mode. Many processes appear to be non-linear ones. Because the process parameters change when operating point changes it is a need to adopt the new parameters to regulators as well, to achieve optimised regulation curves. For such purposes the regulation loop is to be extended with the reference unit or estimator. Another aspect of regulation is disturbances. If the level of disturbances can be measured or can be estimated or the mathematical model of it is known than the regulator

Table 1. Fuzzy logic hardware

Standard circuits	Computers i.e. PC
	Processors: μP, μC, DSP
	Look-up tables (memories)
	FPGA circuits
Special circuits	FLC
	FP: μP with add-on
	FμC: μC and FLC
	AFC: analogue circuits

parameters can be modified that regulation is done satisfactorily. Both described items lead to hybrid architectures in control. The fuzzy controller can be applied to the process directly or it can be applied as a tune-unit to achieve optimal regulation. The expected results of such solutions are: speed up the system reaction, the system behaviour improvement in selected domains of interest, supervision of a classical controller by merging information from controller and system together to improve overall performance.

All presented aspects of fuzzy algorithm implementations and their application play important rule during design and develop phase of project. Simplicity, processing speed, availability of the hardware and user tools, test and simulate are to be taken into consideration. The choice depends on specifications, parameters required and experience. It is relatively easy to program standard element with control algorithm. To design fastest devices the building of a special processor is to be considered. The interesting proposal is utilisation of PLC in fuzzy control field. For this reason authors have decided to develop the fuzzy module which can work as an independent MISO (Multiple Inputs Multiple Outputs) /MIMO (Multiple Inputs Single Output) unit or communicate with PLC as I/O module. To keep the fundamental attributes of PLC system (free configuration and programming) and to design the fuzzy coprocessor the prototype is based on the standard processor. Additional assumption was that fuzzy module is to be an I/O module of the PLC developed in Institute of Electronics, Technical University, Gliwice.

4. PROTOTYPE SPECIFICATION

Prototype features are as follows:
- 4 inputs and 2 outputs
- input signals: 12-bit resolution, 8 linguistic terms
- output signals: 8-bit resolution, 7 linguistic terms
- membership function to be freely programmed
- inference method: MAX-MIN and MAX-PROD
- defuzzyfication method: COG, MoM
- operating modes: independent regulator or I/O module in PLC configuration

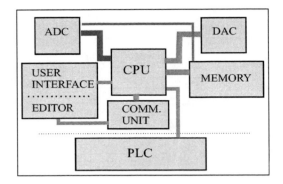

Fig.3. Block diagram of designed fuzzy controller.

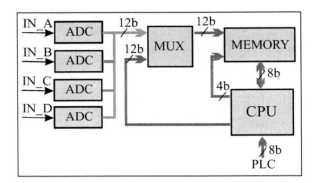

Fig.5. Block diagram of fuzzyfier unit.

Very important aspect of the controllers is processing time. Nowara and Zając (1998) present untypical implementation of the inference unit that ensures high speed processing. Such solution requires extended hardware design. The other (cheaper and slower) approach is based on the inference method related to the single rule. The standard electronic components can be applied. To speed up the processing phase the special memory organisation is proposed. It was noticed during development phase that only software modifications are required to obtain additional operating mode of controller. The controller is able to act as 4-input, 2-otput MIMO controller or two independent 2-input, 1-otput MISO units.

4.1. Knowledge base

The knowledge base is programmed using special software installed on the PC computer. The software is an integrated part of the fuzzy system and allows:

- to define membership functions
- to chose operating mode
- to set the rule base,
- to determine the inference and defuzzyfication method.

4.2. PLC-fuzzy controller interface

Fuzzy controller accesses input data from system bus

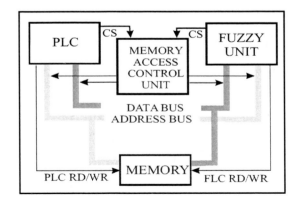

Fig. 4. Communication fuzzy controller-PLC

of PLC directly and PLC accesses to the processing results. The control signals ensure proper synchronisation between those operations and to define the operating modes ('continuous' or 'on request' data acquisition from PLC or from ADC).

4.3. Fuzzyfication

Taking into account the project assumption the memory demand of the fuzzyfier unit can be calculated as:

$$SIZE\_RAM = m*k*2^n \qquad (3)$$

where m is number of inputs (4), k is number of bits required to present membership grade (8) and n is input signal resolution (12). Hence the memory demand is 48kB. No restriction to the shape of the membership function exists. The memory size can be limited by introducing only some kinds of shape of membership functions and mathematical formulas to get the grade. However this makes the processing time longer.

The fuzzyfier unit is addressed with the n-bit crisp value and the memory cells contain:

- 8-bit membership grade of the term with the lower number
- 8-bit membership grade of the term with the higher number
- term numbers (0...7) written in the single memory cell

The further assumption is that the overlap degree of the membership function is 2.
The input data can be acquired from the analogue to digital converter (ADC) or from the PLC. The central unit of the fuzzy controller manages this process and sets the muliplexer properly.

4.4. Inference methods – MIMO mode

The inference mechanism is referred to the single rule, it means the fire grade value antecedent part is used to determine the fuzzy set for the corresponding output term. This action is repeated for each rule separately and finally the result fuzzy set is determined. Assuming number of inputs (m) and number of terms (t) equal to 4 and 8 respectively the total number of rules is 4096 according to statement:

$$R = t^m \qquad (3)$$

To reduce the processing time (no need to take into account number of output variable) the rule base is written in separate 4kB memory blocks for each output variable. The overlap factor implies that the number of rules to be looked up is reduced to 16.

The rule base memory is organised in such way that the address combined with the term numbers of input variables determines the term number of the output variables respectively:

A2A1A0 B2B1B0 C2C1C0 D2D1D0=>X2X1X0
A2A1A0 B2B1B0 C2C1C0 D2D1D0=>Y2Y1Y0

where A, B, C, D are input variables X, Y – output variables. Index 2, 1, 0 identifies bit number.
The fire grade of the rule is computing in the same processing cycle, this grade modifies the output fuzzy set, and its temporary image is stored in the internal memory of controller. The calculations are realised in reference to the inference method MAX-MIN or MAX-PROD defined during parameterisation of fuzzy controller.

4.5. Inference methods – MISO mode

In case of such controller configuration the total number of rules is equal to 64. The inference is performed similar to the MIMO configuration:

000 000 A2A1A0 B2B1B0=>X2X1X0
000 000 A2A1A0 B2B1B0=>Y2Y1Y0

The not used bits are set to value 0.

4.6. Defuzzyfication

The mathematical formula of COG and MoM defuzzyfication methods requires computing two mathematical expressions and then their quotient. The multiplier and divider units are to be used which are time consuming units. The alternative approach is to memorise auxiliary results in the look-up tables, whereas the multiply and divide results are stored. In the fuzzy controller presented, the multiplier subcircuit is based on the memory module, the divider subcircuit is implemented as a software instructions.

The COG expression is modified to formula:

$$u_c = \frac{\sum_{i=1}^{7} S_i * \tau_i * COG_i}{\sum_{i=1}^{7} S_i * \tau_i} \qquad (4)$$

where S_i is surface under membership function representing term i, τ_i is modified factor, COG_i is crisp value in i term range. The nominator and denominator are calculated summarising the data stored in the memory and accessed by addressing word

n T2 T1 T0 $\tau 7$ $\tau 6$ $\tau 5$ $\tau 4$ $\tau 3$ $\tau 2$ $\tau 1$ $\tau 0$

T2 T1 T0 means actual read term number, $\tau 7$ $\tau 6$ $\tau 5$ $\tau 4$ $\tau 3$ $\tau 2$ $\tau 1$ $\tau 0$ is modify factor, for n=0 the $S_i * \tau_i$ value is read, for n=1 - $S_i * COG_i * \tau_i$ value. The final crisp value of fuzzy algorithm is derived from the dividing algorithm. The same procedure is to be implemented for the second output variable.
In case of MoM method the modification factors τ_i are compared to all terms of output variables. The term with the maximum factor is chosen and than the memory value referenced to the address

T2 T1 T0 $\tau 7$ $\tau 6$ $\tau 5$ $\tau 4$ $\tau 3$ $\tau 2$ $\tau 1$ $\tau 0$

represents the crisp output value.

4.7. RAM memory, control bits

The designed controller occupies 64kB internal memory RAM. Besides the memory blocks recognised above the control bits are to be defined additionally. The control vector comprises information about the inference mechanism, defuzzyfication method, operating modes MIMO/MISO and operating modes regarding PLC communication.

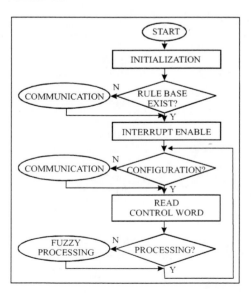

Fig.6. Operation principle flowchart.

4.8. The operation principle.

Fig.6 presents operation flowchart of program realised by fuzzy controller written in assembler language. The program consists of main part including initialisation of CPU and external components and main loop where suitable subprograms run depend on control bits. The subprograms are referred to interrupt regulation communication and auxiliary procedures. The program realises the operation principle describing in the paper.

5. SOFTWARE TOOL, EDITOR

The last described element of the system is editor. The software tool is written in the C language runs in graphical environment under Windows operating system. The software supports defining input/output variables, their membership functions, to define and pre-process rule base, to parameterise the fuzzy controller configuration, to compile and transfer data to controller. The very helpful option is simulation function, which allows testing the behaviour of controller offline.

6. SUMMARY

The presented fuzzy controller was designed based on DS80C320 microcontroller (Dallas Semiconductors, 1995). This microcontroller represents high processing speed comparing to controllers from other manufactures.

The functionality of the fuzzy controller was proved by simulation of identical rule base applying FuzzyCAT ver. 2.0 software from CePlus Gmbh.

The inference time is independent from the chosen method what seems to be advantage of designed unit.

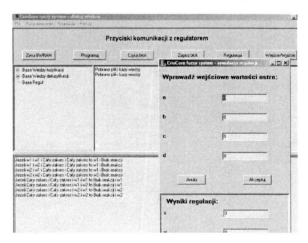

Fig.7. Software tool – application screen

The results achieved are comparable to the Omron

Table 2. Processing times of fuzzy controller.

	MIMO	MISO
Fuzzyfication		
• inputs – PLC	327µs	327µs
• inputs - ADC	277µs	277µs
Inference		
• MAX-MIN	432µs	411µs
• MAX-PROD	414µs	406µs
Defuzzyfication		
• MoM	16,8µs	16,8µs
• CoG	164µs	164µs

FZ001 inference unit for Omron C200 PLC (Omron,1994), but presented unit deals with larger number of rules.

The way to reduce the processing time is to design the special fuzzy processor. It is common, not only in case of fuzzy controllers, that extended hardware allows speed up the data processing significantly. The examples of SAE81C99A (Siemens 1996) or fuzzy processor presented by Nowara and Zając (1998) confirm such approach.

Applying the standard components the simplicity and cost are the important win. Such controller with the configuration and parameterisation software as a integral part of product can became interesting solution for industry applications.

REFERENCES

Eichfeld H., Kuenmund T., Menke M. (1996). *A 12b General-Purpose Fuzzy Logic Controller Chip.* IEEE Transactions on Fuzzy Systems, vol.4., no.4. p.460

Nowara A. and Zając R. (1998). *Fuzzy Logic Processor based on FPGA Circuit.* International Conference Programmable Devices and Systems PDS'98, Gliwice, Poland. p.43.

Nowara A. and Chmiel M. (1998). *Fuzzy Logic Control Implementation in PLCs,* International Conference Programmable Devices and Systems PDS'98, Gliwice, Poland. p17.

Ross T.J. (1995). *Fuzzy Logic with Engineering Applications.* McGraw-Hill, Inc., International Edition.

Yager R. R. and Filev D. (1995). *Podstawy modelowania i sterowania rozmytego.* WNT, Warszawa.

Omron (1994). *SYSMAC C200H-FZ001 Fuzzy Logic Unit.* Operation Manual.

Siemens (1996). *The SAE 91C99A Fuzzy Logic Coprocessor.* Semiconductor: Technical Product Information Edition 8.

Dallas Semiconductor (1995) *High-Speed Microcontroller Data Book.*

IFAC
Publications
www.elsevier.com/locate/ifac

STANDARD RESISTOR DIVIDER FOR PRECISE ADC CALIBRATION AND NONLINEARITY CORRECTION

Maciej Nowiński, Władysław Ciążyński

Institute of Electronics, Silesian University of Technology, Gliwice, Poland

Abstract: The paper presents an idea of the precision voltage dividers built of standard metal-film resistors, but obtaining the accuracy of a few ppm as a result of switching over and averaging process. The division factor error analysis is presented for the worst case. The way of that error estimation based on the real case measurements is described. An application example is shown concerning the precise ADC calibration and its residual nonlinearities correction. Some measurement results confirm the described idea. Its potential application to the precise DVM self-calibration is indicated. *Copyright © 2001 IFAC*

Keywords: A/D converters, electronics, microprocessors, nonlinearity, precision measurements, self-adjusting systems, voltmeters.

1. INTRODUCTION

The simple series connection of two resistors/impedances is no doubt the most commonly appearing stage in all electric and electronic circuits. In the DC circuits it allows the DC voltage to be divided in the proportion determined by the two resistor values. The voltage divider of that type is e.g. used as the input stage of every multirange ADC or DMM, their resistors being switched by the range selection switch or the autoranging circuit. The accuracy of the resistors used determines the voltage division factor error, i.e. affects the overall voltage measurement accuracy.

The simplest two-resistor DC voltage divider is shown at Figure 1. The resistors' nominal resistance values are equal to R_A and R_B, and have the same relative bilateral tolerances equal to $\delta_R = \Delta R / R_0$. The circuit division factor K (and its reciprocal $A = 1/K$ i.e. the voltage suppression factor) has nominal value of K_0 (and A_0 correspondingly). The biggest K

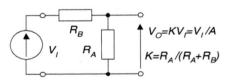

Fig.1. Simple voltage divider

(or A) value deviation from its nominal K_0 (or A_0) value will appear when the both real resistance values deviate correspondingly by the maximum values of δ_R in opposite directions. Let us consider an example when the R_A and R_B are both the 0.1-% resistors of the same nominal value of R_0. The nominal value $K_0 = 0.5$ and maximum deviation of that value is $\delta_K = 0,1$ %. It means the actual division factor (for the two freely chosen resistors meeting given δ_R tolerance) is obtained in the range of $0.4995 \div 0.5005$. The idea of precise calibration using the modest accuracy resistors can be first presented already on that very simple example:

- first for the two freely chosen resistors we are obtaining the value of the voltage division factor K_1;
- then after **swapping** these two resistors (resistor used before as R_A will now play the role of R_B and vice versa) we are obtaining the value K_2.

We can easily find that always $K_2 = 1 - K_1$ and the arithmetic mean value of both K_1 and K_2 factors is equal exactly to 0.5000. So in every real voltage divider we can assume that mean value as accurate half voltage of the input reference and use it e.g. to check linearity of the ADC transfer characteristic in its half-range point with the voltage divider causing no loss in accuracy.

Certainly the presented method, inspired somehow by the early work of (Hamon, 1964) mentioned also in (Oliver and Cage, 1971), can be extended to the division factors other than ½. It is always advantageous to use resistors of the same nominal resistance value (especially at high tolerances). We could then build both R_A and R_B resistances of identical resistors taken from the same box (i.e. the same production lot), in the hope that they show the same or very close other parameters. The result is better temperature and long term stability of the obtained K value. The maximum K factor error is certainly not better, but statistical expected value of that error can be greatly improved that way.

2. DIVISION FACTOR ACCURACY ANALYSIS FOR THE WORST CASE

Below is thoroughly investigated the most interesting case of the mentioned voltage divider,

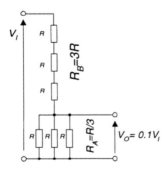

Fig. 2. Voltage divider considered (for $n = 3$)

when both R_A and R_B resistances are composed of the resistors having the same nominal resistance values in such a way, that n resistors connected in parallel constitute the R_A resistance, and another n resistors connected in series constitute the R_B resistance (see Fig. 2). The obtainable division factor K_0 nominal values are then discrete and depending on the number n of resistors used:

$$K_0 = \frac{V_O}{V_I} = \frac{1}{n^2 + 1} \tag{1}$$

The nominal suppression factor A_0 is then determined as: $A_0 = n^2 + 1$, what for the n value of 1, 2, 3, 7 gives interesting and widely used A_0 suppression values of 2, 5, 10, 50.

The biggest deviation of K from its nominal K_0 value will appear for the worst case when:
- all n resistors connected in parallel (constituting R_A) deviate from their nominal R_0 resistance value by the maximum values of δ_R, and (at the same time),
- all n resistors connected in series (constituting R_B) deviate from their nominal R_0 resistance value by the same maximum value of δ_R but in opposite direction.

Now, when **the two groups of resistors are swapped** (i.e. connected in parallel resistors of R_A will be connected in series and playing the role of R_B, and vice versa) the deviation appears of opposite sign. Although the deviation is exactly the same in value only for for $n = 1$, the mean value of both cases (before and after the resistor groups swap) assumed as the accurate value of K_0 will be loaded with the substantially smaller error (and zero error for $n = 1$).

The general expressions for the voltage division factor values of K_1 (before the swap) and K_2 (after the swap) are given below:

$$K_1 = \frac{V_{O1}}{V_I} = \frac{(R_0 + 2\Delta R)/n}{(R_0 + 2\Delta R)/n + nR_0} \tag{2}$$

$$K_2 = \frac{V_{O2}}{V_I} = \frac{R_O/n}{(R_0 + 2\Delta R)/n + nR_O} \tag{3}$$

The above expressions has been obtained in (Nowiński, 1998 and 1999) for the case when all resistors have nominal values of R_0 in one group, and values of $R_0 + 2\Delta R = R_0(1 + 2\delta_R)$ in the second group. This case at small δ_R is equivalent to the worst case. The mean value of K_1 and K_2 is then equal to:

$$K_{AV} = \frac{K_1 + K_2}{2}$$
$$K_{AV} = \frac{1}{2}\left(\frac{(1 + 2\delta_R)/n}{(1 + 2\delta_R)/n + n} + \frac{1/n}{1/n + n + 2n\delta_R} \right) \tag{4}$$

The relative deviation δ_K, i.e. the relative difference between K_0 and the mean value of $K_{AV} = \frac{1}{2}(K_1 + K_2)$ is equal to:

$$\delta_K = \frac{K_{AV}}{K_0} - 1 = K_{AV}(n^2 + 1) - 1 =$$
$$= \frac{n^2 + 1}{2}\left(\frac{(1 + 2\delta_R)/n}{(1 + 2\delta_R)/n + n} + \frac{1/n}{1/n + n + 2n\delta_R} \right) - 1 \tag{5}$$

After some calculations one can obtain the accurate relation in the form of:

$$\delta_K = \frac{2n^2(n^2-1)\delta_R^2}{(1+2\delta_R+n^2)(1+n^2+n^2\delta_R)} \qquad (6)$$

This for the small δ_R values can be simplified to the form of (7), better suited for practical calculations.

$$\delta_K \approx \frac{2n^2(n^2-1)}{(n^2+1)^2}\delta_R^2 \qquad (7)$$

Table 1 below contains the worst case voltage divider error δ_K estimates, derived from the equation (7) for some selected n and δ_R values.

Table 1. Voltage divider's worst case accuracy after arithmetic averaging

	δ_K [ppm]			
Number of resistors in the group: n	1	2	3	7
Nominal voltage division factor: K_0	0.50	0.20	0.10	0.02
Nominal voltage attenuation: $A_0 = 1/K_O$	2	5	10	50
Resistors tolerance δ_R [%] — 1.0	0	100	150	200
0.5	0	25	35	50
0.2	0	4	6	10
0.1	0	1	1.5	2

For $n = 1$, i.e. for $K_0 = 0.500$ there is no error for any tolerance of both resistors. With typical 0.5% metal-film resistors error never exceeds 50 ppm, and when even the simplest ohmmeter is used to select a few resistors that differ no more than 0.1%, the error decreases to the astounding level of 1 ppm.

The implementation of the idea described here at the extreme accuracies obviously requires the proper measurement methods to be applied and, in particular, adequate thermal conditions to be ensured and maintained. It should be noted e.g. that:

- the temperature resistance coefficients should be small and as close to each other as possible;
- the thermoelectric effects (thermo-EMFs) should be avoided or minimised;
- for $n > 1$ small difference appears in the individual resistors self-heating process;
- the divider output resistance after the resistors switching slightly changes and it does make difference when we take into account the measuring device input resistance.

3. VOLTAGE DIVIDER ACCURACY ESTIMATION FOR REAL RESISTOR TOLERANCES DISTRIBUTION

The relative output voltage difference in both configurations for the worst case can be calculated basing on equations (2) and (3) as:

$$\delta_V = \frac{\Delta V}{V_O} = \frac{V_{O1}-V_{O2}}{V_O} = \frac{K_1-K_2}{K_O} = $$
$$= \left(\frac{1+2\delta_R}{1+2\delta_R+n^2} - \frac{1}{1+2\delta_R n^2+n^2}\right)(n^2+1) \approx \frac{4n^2\delta_R}{1+n^2} \qquad (8)$$

From the equation (8) we could find for example that for $n = 3$ and $\delta_R = 0.1\%$ the output voltage in the worst case would deviate 0.36% from the nominal value of $V_O = 0.1V_I$ (approx. symmetrical 0.18% in both directions).

In real applications we certainly do not know the individual resistance deviations from the nominal R_0 value. Composing the voltage divider we would like to select individual resistances freely if only each of them meets the given δ_R specification. It is only the ΔV voltage difference effect after the resistor groups swapping what is in fact observed as a result of the resistors' values deviation. Normally the ΔV is much smaller than resulting from the equation (8) due to the individual resistor tolerances smaller than nominal and causing the output voltage to deviate in opposite directions, thus partially compensating.

So knowing the value of δ_V, we can from (8) estimate the measure of the equivalent real resistance deviation:

$$\delta_{RAV} \approx \frac{n^2+1}{4n^2}\frac{\Delta V}{V_O} = \frac{n^2+1}{4n^2}\delta_V \qquad (9)$$

The value obtained has been denoted as δ_{RAV} because it has the meaning of arithmetic average tolerance value. It is obvious that the smaller ΔV voltage difference observed results in the equivalent δ_{RAV} tolerance smaller than nominal δ_R value.

To make the definition of δ_{RAV} quite clear, let us assume that in the above example divider of $n = 3$ and $\delta_R = 0.1\%$ the output voltage (which in the worst case could deviate 0.36% from the nominal value of $V_O = 0.1V_I$) really deviates ten times less, i.e. only by $\delta_V = 0,036\%$. It corresponds to the mean resistor tolerance in both groups calculated from the equation (9) as low as $\delta_{RAV} = \pm0.01\%$. Such a mean value of δ_{RAV} can appear at the continuum of very many possible distributions of the resistance deviations. With no full proof we can probably assume that the extreme two cases leading to the same ΔV voltage difference are:
- case A - all three resistors in the R_A group have the values of $R_0(1 - 0.01\%)$ and at the same time all three resistors in the R_B group have the values of $R_0(1 + 0.01\%)$;
- case B - all three resistors in the R_A group and two of resistors in the R_B group have the values of $R_0(1 - 0.01\%)$, and at the same time, the third resistor in the R_B group has the value of $R_0(1 + 0.05\%)$.

In general terms we can say that case A corresponds to the uniformly worst case of the resistance value distribution analysed in paragraph 2, with the only difference in the resistors' values deviating much less than δ_R. The accuracy of the voltage divider after arithmetic averaging can certainly be obtained from the equation (7) with δ_R substituted by δ_{RAV}. Having δ_{RAV} expressed in terms of observed $\Delta V/V = \delta_V$ (see equation 9) we can easily get:

$$\delta_K^A = \frac{(n^2-1)}{8n^2}\left(\frac{\Delta V}{V}\right)^2 = \frac{(n^2-1)}{8n^2}\delta_V^2 \qquad (10)$$

The case described above as B is equivalent to the situation when we have one resistor of $R_0(1+2n\cdot\delta_{RAV})$ value and all remaining of the nominal R_O values. For general n the analysis of that case is quite complicated and leads to the following expression (Nowiński, 1999):

$$\delta_K^B = -\frac{(n-1)^2}{8n}\left(\frac{\Delta V}{V}\right)^2 = -\frac{(n-1)^2}{8n}\delta_V^2 \qquad (11)$$

The obtained division factor error for that case not only differs in value but also has the opposite sign. It is very likely that real error corresponding to the δ_V observed will be somewhere in between of the values resulting from equations of (10) and (11).

$$-\frac{(n-1)^2}{8n}\boldsymbol{\delta}_V^2 \le \boldsymbol{\delta}_K \le \frac{(n^2-1)}{8n^2}\boldsymbol{\delta}_V^2 \qquad (12)$$

4. IMPROVING OF ACCURACY ESTIMATION USING GEOMETRIC AVERAGING

It may be interesting to note that using the geometric rather than arithmetic average can produce even better results. The rigorous error calculation leads to very complicated expressions. With some approximations it is possible however to achieve the following simplified formula (Nowinski, 1999):

$$\delta_K^G \approx \frac{2n^2}{(n^2+1)^2}\delta_R^2 = \frac{\delta_K}{n^2-1} \qquad (13)$$

The corresponding numerical values are (n^2-1) times smaller than δ_K obtained from equation (7). E.g. for a 1:10 voltage divider ($n = 3$) made of 0.5% resistors it gives the δ_K^G division factor error as small as 5 ppm. To calculate the geometric mean value we need to know the accurate V_1 and V_2 values in both configurations [1]. Nevertheless, the realisation that the geometric mean value gives better results allows below introducing ε, the correction component of δ_V, decreasing the error estimate resulting from arithmetic averaging.

---

[1] For the arithmetic averaging only the difference between the divider's output voltages has to be known rather than their global values of V_1 and V_2. It is possible to take very accurate measurement of ΔV at the DVM's low range (high resolution) with the appropriate reference voltage used.

The difference between the arithmetic and geometric mean values of the two voltages that differ by ΔV is equal to:

$$\varepsilon = \frac{V_{O1}+(V_{O1}+\Delta V)}{2} - \sqrt{V_{O1}(V_{O1}+\Delta V)} = \qquad (14)$$
$$= V_{O1}(1+\frac{1}{2}\frac{\Delta V}{V_{O1}}) - V_{O1}\sqrt{1+\frac{\Delta V}{V_{O1}}}$$

Expanding the square root function into a power series and taking into account the first three terms we get:

$$\varepsilon = V_{O1}(1+\frac{1}{2}\frac{\Delta V}{V_{O1}}) - V_{O1}[1+\frac{1}{2}\frac{\Delta V}{V_{O1}} - \frac{1}{8}(\frac{\Delta V}{V_{O1}})^2 + ...] \qquad (15)$$

Finally taking the accurate value of V_O for the V_{O1} we get the correction:

$$\varepsilon = \frac{1}{8}V_O(\frac{\Delta V}{V_O})^2 = \frac{1}{8}V_O\delta_V^2 \qquad (16)$$

An arithmetic mean yields always a value that is greater than one for a geometric mean. Thus the correction ε of equation (16) should always be (if only it is justified enough by the high measurement accuracy) subtracted from the $\delta_V = \Delta V/V_O$ observed. It ensures improvement of the divider accuracy estimation both for the worst case and real case analysis.

5. APPLICATION EXAMPLE

When the described idea is applied to the ADC calibrating purpose we know the two values of V_{O1} and V_{O2} (see equations 2, 3 and 8) and can assume that their mean value (possibly corrected by ε resulting from equation 16) corresponds to the accurate value of V_O. The error of such an assumption is determined by the values of inequality (12). If the error observed falls outside we can say it is caused by the ADC nonlinearity at that point of the conversion characteristic. Taking the reference input voltage V_I corresponding to the range of ADC under test:

- for $n = 1$ we can estimate half-range nonlinearity;
- for $n = 2$ we can estimate nonlinearity at 20% of the range, and (after inverting positions of the R_A and R_B groups) at 80% of the range;
- for $n = 3$ we can estimate nonlinearity at 10% of the range, and (after inverting positions of the R_A and R_B groups) at 90% of the range; and so on (but the practical problems are very quickly rising for higher n).

Figure 3 shows the example configurable resistance divider built of four $10\,k\Omega/0.05\%$ resistors, and some manual switches:

- SW1 - SPDT switch to control the unipolar/bipolar reference voltage V_{REF};
- SW2 - DPDT switch to change V_{REF} polarity;

- SW3 - DPDT switch to reconfigure voltage divider, i.e. to swap the resistors; and
- SW4 - 4-section 4-position rotary switch to select the division factor.

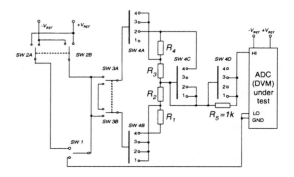

Fig.3. Circuit for ADC nonlinearity testing.

That test circuit has been first proved using the high accuracy HP3458 digital multimeter configured to the ±10 V measuring range, with 7.5-digit resolution and the measurement period of approx. 1 s. The instrument at the specified range shows extremely low nonlinearity of 0.05 ppm.

For SW1 & SW3 switched off and SW4 in position 1 the DVM input has been connected to the circuit common through the two resistors in parallel and the DVM zero check procedure has been followed. Next at SW1 switched on, the $-V_{REF}$ and $+V_{REF}$ reference voltages symmetry has been verified by SW3 being switched over and the DVM readings adjusted to show to show ±10 000.000 mV. During the test the V_{REF} stability was better than ±0.005 mV, and the multimeter reading uncertainty was at the level of ±(0.002....0.003) mV.

After zero has been corrected and the reference voltages adjusted symmetrical, the linearity verification of the DVM conversion characteristic could start. Proper manipulation of the switches enabled application to the DVM input voltages of the nominal values of: $+V_{REF}$, $+0.8V_{REF}$, $+0.5V_{REF}$, $+0.2V_{REF}$, 0(V), $-0.2V_{REF}$, $-0.5V_{REF}$, $-0.8V_{REF}$ and $-V_{REF}$. Each measurement was made twice, for both positions of the SW3 switch, which reconfigured the four resistors constituting the voltage divider as described in previous paragraphs. For each input voltage the mean value of the two measurements in ideal situation equals to the input voltage applied.

The voltage division error for every checked point of the conversion characteristic can be estimated from the equation (12) for $n = 1$ or 2 correspondingly. Possible bigger deviations would give evidence of the DVM nonlinearity in the examined point. In the described case the voltage divider mean value errors shown in Table 2 were found at the noise level and much better than 0.5 ppm.

The examined HP3458A DVM has very high input resistance, but in case of the other DUT the input resistance, usually equal to 10 MΩ, can negative influence the accuracy obtained. To eliminate that effect we have to consider the divider output resistance. It can be easily found that for the division factor of 0.0, 0.5 and 1.0 the divider's output resistance is equal to 5 kΩ, and for the division factor of 0.2 and 0.8 is equal to 4 kΩ. The additional resistor of 1 kΩ/0.1% has been introduced to the circuit by the SW4D switch for the division factor of 0.2 (switch position 2) and 0.8 (position 4) and shorted for the other factors to clear that error.

Table 2. Voltage divider accuracy verified with HP3458A digital voltmeter

Nominal Voltage [mV]	First reading [mV]	Second reading [mV]	Mean value [mV]
0.00	0.001	–	–
–2 000.00	–1 999.938	–2 000.059	–1 999.999
–5 000.00	–4 999.409	–5 000.593	–5 000.001
–8 000.00	–7 999.945	–8 000.062	–8 000.004
–10 000.00	–	–10 000.000	–10 000.000
+2 000.00	+1 999.942	+2 000.063	+2 000.003
+5 000.00	+4 999.412	+5 000.595	+5 000.004
+8 000.00	+7 999.944	+8 000.061	+8 000.003
+10 000.00	–	+10 000.000	+10 000.000

In the considered test circuit the DUT conversion characteristic could also be verified for the $+0.6V_{REF}$ and $-0.6V_{REF}$ when the voltage divider having the suppression factor of 5 (i.e. all four resistors used, SW4 in position 2 or 4) is connected rather across the $+V_{REF}$, $-V_{REF}$, not across $+V_{REF}$ (or $-V_{REF}$) and the common.

The test circuit of Figure 3 has also been used to estimate the linearity error of the precise 6-digit DVM built according to the new idea presented in (Nowiński, 2000). The internal reference voltages of the device under test obtained from the MAX670 integrated circuit and showing the temperature coefficient below 3ppm/K, have been used as the divider input voltages. The nonlinearity has been observed at the level not higher than 1.5 ÷ 2 ppm, while the values resulting from the inequity (12) are clearly smaller.

All results obtained in the example circuit confirm the relations derived in the previous paragraphs. The idea presented in the paper proved to be useful, although the practical problems connected with its application limited the example divider to $n = 1$ and 2. The microprocessor-controlled circuit built using the electromagnetic relays should give similar results. Application of the semiconductor switches seems to be much more difficult for high accuracy, due to their on-resistance tolerances.

6. CONCLUSIONS

The presented idea can find many applications. The most interesting is testing of the conversion characteristic nonlinearity for the microprocessor controlled ADCs. High accuracy can be obtained here with resistances of modest tolerances. Using the internal reference voltage of such a converter and the configurable voltage divider, built as described above of some nominally identical resistors, makes possible to find deviations of the real conversion characteristic from the nominal one in some characteristic points. The accuracy of the linearity estimation in practice is limited only by the resolution of the device under test.

Knowing the true conversion characteristic we can properly correct it in the software way and even possibly provide means for the approximate self-calibration.

True ADC calibration would require access to the external reference voltage of standard quality. The method presented in the paper allows for the significant simpler and faster calibration procedures in such a case too, and the final calibration accuracy obtained is greatly improved. For the best accuracy the voltage division reconfiguration needs for the manual switches or the electromechanical relays. The bigger on-resistance values of the semiconductor switches and especially their high allowances make them unsuitable in these applications.

The method described here can also be directly applied to the selection of the two resistors whose ratio of resistance values is accurately determined. A voltage divider with the attenuation equal to $A = n^2 + 1$ is made of resistors whose resistance values differ n^2 times. So the precise selection of the two resistors whose ratio of resistance values is in the sequence of 1:4:16:64. Such resistors further connected in series and in parallel allow getting also other ratios useful in high accuracy DACs.

REFERENCES

Hamon B.V. (1964). A 1-100Ω Build-up Resistor for the Calibration of Standard Resistors. *Journal of Scientific Instrumentation,* **vol. 31**, no.12.

Nowiński, M. (1998). Standard Resistors in Precision Voltage Dividers (in Polish). In: *Podstawowe Problemy Metrologii, Prace Komisji Metrologii PAN,* pp. 207 - 217. Gliwice – Ustroń.

Nowiński, M. (1999). Design and calibration of precision voltage dividers using modest tolerance resistors (in Polish). *Elektronizacja,* nr 1, pp. 8 - 11.

Nowiński, M. (2000). Precision voltage-to-period and voltage-to-frequency converters. In: *Programmable Devices and Systems, A Proceeding from the IFAC Workshop, Ostrava, Czech Republic, 8-9 February 2000,* pp. 229 - 234. Pergamon.

Oliver B.M. and J.M. Cage (1971). *Electronic Measurements and Instrumentation.* McGraw-Hill Inc. (Polish edition, WKŁ, Warszawa, 1978).

IFAC

Publications
www.elsevier.com/locate/ifac

PH MODEL APPROXIMATION FOR PURPOSE OF INDUSTRIAL PROGRAMMABLE CONTROLLERS

Krzysztof Stebel

*Institute of Automatic Control,
Silesian University of Technology
ul. Akademicka 16, 44-100 Gliwice,
Poland
FAX: (+48) 32 237 21 27;
E-mail: kstebel@terminator.ia.polsl.gliwice.pl*

Abstract: The aim of this paper is to formulate advanced algorithms as GMC and PFC in the way applicable in programmable industrial controllers. The highly nonlinear dynamical properties of the pH value to the addition of acid or base makes the pH control very difficult. Changes in process sensitivity with pH makes difficult to design conventional PID controllers. Several, advanced, non-conventional control strategies based on model of process, have been developed recently for attack this problem. Although good results are obtained, the main disadvantage of them is complexity and they cannot be applied in industrial controllers. Proposed pH model approximation allows formulating advanced control algorithm ready to use for most of programmable controllers. *Copyright © 2001 IFAC*

Keywords: programmable controllers, pH control, control algorithms, model approximation, model-based control, polynomial models.

1. INTRODUCTION

A main objective when designing a control scheme for a given plant is often to achieve good control performance which is insensitive to process uncertainties. That is, even if the model used for design differs from the actual process, the controller should provide as good performance as possible (Nystrom, et al., 1998). Although good results are obtained, the main disadvantage of them is complexity hence part of them has only value for simulation and theory purpose. Some studies present nonlinear IMC algorithms for SISO system applicable for very fast computer-based pH measurement and control system (Kulkarni, et al., 1991). Fuzzy- logic algorithm can be also

successfully applied using computer-based control system (e. g. Menzl, et al., 1996, Ylen 1997). Although large progress in computer based system was achieved in recent years there is still need for new advanced algorithms in a commercially available industrial controllers. Self-tuning algorithms based on traditional PID algorithm or different kinds of gain scheduling can be found in industrial controllers (Jutila and Jaakola 1986, Piovoso and Williams 1985).

Nowadays most of controllers offer free programmable block functions according to international IEC II3I-3 norm (Lewis 1996) e.g. as a possible way of nonlinear control algorithm implementation. It cannot be forgotten that even though possibilities of control algorithm implementation are growing there are still strong limits in complexity of algorithm and sampling time assured is specific conditions. Applications of nonlinear control strategy usually require that the model differential equations will be solved sufficiently fast so as to give the model output in less than one sampling period (Kulkarni, et al., 1991). In fact it is considered real-time control as well as real-time simulation requirements e.g. fixed computing time in which the integration must be performed in each control step. Hence, in the case of real time control, the integration step size is fixed, bounded and determined by the real-time clock. It should be also noted that the computing time necessary to perform calculation for an integration interval of the model must be constant and should be as small as possible. Hence, the explicit single-pass methods are most suitable for real-time control and simulation without any modification or approximations (Metzger 2000).

Mathematical models of pH in well-stirred tanks are widely discussed in the literature see for example: McAvoy (1972); Gustafsson and Waller (1983); Gustaffson et al., (1995). In this paper the model proposed for process simulation is similar to model used by Wright and Kravaris (1991). This model consists of four bilinear ordinary differential equations, in accordance with the mixing characteristics of the reactor, coupled with two strongly nonlinear algebraic equations. Simulation is based on coefficients taken form pilot plant installation. Although such model is good for simulation it cannot be used as internal model. Numerical computation can be very time consuming, even if analytical way of computation would be chosen this model is to complex to tune and control algorithm is very sensitive to parameters changes (Stebel 2001).

Proposed pH model approximation was applied to input-output linearization, process model based algorithm or GMC (Generic Model Control) (Lee and Sullivan, 1988; Henson and Seborg, 1990, Riggs and Rhinehart, 1990).

Second algorithm considered in this paper is Predictive Functional Control (PFC) (Richalet et al 1978; Richalet, 1993). This algorithm is not common for pH process but is known for other nonlinear processes and gives good results. PFC algorithm for pH processes uses the linearization of pH model approximation in digitized form with fixed time constant and changeable gain.

2. PH MODEL

Models for multi species systems are very complicated. To simulate such multi species systems two hypothetical weak acids, strong base and weak base are combined in model (1-4) to simulate known and unknown reagents. Such model cannot simulate accurately every multi species system. However it can be treated as a kind of approximation that allows

us to observe main feature of such system. Following form will be considered (see e.g. Wright and Kravaris 1991):

$$\frac{V d\,x_1}{dt} = F_1 C_1 - (F_1 + F_2) x_1 \tag{1}$$

$$\frac{V d\,x_3}{dt} = F_1 C_{31} + F_2 C_{32} - (F_1 + F_2) x_3 \tag{2}$$

$$\frac{V d\,x_2}{dt} = F_2 C_2 - (F_1 + F_2) x_2 \tag{3}$$

$$\frac{V d\,x_4}{dt} = F_1 C_{41} + F_2 C_{42} - (F_1 + F_2) x_4 \tag{4}$$

And pH equation

$$[H^+]^5 + A \cdot [H^+]^4 + B \cdot [H^+]^3 + C \cdot [H^+]^2 + \\ + D \cdot [H^+] + E = 0 \tag{5}$$

$$pH = -\log_{10}(H^+) \tag{6}$$

Where:
$A = x_2 + x_4 + K_a + K_b + K_c$

$B = K_a(x_2 - x_1) + K_c(x_2 - x_3) + K_b x_2 + (K_a + K_c) K_b + \\ + K_a K_c + (K_a + K_c) x_4 - K_w$

$C = K_a K_c(x_2 + x_4 - x_1 - x_3) + K_a K_b(x_2 - x_1) + \\ + K_b K_c(x_2 - x_3) + K_a K_b K_c - K_b K_w - (K_a + K_c) K_w$

$D = K_a K_b K_c(x_2 - x_1 - x_3) - (K_a + K_c) K_w K_b + \\ - K_a K_c K_w$

$E = -K_a K_b K_c K_w$

x_1- acetic acid concentration in reactor
x_2- potassium base concentration in reactor
x_3- carbon acid concentration or other weak acid in reactor
x_4- calcium base concentration in reactor
V – volume of reactor (2 [liter])
$[H^+]$ – hydrogen ion concentration
K_a – acetic acid equilibrium constant ($1.8*10^{-5}$)
K_w – water equilibrium constant (10^{-14})
K_c – carbon acid equilibrium constant ($4.2*10^{-7}$)
K_b – calcium base equilibrium constant ($4.3*10^{-2}$)
$C_{31}=0.0037$[mole/l], $C_{32}=0.0015$[mole/l] - carbon acid or other weak acid inlet concentration in F_1 and F_2 stream. $C_{41}=0.0015$[mole/l], $C_{42}=0.0066$[mole/l] – calcium base or other weak acid inlet concentration in F_1 and F_2 stream. Following coefficients ware taken from pilot plant installation (CSCE group, Institute of Automatic Control, Silesian Technical University, Poland): acetic acid inlet concentration $C_1=$ 0.0085 [mole/l], potassium base inlet concentration $C_2=$ 0.0089 [mole/l], acetic acid inlet flow $F_{10}=$ 0.4 [l/min]= const., range of potassium base inlet flow $F_2=$ 0÷0.8 [l/min].

Figure 1 shows how accurately it is possible to simulate real world titration curve using model 1-6. The problem is that even small changes in any reagent concentration or equilibrium constant has significant influence on titration curve not only during simulation but similar behavior can be observed during experiments on pilot plant installation. Figure 2 shows titration curves taken in the same conditions (same reagent concentrations) during three different experiments.

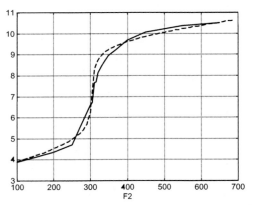

Fig. 1 Titration curve a) taken from pilot plant installation – solid line, b) model (1-6) dashed line

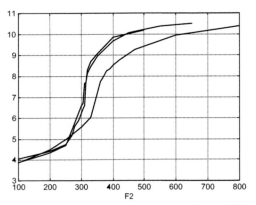

Fig. 2 Titration curves taken from three different experiments on pilot plant installation.

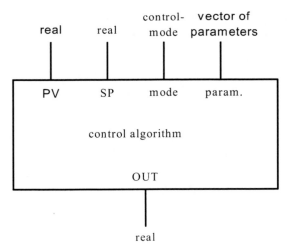

Fig. 3 Schematic function block

Model 1-6 can not be used for control purpose because of complexity and difficulties in parameter estimation in presence of disturbance in real world installations. New model should be easy adjustable to

titration curve taken experimentally. Finally desirable control form was shown on figure 3.

3. APPROXIMATION FOR PURPOSE OF CONTROL

Instead of four differential equations (1-4) in model one differential equation can be proposed

$$\frac{V d_{x_w}}{dt} = F_2(C_2 + C_{42} - C_{32}) + \\ - F_1(C_1 + C_{31} - C_{41}) - (F_1 + F_2) x_w \tag{7}$$

Where new: $x_w = x_2 + x_4 - x_1 - x_3$ is a difference between reagent concentration in reactor. Analyze of equation 5 and 6 suggests that pH value depends mainly on reagent concentration difference x_w in reactor that is why this variable was chosen as a new state variable. Equation 5 and 6 cannot be effectively approximate using one polynomial function, even if high order polynomial function is used. It is not easy problem to balance between high accuracy and simplicity of model. The idea was to find three simple functions with restriction that in case of switching between them continuity of value and first derivative is assured. Finally third order polynomial functions were chosen. First function responsible for low part of titration curve

$$pH_1 = a_0 + a_1 \cdot x + a_2 \cdot x^2 + a_3 \cdot x^3 \tag{8a}$$

Second function responsible for part in the middle of titration curve

$$pH_2 = b_0 + b_1 \cdot x + b_2 \cdot x^2 + b_3 \cdot x^3 \tag{8b}$$

Third function responsible for upper part of titration curve

$$pH_3 = c_0 + c_1 \cdot x + c_2 \cdot x^2 + c_3 \cdot x^3 \tag{8c}$$

In order to switch among functions value x_{01} and x_{02} were chosen as a kind of threshold value, where function (8) have to be switched.

After recalculation using equations (7) and (8) following form was obtained:

$$V \cdot \frac{d(pH)}{dt} = -(pH + w(x)) \cdot (F_1 + F_2) + z(x) \cdot \\ \cdot (F_2 \cdot (C_2 + C_{42} - C_{32}) - F_1 \cdot (C_1 + C_{31} - C_{41})) \tag{10}$$

Where
$w(x) = -b_0 + b_2 \cdot x^2 + 2 \cdot b_3 \cdot x^3$
$z(x) = b_1 + 2 \cdot b_2 \cdot x + 3 \cdot b_3 \cdot x^2$

Coefficients of polynomial functions
a0 = 5.760728;
a1 = 9.102692e+002;

a2 =   1.861139e+005;
a3 =   1.514263e+007;
b0 =   7.240765;
b1 =   7.595050e+003;
b2 = -2.555149e+010;
b3 = -2.555149e+010;
c0 =   8.442167;
c1 =   9.181822e+002;
c2 = -1.955329e+005;
c3 =   2.154936e+007;

Threshold values, where functions have to be switched

gd = -3.188405e-004;
gg =   2.777778e-004;

Model approximation results are shown on figure 4. Reverse model approximation was not shown in this work but it would be useful to find suitable value of x_w for set point value. Both approximations are accurate and difference between model and his approximation is not visible. Obtained approximations were applied to GMC and PFC control algorithm

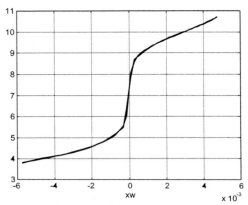

Fig. 4 Approximation of titration curve from model equations 1-6

4. GMC CONTROL LAW

According to Lee and Sullivan (1988)

$$\dot{y} = \ddot{e}\left(y - y_{sp}\right) + \ddot{e}_0 \int_0^t \left(y - y_{sp}\right) \qquad (11)$$

λ , λ_0 - Tuning algorithm parameters

Using equation (10) and (11) and control law (12) is obtained:

$$F_2 = \frac{\left(\lambda\left(y - y_{sp}\right) + \lambda_0 \int_0^t \left(y - y_{sp}\right)\right) \cdot V}{(C_2 + C_{42} - C_{32}) \cdot z(x) - w(x) - y} +$$

$$\qquad (12)$$

$$+ F_1 \cdot \frac{(C_1 + C_{31} - C_{41}) \cdot z(x) + w(x) + y}{(C_2 + C_{42} - C_{32}) \cdot z(x) - w(x) - y}$$

Presented algorithm was tested via simulation for different set point values and accessible information assuming 3 second sampling time. In case of full accessibility best results were obtained for $pH_{sp}=6$

were AES (absolute error sum)< 0.22 (see fig. 5). For $pH_{sp}=7$ and 8 results are much worse (for $pH_{sp}=7$ AES=54, for $pH_{sp}=8$ AES=225) It was mainly because long transition time (see fig 6 and 7). Algorithm was tuned to be stable in each operation point, hence in some regimes worse performance was obtained.

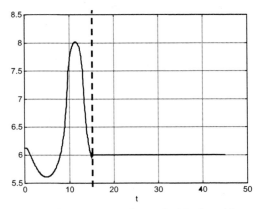

Fig. 5 pH value response for t= 0÷15min without any regulation, for t>15min GMC control algorithm with assumption of full information accessibility – ideal case.

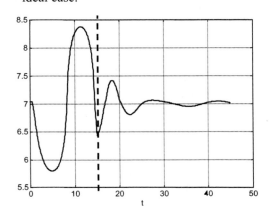

Fig. 6 pH value response for t= 0÷15min without any regulation, for t>15min GMC control algorithm with assumption of full information accessibility – ideal case.

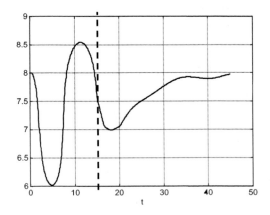

Fig. 7 pH value response for t= 0÷15min without any regulation, for t>15min GMC control algorithm with assumption of full information accessibility – ideal case.

In case of not complete information accessibility it was not possible to assure stability of control and satisfying results for one set of tuning parameters. For each set point algorithm was tuned separately that is why results shown on figure 8 and 9 (for $pH_{sp}=7$ AES=2, for $pH_{sp}=8$ AES=24) are better then results on figure 6 and 7.

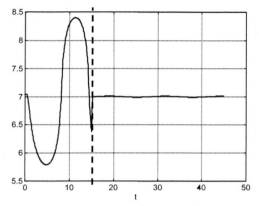

Fig. 8 pH value response for $t=0\div15$min without any regulation, for $t>15$min GMC control algorithm with assumption of partial information accessibility – realistic case.

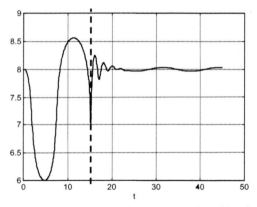

Fig. 9 pH value response for $t=0\div15$min without any regulation, for $t>15$min GMC control algorithm with assumption of partial information accessibility – realistic case.

5. PFC CONTROL LAW

Using equation (10) after linearization and digitalization following equations are obtained:
Internal model

$$\ddot{A}pHm\ (n+1) = \acute{a}\cdot\ddot{A}pHm\ (n)+$$
$$+\left(1-\acute{a}\right)\cdot k(x)\cdot\ddot{A}F_2(n) \tag{13}$$

Control law (Richalet 1993)

$$\ddot{A}F_2(n) = \frac{\left(1-\overset{\cdot h}{\acute{e}_R}\right)\cdot\left(pHs(n)-\ddot{A}pH(n)\right)}{k(x)\cdot\left(1-\acute{a}^h\right)}+$$
$$+\frac{1}{k(x)}\cdot\ddot{A}pHm(n) \tag{14}$$

Where

$$\hat{o}= \frac{V}{F_1+F_{20}}$$

$$k(x) = \frac{(C_2\,C_{42}\,C_{32})\cdot z(x)-w(x)-pH_0}{F_1+F_{20}}$$

$$\acute{a}= e^{-\frac{tsamp}{\hat{o}}}$$

$$\overset{\cdot h}{\acute{e}_R}= e^{-\frac{tsamp}{t_R}}$$

$$t_R = \frac{t_{RBF}}{3}$$

$\ddot{A}pHm(n)$ -from model
Tuning parameters
t_{RBF} -expected response time
h – prediction horizon

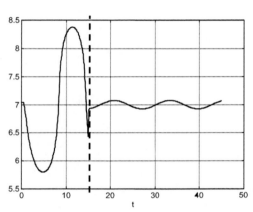

Fig. 10 pH value response for $t=0\div15$min without any regulation, for $t>15$min PFC control algorithm with assumption of full information accessibility – ideal case.

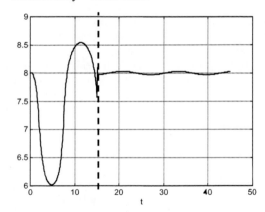

Fig. 11 pH value response for $t=0\div15$min without any regulation, for $t>15$min PFC control algorithm with assumption of full information accessibility – ideal case.

PFC algorithm was tested in similar way as GMC, but in ideal case when full information was accessible as well as partial information accessibility individual tuning was required. Performance of control algorithm when full information was accessible (fig. 10 and 11) was similar to performance when partial information was accessible (fig. 12). Algorithm utilizes linear model that is why it uses only part of accessible information and is less sensitive to parameters change.

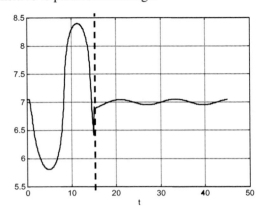

Fig. 12 pH value response for t= 0÷15min without any regulation, for t>15min PFC control algorithm with assumption of partial information accessibility – realistic case.

6. CONCLUSIONS

This paper presents algorithms applicable in industrial programmable controllers. Key issue was to find simple pH model suitable for control algorithm utilized directly in algorithm or after linearization. In both cases simple form of control law was obtained. In ideal case it was possible to assure stability and sensible good performance only for GMC algorithm. Considering partial lack of information individual tuning was required but also good results were obtained. Tested algorithms are simple and do not require short sampling time. It is well known that control quality strongly depends on measurement accuracy and delays in control loop. Propose approach requires model which can be easy fitted to titration curve taken from real world process. In present stage of researches results are very promising and fully applicable to programmable commercial available controllers.

7. ACKNOWLEDGEMENTS

This research was supported by the National Committee for Scientific Research (KBN) under Grant No. 8-T11A-001-16

8. REFERENCE

Gustafsson T. K. and Waller K. V., (1983) Dynamic Modeling and Reaction Invariant Control of pH, *Chemical Engineering Science* Vol. 18, No.1, pp. 389-389.

Gustafsson T. K., Skrifvars B. O., Sandstrom K. V. and Waller K. V., (1995) Modeling of pH for Control, *Ind. Eng. Chem. Res.*, Vol. 34, pp. 820-827

Henson, M.A. and Seborg, D.A., 1990, Input-output linearization of general nonlinear process. *AICHE J.* 36, pp. 1753-1757.

Jutila, P.and P. Jaakola, (1986) Tests with five adaptive pH-control methods in laboratory. Helsinki University of Technology, Report 65.

Kulkarni, B. D., S. S. Tambe, N. V. Shukla and P. B. Deshpande (1991), Nonlinear pH control. *Chemical Engineering Science*, Vol. 46, No.4, pp. 995-1003.

Lee P.L. and G.R. Sullivan. 1988. "Generic model control (GMC)." *Comput. Chem. Engng.*, 12, 573-580.

Lewis R.W. (1996) *Programming industrial control systems using IEC II3I-3*, The Institution of Electrical Engineers Michael Faraday House, United Kingdom

McAvoy T. J., (1972) Time Optimal Ziegler-Nichols Control, *Ind. Eng. Chem. Process Des. Develop*, Vol. 11, No. 1 (1972)

Menzl, S., M. Stuhler and R. Benz (1996), A self-adaptive computer-based pH measurement and fuzzy-control system. Wat. Res. Vol. 30, No. 4, pp. 981-991.

Metzger M. (2000) A comparative evaluation of DRE integration algorithms for real-time simulation of biological activated sludge processes. *Simulation Practice and Theory*, vol. 7, No.7, pp.629-643.

Nystrom, R. H., K. V. Sandstrom, T. K. Gustafsson and H. T. Toivonen (1998), Multimodel robust control applied to pH neutralization process. *Computers chem. Engng* Vol.22, Suppl., pp. S467-S474

Piovoso, M. J. and J. M. Williams (1985) Self-tuning pH control: a difficult problem, an effective solution. *InTech.*

Richalet J., Rault A., Testud J.L., Papon J., (1978), "Model Predictive Heuristic Control: Applications to Industrial Processes", Automatica, Vol. 14, pp. 413-428.

Richalet J., (1993), "Pratique de la commande predictive", Hermes, Paris

Riggs, J.B. and Rhinehart, (1990), Comparison between two nonlinear process-model based controllers. *Comput. Chem. Engng.* 14, 1075-1081.

Stebel K. (2001), Input-output linearization and PI control algorithms applied for pH process, *Proc. Of the 7th IEEE International Conference*, Vol 2 Poland Miedzyzdroje

Wright R. A. and C. Kravaris, (1991), Nonlinear control of pH processes using strong acid equivalent. *Ind. Eng. Chem. Res.*, Vol. 30, No. 7 pp. 1561-157

Ylen, J. P. (1997) Practical aspects of self-organizing fuzzy controller (SOC) implementation. *6th IEEE International Conference on Fuzzy Systems.* Barcelona, Spain.

ANALYSIS, MEASURING AND MODELING OF LOWER ARTIFICIAL LIMB MOTION

Tiefenbach, P., Soušková, H., Václavík, L., Novotný, J.

Vysoká škola báňská - Technical University of Ostrava
Faculty of Electrical Engineering and Computer Science
Department of Measurement and Control
17. listopadu 15, 708 33 Ostrava - Poruba
Czech Republic
tel: +420-69-6994290
e-mail: { petr.tiefenbach | hana.souskova }@vsb.cz

Abstract: The project answers the purpose of design of electronically controlled lower peg-leg that will make it possible to respond to change of handicapped person's pace of walking. In this project we deal with kinematics of lower limb when walking, also with angular deflections of individual joints, with reaction forces distribution and pressure effects within knee joint and on sole in particular phases of the step. Knee joint of the limb control is carried out using a pneumatic cylinder. Within this part speed and quantity of flowing media is regulated by means of needle valve. Mobility of needle valve is regulated by step motor using servo unit. The design of mathematical model stems from laws of gas flow, from law of thermomechanics and equation of motion of pneumatic cylinder piston. Simulation of system model is implemented in Matlab & Simulink environment. The simulation results in determination of such a system parameter that defines valve permeability within pneumatic cylinder in pursuance of measured data. *Copyright © 2001 IFAC*

Keywords: mathematical model, analysis, measuring transducer, measured values, hydraulic piston.

1. INTRODUCTION

For lower limb replacement are used prosthesis, that make to possible of handicapped persons to walk, to run, to ride bicycle and so to include to the standard life. Every handicapped is not possible to perform the whole activities. The limitations are given of patients' possibility and demands and potentialities of prosthesis.

We can separate the prosthesis intended to the patients with above knee amputation according to type of used knee joint:
- uncontrolled
- mechanical joints
    - combined (pneumatic, hydraulic)

- controlled (microprocessor)
  - pneumatic actuator
  - hydraulic actuator

We are got to the two levels by the controlled prosthesis. There are knee joints that are possible to control the process of walking by the very fine mechanism. Practically it means that the prosthesis not stretched out wildly but the whole process is nonlinear (damped – controlled). The next options are the controlled prosthesis that has got it electronic kernel with sensors, controlled hydraulic or pneumatic damper. The sensors pursue recognition of walking style and state. The passing speed of media by inflexion and extension is controlled by the semi-permeable valves. The valves are controlled by actuator in the electronic kernel. As the controller is used stepping motor or direct current motor or an electromagnet.

2. THE MOTION ANALYSIS OF THE LOWER LIMB

By the controlled above-knee prosthesis it is necessary to copy, or to reach a process with a standard walking character. An example of an ideal knee motion character is presented on the Fig. 1. In this picture the x-coordinate describes a part of one step in percents and a y-coordinate describes the knee inflection angle from the stretched position.

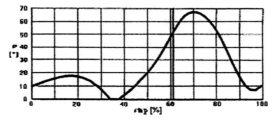

Fig. 1. The kinematics of the knee joint motion.

The basic split and walking check is deduced from a double step. It is possible to divide each step in two parts:
a) standing phase
b) swinging phase
The standing phase of a step is that one, when the sole is in contact with a bed. That means the moment when the heel becomes contact with the bed to the moment of a land return of a sole tip. The rest of the time, when the leg is moving in the air, is the swinging phase. By standard walking the standing phase represents 60 % share, and the swinging phase 40 % share of the whole step time.

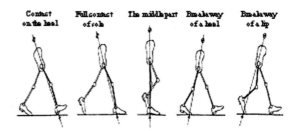

Fig. 2. The standing phase process.

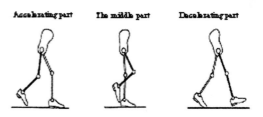

Fig. 3. The swinging phase process.

2.1 Detailed description of one walking step.

From the undermentioned graphic dependencies the positions of each joint during walking are evident very good. In the graphs the angular turning dependence of each joint on the actual position, let us say on the partial walking point, is shown.. The graphic dependence matches one's period of walking that means one's double step. This double step is repeating on both legs, but with a time drift. When the standing phase begins at the left leg, then the right leg is in the swinging phase.

2.2 The bed reaction.

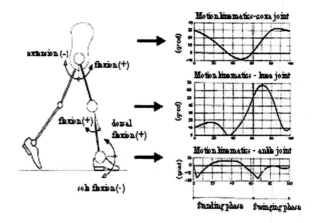

Fig. 4. Motion kinematics of a lower limb.

At walking or running happens, owing to physical laws, to a phenomenon that is named bed reaction on sole and a whole leg. Essentially it is the weight transfer on one leg or on the other leg. At a full tread onto the bed then happens absorbing of those reaction forces in joints of the leg. Out of this force also the inertial forces express oneself that result from the man's movement.

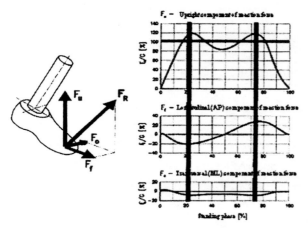

Fig. 5. Bed reaction.

3. WALKING SPEED ANALYSIS BY SCAN FROM THE SOLE

The whole metering is found on placing two groups of sensors. The sensors inform about the force of which the body works onto the bed. The most important moments for the metering are: the heel touch on the bed at the beginning of the standing phase and the tip breakaway by a passing to a swinging phase. This time interval indicates, after recounting to the sole length, at what speed a man is moving. Among others, that gives a possibility to analyze, without problems, whether the moving is pointed, if forward or to the back.
The placing of the scanpoints follow from the Fig.6. The used sensor type: FSR - 151

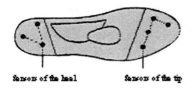

Fig. 6. The sensor lay-out on the sole.

It was necessary to put together the sensor operating system, data reading and their plotting. For this object necessary software was programmed

and hardware designed (see Fig. 7.). On this picture follow the block diagram of the whole measurement system. The software itself was programmed in Delphi environment, version 5.

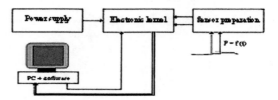

Fig. 7. Measurement system.

3.1 Measured data.

Several series of measurement has been done:
a) according to the walking type (quickly, slow, very slow)
b) according to the sensor placing (on the shoeless foot, on the shoe with rubber sole)
c) according to the bed type (linoleum, carpet)

In the Fig. 8. and 9. you see the measured data with the sensor placed in a shoe by walking on linoleum.

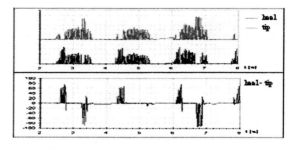

Fig. 8. Very slow walking.

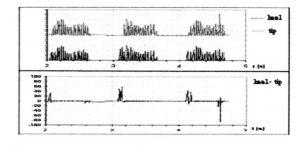

Fig . 9. Quickly walking.

Description of the graphic processes:
1) The processes heel and tip are a time record of the pressure difference onto the given sole points. The values on the y-coordinate agree with the bit-number after AD transmission, when 2 V = 128.

For this metering are the amplitude values unsubstantial and further they will not be mentioned. Important is the timing process on the heel and tip of the sole, eventually the difference between them.

2) The heel-tip process is a projection of load differences between heel ant tip of the sole. Such graphic process is more preferable for a rating of the standing phase. It is possible to separate the values on:

- positive values - fully treading on the heel, heel reaction
- negative values - the weight transfer on the tip of the foot, tip reaction.

3.2 Method evaluation.

According to staging graphic processes it is perceptible, that the speed change of walking will express itself in a time response among tread fully onto heel-heel, if need tip-tip. For more accurately metering that is more suitable to use comparison heel-heel, because here happens a sharper load growth.

A second variation how to determine the walking speed is to find out a time-lag among tread fully onto heel and that moment, when all load devolves on the sole tip. Resembling problem solution like timing the whole standing phase time is possible.

This data it is possible to compare only behind conditions, where the same parameters are preserve.

4. WALKING ANALYSIS BY METERING OF KNEE-JOINT BEND

For transmission of angle change or angle size, it is possible to use a lot of sensors that might be of analog or directly digital exit in standard form. In this proposal a solving with an analog output is used.

For this type of angle transmission and angle plotting in an interval $0° - 360°$ is very acceptable to use as an inverter a classical potentiometer with linear course of resistance changeover. With a potentiometer, that is connected up like a knee angle bend sensor, it is possible in every moment to find out instantaneous angle value and without problem determine, whether the leg in knee is bending or straightening. Then it is possible to say, that to every value of resistance it is possible assign a size of angle, that indicates, how the knee is flexed. Block diagram of an angle displacement

potentiometer sensor with transmission onto a digital signal, is displayed on following Fig. 10.

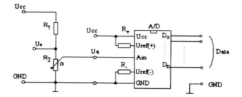

Fig. 10. Wiring diagram of angle turning sensor.

Voltage at the max. deviation:

$$Ua_{max} = Us = \left(\frac{R_2}{R_1 + R_2} \right).Ucc \ [V]$$

5. WALKING ANALYSIS BY PRESSURE MEASURING BELOW THE DAMPER PISTON IN KNEE

For the pressure measuring a sensor PX71-060AV was used. It is founded on a bridge measurement, when in its two branches they are wiring resistive tensiometers that balance the bridge. The sensor is placed in the mantle of the pneumatic cylinder so that it can measure the pressure below the piston, namely at small stroke too.

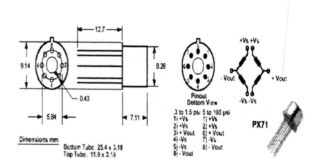

Fig. 11. Wiring of sensor series PX70.

Catalog values are mentioned in (Václavík, 2001).
For pressure control below the piston in the pneumatic cylinder a throttle-valve is used, that is of service for passing as from the space below the piston into space behind the piston. The help of stepping motor handles the valve himself. To further signal processing that agrees with the pressure on the sensor entrance, it is necessary at first amplify the analog value and then transfer it to digital. The plotting himself transacts already the single-chip microprocessor.

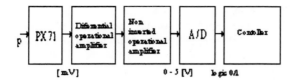

[mV]   0 - 5 [V]   logic 0,1

Fig. 12. Wiring block diagram for signal
modification from sensor PX71-060AV.

6. CONTROL CIRCUIT OF THE LOWER LIMB PROSTHESIS

The control circuit is the most important part of the
whole electronic proposal, because in it all the enter
data from the sensors are plotted. As an actuator
serves a needle valve, operated by the help of a
stepping motor. This needle valve is an instrument
that supports the controller to hit into the controlled
system. The whole regulation works in a pneumatic
cylinder. The schematic drawing of the cylinder is
on the Fig. 13. Here, by the help of needle valve,
the speed and the quantity of the passing medium is
regulated. Mobility of a needle valve is regulated
by a step motor using a servo unit.

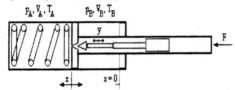

Fig. 13. Schematic drawing of the pneumatic
cylinder.

p - is the pressure value in the given space
V - is the actual volume of medium
T - is the medium temperature
x - position is given by the leg bend in the knee
joint and is characteristic for the piston locality

6.1 Operating model in the user-own Matlab environment.

A mathematical system that goes out from
application of laws describing gas flow and
thermomechanics was set together. The simplified
model of this system goes out from the fundamental
motion equation. This equation is set together on
the basis of force incidence inside the system and
the forces operating external. Schematic drawing of
the system is on the Fig. 14.
The motion equation has the following form

$$F = m . \ddot{x} + b . \dot{x} + k . x$$

F [N]   = the prosthetic force against the
                    pneumatic cylinder
m [kg]   = piston mass
b [Nsm^{-1}] = damping factor
k [Nm^{-1}] = stiffness of spring
The external force operates against three forces,
that are Fk = spring force, Fs = inertial force, and
Fb = damping, evoked through speed control of the
passing medium from the space in front of piston to
the space behind it.

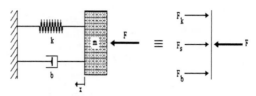

Fig. 14. Physical system description.

The resulting relation among the forces has the
following form:

$$F = F_s + F_b + F_k$$

The most interesting force for the regulation is the
force Fb that is controlled by throttling the valve.
From here results, that it is a change of the damping
parameter b. Transmission of such system then is
written in a form:

$$G(p) = \frac{X(p)}{F(p)} = \frac{1}{m.p^2 + b.p + k}$$

During the regulation the parameter "b" is
changing.   The parameter b depends on the
throttling size of the transfer valve. It is possible to
write: b = a / f(y), where a = constant, y = needle
insertion into the valve slot. The smaller y and the
passing quantity of medium, the greater the
damping. The proposal spring is defined by the
stiffness. The stiffness parameter "k" is used.   In
the proposal case the spring parameter has the value
k = 2130 Nm^{-1} . The parameters "m" and "b" are to
be preset according to the prosthesis parameters
and according to the man for whom the prosthesis
is prepared. The determination of the parameters is
performed for elected values, as it were a virtual
patient.   In a model case the counted relations
express single invariable, after the system has to
behave itself. By the help of the aperiodic test
computation, the constants can be proposed in such
way, that the system has a minimal overshoot, and
smallest damping. By setting up of this parameters
a standstill time of the system can be changed to a
fixed value. For the system simulation in the user-
own environment Matlab&Simulink, is the system
set together in this form:

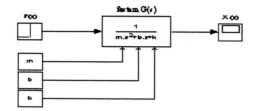

Fig. 15. Model of the system for transmission response assignment.

For an aperiodic system is accepted:

$$k > 0, \ b > \sqrt{4.m.k} \ .$$

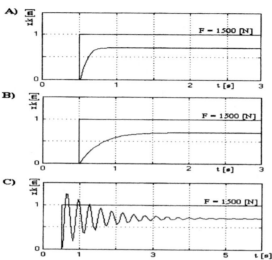

Fig. 16. Step responses of the system.

Description of graphic relations:
A) system on aperiodicity limit

$$b = 92,32 \ . \ \sqrt{m} \ , \text{ set for } b = 206,41 \text{ Nsm}^{-1}$$

B) system is aperiodic for $b > 92,32 \ . \ \sqrt{m}$, elected $b = 1000$ Nsm^{-1}

C) system is damped oscillate for $b < 92,32$

$\sqrt{m}$ elected $b = 10$ Nsm^{-1}

Adjusted parameters of a virtual patient and of the system are:

- weight ... m = 5 kg
- spring stiffness ... k = 2130 Nm^{-1} (correspond with the used spring)
- input power ... F = 1500 N
- damping ... b is elected for single causes

The characteristics are projection of system step response. The unit step is created by the help of power F. Its amplitude in steady state is 1500 N. It is the unit step from 0 N up to 1500 N. For a better illustration the unit step is moved to the time 0,5 sec. The coefficients are proposed for such regulator, that regulates medium transmission from the space under the piston to the space behind it only by flexion. By extension it is a matter of regulation based on the angle of knee bend.

7. CONCLUSION

In this project the motion analysis of the lower limb is maintained. Some measuring methods for walking speed were described: walking speed measuring by the help of sensors on the sole, the measuring way of the knee joint bend and the measuring way of the pressure under the damper piston in the knee joint of the prosthesis. The first method is for the prosthesis operating not usable. The reason is the great dependence of the measured values on the outside conditions. This project recommends to use for the regulation of speed flexion the method of pressure measuring directly in the knee joint. Therefore this project solves simulation of such system. The system is created as a pneumatic cylinder. Within this part speed and quantity of flowing media is regulated by a step motor.

This project will serve as a theoretical groundwork for construction of a functional prosthetic model for laboratory purposes.

Support for GACR project "Pilot extensible assistance framework for control and information systems development" (No 102/01/0803) is gratefully appreciated.

REFERENCES :

Valenta, J.(1985). *Biomechanika*, Academia Praha
Noskievič, P.(1999). *Modelování a identifikace systémů*, Montanex a.s., Ostrava .
Václavík, L. (2001). *Diplomová práce*, VŠB-TUO, Ostrava.
In: *Ortopedická protetika*. ING corporation, s.r.o. Year 2, 5/2000 and year 3,9/2000. Ostrava.

IFAC

Publications
www.elsevier.com/locate/ifac

HC11 CONTROL UNIT APPLICATIONS

Vladimír Vašek, Manh Thang Pham

Faculty of Technology Zlín, Tomas Bata University in Zlín
76272 Zlín, Czech Republic

Abstract: The simply control unit on the basis of the single chip microprocessor Motorola
HC11 was created. This unit is applicable especially for the small technological processes
and in the laboratory conditions. The control unit is equipped by the real time running
software with the possibilities to display important information either on the small LCD
display (two rows, 16 columns) or by the help of the serial communication to send the data
to the high level computer system to display them there. Several control algorithms is
possible to use for the automatic control. The practical applications are shown on the control
system for the injection machines heating unit and the units used in the distribution control
system of the tannery wastes processing technology. *Copyright © 2001 IFAC*

Key words: Control unit, computer applications, control algorithms, hardware, software,
real time.

1. INTRODUCTION

Both described applications are based on the
hardware and system software, which were presented
on the last year workshop. There is used hardware
with the possibility to work with 8 digital inputs in
the area 0 - 5V, 8 digital outputs in the any area
(relays) and 8 analogue inputs several kinds. In the
special cases it is possibility increase the number of
digital inputs (by 3) and outputs (by 5). The
microcontroller Motorola HC11 in the minimal
configuration equipped by the small keyboard and the
two rows display is used as the basic part of the
control unit. The microcontroller is running in the
expanded mode with the 32kB RWM and 32kB
EPROM memory including 512 bytes EEPROM
memory. Discrete outputs are used as the
manipulated values generators with the wide impulse
modulation. The special attention is given to the
analogue inputs. They are solved generally and by the
help of switches it is possible to change the mode of
their activity. There are following possibilities

♦   analogue input in the range 0 – 5V

♦   analogue input in the range 0 – 10V
♦   analogue input for temperature sensor KTY
♦   discrete input TTL

The software system was built in the HC11
assembler. For the real time running of the program
system there was used special pre-emptive real time
operating system RTMONHC11 (Vašek, V., 1994),
which was built for the using of the monitoring and
control system HC11 application for not too large
technological processes. (Klofáč, 1995). It allows
multitasking of defined number of processes. User's
programs are structured on the basis of the priority
hierarchically. The choosing of the program, which
will be running on the processor, is carried out on the
basis of its priority level. If there are more programs
with the same priority, the processor is assign
periodically for the time of defined account of time
quantum. There is used one time quantum as 200ms
in this system. By the help of only one interrupt
signal - timer interrupt - this simple priority system
allows to fill many requirements for the real time
processes running. The special services are used for
program modules programming to run in the real

time. For the correct data moving between several processes in this system there are used special means - the boxes and the reports.

2. HEATING UNIT APPLICATION

There was created the simple and cheapen heating and cooling system for using in the field of plastic's production by the injection technology and other industrial applications. For the high quality of the final products there is necessary to keep both productions' parameters and exact sequence of production steps. One of the most important production parameter in the field of plastic production processes is often the temperature. Mostly there is used medium heating and cooling system in the praxis, where the heat transfer is carried out by the help of the water or the oil. In the case of the water using especially in the case of the cooling most of used systems are open, e.g. the water after the way through the form leaves the system. In this time the price of the water is increased and that is why was built closed system.

Principle of the heating and cooling unit there is shown on the Fig.1. (Vašek, V., 1996). The basic part of this system is the reservoir 1 fulfilled by the water or the oil as the warmexchanging medium. Medium in this reservoir is heated up by the help of electrical heating element 2 and cooled by the help of water cooling system 3. The pump 12 draws the medium to the hydraulic circuit with the injection form 11. For the temperature and level of the medium control there was built the microprocessor control unit 10.

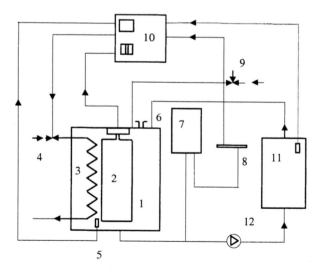

Fig.1 Scheme of the heating and cooling unit

In the case of the loss the medium level, there is opened the valve 9 and the level is increased.
The main tasks of the control unit are as follows:

- the temperature in the range 20 - 140 °C measurement and control
- the level of the heat transfer medium measurement and control
- temperature sensors signal processing
- running in the real time
- the basic user communication
- the basic information monitoring

The linear discrete PID algorithm, which can be written in the form (2) is used for the calculation of the manipulated values. The right function of the temperature control program system was verified on the real heating process. The mathematical description of this process is in the form of the transfer function

$$G_s(z^{-1}) = \frac{0,555z^{-1} + 0,204z^{-2}}{1 - 0,812z^{-1} + 0,044z^{-2}} \qquad (1)$$

The following controller transfer function was calculated by the dynamic inversion method

$$G_R(z^{-1}) = \frac{0,290 - 0,228z^{-1} + 0,014z^{-2}}{1 - z^{-1}} \qquad (2)$$

Time period of the discrete control circuit was used 30 sec. On the Fig.2 there is shown the temperature course as the required value change response.

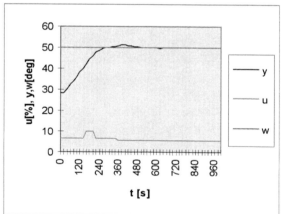

Fig.2 Real temperature control course

Very good results in the area of the temperature control we obtain also with the penalised three point controller, because of the temperature unit has the possibility to generate negative manipulated value – water cooling.

The software system was built in the assembler HC11. For the real time running of this application there was used real time operating system RTMONHC11. The four bytes mathematical function library there was used for the real number calculations.

Program system includes the basic part of the real time operating system and the next 17 processes, which are running in the multitasking mode. The structure of the main processes is shown on the Fig.3.

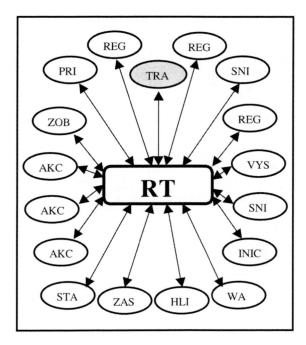

Fig. 3 Program system scheme

Initialisation process INIC defines data structure of the real time operating system and fills the program variables by the initial values. The process SNI for the binary and analogue inputs periodically reads the actual values of the continuous and discrete input values. The control algorithms processes REG calculate the manipulated values for heating or cooling and manipulated members processes AK realise them by the help of the wide modulated binary output. The visualisation process ZOB allows to write the most important information about control running to the small display. The user commands PRI, start STA and stop ZAS processes allow the hand inputs for the choice of the control system running mode. Process control supervisor HLI is the program for the periodical watching of the main technological parameters of the temperature control unit (temperature range, heating medium level, timeouts after the system start, cables connection etc.). Binary outputs VYS program allows periodically to send any combination of the logical „0" or „1" to eight bits of the output port and this way to control two level power signals (except the heating or cooling). WA process realises in the co-operation with the Watch Dog system of the HC11 hardware supervising of the right program system running. To this program system was in this year included one new program process called TRANSMIT (TRA), which enables to communicate in both directions

with the up level injection machine control system. This process is started periodically with the time period 0,5s. After the each starting process TRANSMIT is generate synchronisation sequence, which is represented by character string "TRANSMITSTART". Then is generated data block in the following structure

temperature - required value	- 4 bytes
temperature - measured value	- 4 bytes
positive manipulated value – heating	- 4 bytes
negative manipulated value – cooling	- 4 bytes

For this data block connected with the synchronisation sequence is generated check sum and it is added to the end of data block. That is the whole message prepared for communication to the up level control system. After the data block generating the own communication is started by the help of the serial line using RS232 interface. Then is started communication in the opposite direction. The message structure is similar. The difference is in the data content. From the up level control computer is send to the HC11 in this application only required value of the temperature. The format of this data has to be again in the 4 bytes mode.

3. TANNERY WASTES PROCESSING CONTROL SYSTEM

It is estimated those up to 25% of solid tanned wastes are arising during production of chrome/tanned leathers. (Kolomaznik, 1994, 1996) In the past, most of these wastes were transported on dumps in determined places. Conditions of enzymatic hydrolysis were determined in the laboratory scale, according to which it was possible to get enzymatic hydrolyzate almost without chromium content (less than 5ppm), while the chromium filtering cake could be recycled. Enzymatic processing of chromium-containing leather waste, chrome shavings as well as splits and trimmings are hydrolyzed enzymatically under alkaline conditions to obtain a high-value, high molecular weight gelable protein fraction, a hydrolyzed protein fraction and a recyclable cake. (Janáčová, D., 1998). Because there is concern that the character and quality of protein products would have adverse effect on marketing of this product, we especially control proportions of alkali agents and the reaction temperature.

The chromium tannery wastes processing technology is complicated and includes three relative independent parts. The first is fermentation reaction, its product is filtered by the special conditions and the last part of this technology is drying to the final product.

Described technology is now realised in laboratory conditions in our department and at the same time is tested in industry conditions in Hradek nad Nisou. In the laboratory conditions there is solved also microcomputers control system. Each part of the technology process has its own control system for the direct digital control of the physical values as a temperature, pressure, reaction velocity etc. The control units for this level are solved on the microcomputers basis with the Motorola HC11. On this computer controllers there is running software in real time conditions, structure of the application programs is as follows in the next part of this paper. In this time there is possible full using of the first level control of the fermentation and filtration processes (Vašek, V., 1999).. For the last part was not solved this level control system when the real technological equipment was not supplied.

The second level of the control system is solved on the basis of the personal computer. This supervisory computer allows

♦ to insert control parameters for the control loops
♦ to watch by the help of the visualisation software the whole technological process
♦ to co-operate with the standard Windows applications products to supervise all parts of the technology (Fig.4) and generate production protocols.

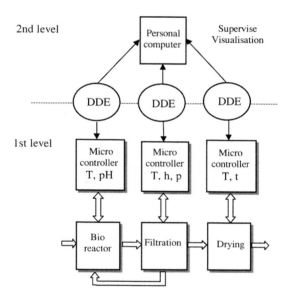

Fig.4 Technology and control system configuration

Bioreactor control subsystem allows to keep required values on the constant level. In our case there are temperature in the reactor and the pH value of the reactor contents. Individual kinds of enzymes have different activation at distinct temperatures. The universal opinion is the fact that their activity is usually obvious in the temperature range of 10 - 80 C. As for that during fermentation there do not occur extreme changes of temperature conditions, the temperature automatic control can be realised by

means of usual algorithms. Because of the existence of the acidophilic microorganisms (pH 0 - 7) and the alkaliphilic ones (pH 7 - 14), the microorganisms used at the actual technology are necessary to be provided with the optimum pH value. It is need to be currently measured and controlled up to the demanded value with the adding of suitable acids and alkalises. For the successfully fermentation processing is mostly necessary to control also next following quantities - oxygen quantity, foam level and mixing speed. Technological equipment includes own bioreactor with temperature unit, converter unit and microcomputer. (Fig.5)

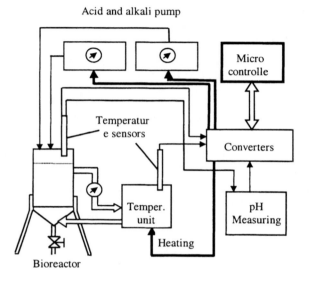

Fig.5 Bioreactor control subsystem scheme

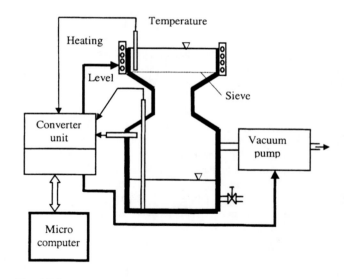

Fig.6 Filtration control subsystem scheme

In the case of the filtration it is necessary to keep the high level of the permeability. That is why the whole filtration equipment is heated and the vacuum degree is very sensitive controlled. The principal scheme of this subsystem is on the Fig.6.

For these subsystems there were built real time software. Application programs of this software system allows the measurement of the physical quantities, counting of the manipulated value and its output back to the controlled system i.e. bioreactor or filtration system. Co-operation rules of the application programs with the real time system RTMON in both aplications – bioreactor and filtration we can see on the Fig.7.

For the second level of the control system has been used commercial software system for monitoring and visualisation InTouch, which allows high level of the user up-to-date facilities for the both commands insert and technological process state watching in the real time. This system is running on the personal computer and for its communication with the first level control system are used the DDE (Dynamic Data Exchange) servers.

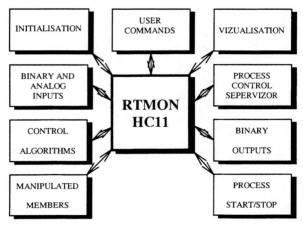

Fig.7 Application programs structure

DDE servers ensure the periodical communication between both levels of the whole control system. (Vašek, V., 1998). System InTouch is running like a Microsoft Windows application and with the help of the DDE protocol is its communication with others Windows applications very effective and automatic. Like a DDE server for the two levels communication was created special program Connect which is using Widows standard function. This program allows initialisation and running of the communication.

4. CONCLUSION

In this contribution there is described the simple and cheapen microcomputer control unit MRS-32 and some its applications in the area of the monitoring and control of the technological processes. There is described heating unit control system and the bioreactor and filtration process control systems. In the microprocessor control system solution there is

included both the hardware proposal on the basis of Motorola microcomputer and the real time running software system pleasant for the user. The described control unit is solved as the multipurpose device. It is possible to use it as the general control unit for maximal eight analogue inputs 0 - 5V, eight binary inputs and eight wide impulse modulated outputs/binary outputs. These possibilities are supported by the developed program system. Using this control unit it is possible to solve the control system as a decentralised system - like the two level system. The first (low) level is based on the microcomputers HC11 and its task is the direct digital control of the continuous variables. The second (high) level works in the Windows environment and is represented by the Wonderware's InTouch professional software system for the visualisation of the technological processes. For the periodical communication between both levels has been developed special program on the basis of the DDE servers. Presented system is running in several types of briefly described applications.

REFERENCES

Janáčová, D., Kolomazník, K.& Langmaier, F., et. al. (1998). Industrial Treatment of Chrome Tanned Solid Waste. In CHISA 98 - CD-ROM of full texts. Prag :

Klofáč, J. (1995). Operační systém pro práci v reálném čase pro mikroprocesor HC11. (Diploma work.), FT Zlín

Kolomazník, K. (1994). The Study of Possibilities Enzymatic Treatment of Chrome Leather Wastes. (In Czech). Research Scientific Paper No. 9404, Zlín, Czech Republic

Kolomazník, K., (1996). Non-Ammonia Deliming of the Cattle Hides with Magnesium Lactate, *J.Am.Leather Chem. Assoc., 89,*

Vašek, V., Langer, P. and Janáčová, D. (1998). Communication in the Distributed Control System. In: *Proc. of 9th International DAAAM Symposium*, TU of Cluj-Napoca, 487-488, ISBN 3-901509-08-9, Romania

Vašek, V. & Pokorný, P.(1999). Tannery wastes processing distributed control system In.: *Proc. Of Process Control 99*, University Bratislava, 400-403, ISBN 80-227-1228-0, High Tatras, Slovak Republic

Vašek, V., Sysala, T. and Navrátil, Z. (1996). Temperature control unit. In: *Proc. of 7th International DAAAM Symposium*, TU of Vienna, 455-456, ISBN 3-901509-02-X, Austria

Vašek, V. (1994). Mikropočítačové monitorovací a řídicí systémy. (Hability work), Zlín.

IFAC

Publications
www.elsevier.com/locate/ifac

DIGITAL CIRCUITS EMULATOR

L. Jacobo Álvarez Ruiz de Ojeda

Dpto. Tecnología Electrónica
University of Vigo
Lagoas (Marcosende) s/n
36280-Vigo (Spain)
Email: jalvarez@uvigo.es

Abstract: The equipment proposed in this paper is an FPGA based Digital Circuits Emulator that permits to test the majority of the standard TTL circuits without the need to mount or wire any of them. This equipment provides an easy way of teaching the Digital Electronics basic concepts, complementing the theoretical explanation of the circuits behaviour. *Copyright © 2001 IFAC*

Keywords: Arithmetic circuits, Combinational circuits, Sequential circuits, Digital circuits, Electronic systems, Programmable.

1. INTRODUCTION

In many countries, the basic Digital Electronics laboratory teaching is still based in the same primitive methods: to mount and wire the old 74xx TTL circuits (Mandado, 1998). Then, when the students are familiarized with these circuits, they start working with Programmable Logic Devices and CAD tools.

The equipment proposed in this paper is an FPGA based Digital Circuits Emulator (see figure 1) that permits to test the majority of the standard TTL circuits without the need to mount or wire any of them. This is because the FPGA is configured to contain every circuit and to select them.

The basic Digital Electronics is covered through the emulation of many common digital circuits, as well combinational, as aritmethic and sequential, without any mounting nor wiring.

Not only the basic TTL functional blocks are included, but some complete synchronous sequential systems that the students must program to do any state graph they want. This help them to understand the operation of the sequential circuits.

In addition to the Emulator function, this equipment is useful when the students are already familiarized with the standard functional blocks because with Xilinx Foundation software tools and a low cost programming cable, the equipment serves as an FPGA based application development platform, with an XC4005E (5000 gate equivalent) FPGA included (Xilinx, 1994b).

2. EQUIPMENT OPERATION

This is an stand-alone equipment that can be used without the need of any additional instrumentation (only a wall-mounted 9 V. transformer or a 5 V. DC power source).

The operation is very simple. Once the equipment is powered up, the FPGA downloads its configuration automatically. Then, the user can select the circuit he wants to test through some switches (X5, X4, X3, X2, X1 y X0). This selection is validated when the SEL button is pushed. The switches X5 and X4 select the type of the circuit being tested among combinational (state 00), arithmetic (state 01) or sequential (state 10). The state 11 of these switches is not implemented. The switches X3, X2, X1 and X0 select up to 16 different circuits of each type.

The selected circuit is shown in the two seven segment displays. In the table 1, included in the following section, some of the emulated circuits are listed, though the Emulator includes many more.

Once the circuit is selected, the user can test its behaviour inmediately, using the switches, push-buttons, LEDs and displays that are mounted in the Emulator board.

It is important to point up that the Emulator includes the possibility of implement some synchronous sequential systems (state machines), both register based and counter based.(Mandado,1998). The user must program the content of the RAM memory, included in the FPGA, to make the sequential system evolve through the desired state graph.

But, besides this autonomous operation, it has been developed a Digital Circuits Emulator software that allows to select the circuits through the parallel port of a personal computer. This software includes basic information about the emulated standard TTL circuits (IEEE symbol, truth table and a brief explanation) that makes the Emulator more complete and adequate for teaching.

At last, it is also important to make the Emulator more useful with the possibility given by the connection of the eight outputs of any emulated circuit to an external system. This is achieved through an output amplifier and the output connector. In this way a simple external system can be controlled by the emulated circuit (i.e. a sequential system).

3. EXAMPLES OF EMULATED CIRCUITS

Table 1 contains a few examples of some circuits of each type that this equipment can emulate.

Table 1. List of some emulated circuits.

COMBINATIONAL CIRCUITS

3 input AND/NAND gates.

1 among 8 decoder with three inhibition inputs and inverted outputs.

8 channel multiplexer with inhibition input and normal and inverted outputs.

8 bits priority encoder with inhibition input and propagation outputs.

ARITHMETIC CIRCUITS

Binary comparator with propagation inputs.

Parity generator/checker.

4 bits adder/substracter with carry/borrow input and output and overflow output.

SEQUENTIAL CIRCUITS

RS, JK, D and T synchronous flip-flops with asynchronous reset.

8 bits register with inhibition input and asynchronous reset.

4 bit synchronous binary up/down counter with synchronous parallel load and inhibition input.

4 bit synchronous left/right shift register with synchronous parallel load, inhibition input and asynchronous reset.

16x4 RAM memory with write enable input.

User programmable, register based synchronous sequential system (state machine).

4. EQUIPMENT DEVELOPMENT

Although almost any expert designer knows how to use any of the basic digital circuits in the context of a particular application implemented with programmable logic, it is not so easy to combine them all in one FPGA to achieve this emulator.

Of course, the first thing to have in mind is the logic capacity of the circuit to be chosen. This is a compromise between logic density and cost. The main restriction in this Emulator was to obtain a low cost equipment, so a capacity of about 5,000 gates was initially selected.

But the internal architecture is also important. All the emulated circuits in this equipment are operated through the same 20 switches and 4 pushbuttons and the state of their outputs are checked through the same 10 LEDs.

So, the matrix-based architecture of the Programmable Logic Devices (PLDs) is not suitable and a Field Programmable Gate Array (FPGA) was chosen.

The XC4005E (5000 gate equivalent) FPGA from Xilinx (Xilinx, 1994b) was the ideal solution. The possibility to implement three-state buses inside the FPGA was definitive to combine the outputs of every emulated circuit into the common outputs of the equipment.

Another challenge was the programmability needed in some sequential circuits. The RAM memory and, mainly, the synchronous state machines must be programmable by the user to implement any graph and so, to teach their internal operation.

The solution is multiplexer based and it leads to two different working modes of the sequential circuits. In the first one, the user only "sees" a memory module, so he can access the address and data inputs to program its content and check it through its data outputs, displayed in the LEDs. In the second one the sequential circuit performs the programmed graph, so the user only has access to the graph inputs and outputs and its internal state through the LEDs.

Finally, it was necessary to implement two different configurations in the FPGA. The first one contains all the basic combinational, arithmetic and sequential circuits and the second one includes the synchronous sequential systems (programmable state machines).

The FPGA was almost used at its maximum capacity in both configurations as the following extract of the implementation report for the first one, confirms:

Number of CLBs: 190 out of 196 96%
CLB Flip Flops: 90
4 input LUTs: 340 (4 used as route-throughs)
3 input LUTs: 60 (20 used as route-throughs)
16X1 RAMs: 4
Number of bonded IOBs: 53 out of   61 86%

Many of the solutions used and the ideas involved in this development are really new, so it was decided to solicitate the equipment patent, that it is now pending.

5. FPGA BASED DEVELOPMENT PLATFORM

One of the inherent benefits of using an FPGA it is the possibility of reprogramming it. So, when the user has learned the basic digital circuits operation, he can design his own applications through a standard CAD tool package, like Foundation Xilinx.

So, if the user has access to the Foundation software (there is a student version (Xilinx, 1998) (Álvarez, 2001)) or the older XACT software and to the Xchecker programming cable (Xilinx, 1994a), he can work with the Emulator board as an FPGA application development board, programming the FPGA through a personal computer standard port.

To test the designs, all the switches, pushbuttons, seven-segment displays and LEDs in the equipment are available, as well as an 8 input connector and the 8 output connector that can be connected to other circuits, including computer ports (i.e. the parallel port).

Fig. 1. Photograph of the Emulator board.

201

6. CONCLUSIONS

As the main conclusion, it is sure that this Digital Circuit Emulator constitutes a complete learning equipment for Digital Electronics, that can emulate many of the basic digital circuits as well as act as an FPGA based development platform.

Although the original idea is really simple, it is a new approximation to the teaching of Digital Electronics that has never been implemented before, so this equipment is being patented.

REFERENCES

Álvarez L.J., *"Diseño de aplicaciones mediante PLDs y FPGAs"* (2001), Tórculo, Santiago de Compostela.

Mandado E., *"Sistemas Electrónicos Digitales 8ª edición"* (1998), Marcombo, Barcelona.

Xilinx, Xilinx Student Edition (1998), Prentice-Hall, New Jersey.

Xilinx, *Hardware and Peripherals Guide* (1994a), San Jose (California).

Xilinx, *The Programmable Gate Array Data Book* (1994b), San Jose (California).

IFAC
Publications
www.elsevier.com/locate/ifac

CASE STUDY : FPGA ACCELERATION OF CRC COMPUTATION [1]

Miloš Bečvář[*] Michal Jáchim[*,**] Martin Jäger[*,**]

*Department of Computer Science and Engineering, Czech
Technical University, Faculty of Electrical Engineering,
Karlovo nám. 13, Prague 2*
** *ASICentrum s.r.o., Novodvodvorská 994, Prague 4*

Abstract: Cyclic Redundancy Check (CRC) computation is widely used in various
network and telecomunication devices. This computation is used as an example
of computing kernel which can be accelerated by FPGA coprocessor. The whole
application was implemented and measured on the common Personnal Computer.
Performance of FPGA implementation was compared to simple and optimized
SW CRC algorithm. FPGA implementation proves to be superior in terms of
speed even with the presence of large overheads of the PCI based coprocessor.
Various components of overhead including the operating system were analyzed
with proposition to improvements. This paper can be used as a case study for
designing of better architectures for reconfigurable acceleration. *Copyright ©
2001 IFAC*

Keywords: Reconfigurable acceleration, Cyclic Redundancy Check, FPGA
implementation, PCI

1. INTRODUCTION

Systems consisting of CPU and reconfigurable
logic (FPGA) have been widely studied as promising architectures for future. These systems balance the efficiency of ASIC implementation together with flexibility of the software. (Hutchings
et al.,1994) It has been proven that for some
specific algorithms the CPU is not always the best
choice. Especially when the CPU instruction set
does not directly support the required operations.
In this case, the FPGA implementation can be
more efficient. Moreover, CPU can be offloaded
and computation can be done by FPGA. With
the growing concern in power consumption the
possibility to power down CPU is also not without
importance.

A case study of one of the possible computing kernels suitable for FPGA implementation
is presented in this paper. Cyclic Redundancy
Check (CRC) is a well-known error-checking code
(Wicker, 1995) that is used in serial data communication systems. Due to hierarchical character
of communication protocols, the CRC is usually
computed multiple times. On the physical level,
Linear Feedback Shift Registers (LFSR) are used.
For the higher levels of protocol stack, CPU is
commonly used for CRC computation. This approach has several disadvantages. The instruction
set of the general-purpose CPU is not well suited
for shift-XOR bit-level operations used in CRC
computation. The second disadvantage is in fact,
that CRC computation needs to process the whole
data stream that may lead to unnecessary data
transfers. This features makes CRC computation
interesting kernel for FPGA acceleration. FPGA

[1] Acknowledgment: This research was in part supported
by grant 102/99/1017 of the Czech Grant Agency.

reconfiguration capability can be used for adaptation to different CRC standard.

The following paper presents comparison of two software algorithms and parallel FPGA implementation. The common Personal Computer is used as a host system for evaluation.

2. CRC ALGORITHM

2.1 CRC background

The theoretical background of CRC is based on division of polynomials over Galois Field $GF(2^n)$. In the common case, systematic CRC codes are used. Transmitted data are divided by the generator polynomial and resulting remainder is attached to the data stream. On the receiving side, the whole data stream is again divided by the same generator polynomial and resulting remainder presents a syndrome of errors that may occur during data transmission.

Various standards exists each of them differing in generator polynomial and interpretation of syndromes (e.g. IEEE Std.802.3-2000, IEEE Std.1596-1992). Usually zero remainder indicates that no error occurred during transmission. The use of non-zero remainder for this purpose is based on adding a constant value to transmitted data. This addition is implemented by presetting of non-zero values in polynomial division unit (e.g. USB, 1995).

Algorithm can be also expressed mathematically

$m(x)$ original message to be send

$g(x)$ generator polynomial of degree n

$t(x)$ transmitted message including CRC

$CRC = m(x) * x^n \% g(x)$

$t(x) = m(x) * x^n + CRC = m(x) * x^n + m(x) * x^n \% g(x)$

$r(x)$ received message

$s(x)$ syndrome

$s(x) = r(x) \% g(x)$

The following study considers generator polynomial g(x) (commonly used in IEEE Std. 802.3) :

$g(x) = x^{32} + x^{26} + x^{23} + x^{22} + x^{16} + x^{12} + x^{11} + x^{10} + x^8 + x^7 + x^5 + x^4 + x^2 + x + 1.$

2.2 Trivial SW implementation

Trivial software implementation corresponds to direct bit-serial implementation of remainder computation in SW. This implementation is not using the capabilities of CPU very well (Boudreau

et.al, 1971). We will denote this implementation as *slowcrc*. The following C code fragment illustrates the main idea of trivial algorithm. Data are shifted through 32-bit register until the MSB is one. When the MSB is one, polynomial is "subtracted" from the register. Subtraction is the simple XOR operation for $GF(2^n)$.

```
reg = 0;
p = array;
cnt = cnt/4;
while(cnt-->0) {
  l = *p++;
  for(i=31;i>=0;i--) {
    reg = (reg<<1) | ((l>>i)&1);
    if((reg&0x8) != 0) reg ^= POLYNOMIAL_WO_MSB;
  }
}
/* flush */
for(i=0;i<SIZE_OF_POLYNOMIAL;i++) {
  reg = reg<<1;
  if((reg&0x8) != 0) reg ^= POLYNOMIAL_WO_MSB;
}
return reg;
```

2.3 Optimized SW implementation

The main idea behind the optimized CRC calculation is the operation on more bits in parallel (e.g. Wecker, 1974) . Depending on the first n bits, register is xored with generator polynomial during the first n shifts. Due to asociativity of XOR operation, these shifted generator polynomials can be precomputed for all 2^n possible bit combinations and stored in the look-up table. During computation data are shifted by the n and first n-bit of the register creates an index to the look-up table. The output of the look-up table is XORed to the content of the register.

Practical implementation is faster when n is a multiple of 8 bits. (Due to fast access to 8-bit quantities in the CPU architecture.) However, for $n >= 16$, the full look-up table can not be stored in the primary Cache and it slows-down the computation. Therefore the $n = 8$ is mostly used in the SW implementation. In our case of 32-bit polynomial, 256 x 32 bit look-up table is used.

Following C-code fragment corresponds to this optimized SW algorithm that we will denote as *fastcrc*.

```
unsigned long table[256];

void FastCrc_Table_Init (unsigned long key)
{
  if (key == 0) return;
  for (unsigned i = 0; i < 256; ++i) {
    unsigned long reg = i << 24;
    for (int j = 0; j < 8; ++j) {
      bool topBit = (reg & 0x80000000) != 0;
      reg <<= 1;
      if (topBit) reg ^= POLYNOMIAL_WO_MSB;
    }
```

```
      table[i] = reg;
  }
}

reg = 0;
p = array;
for (i = 0; i < cnt ; ++i) {
 top = reg >> 24;
 top ^= *p++;
 reg = (reg << 8) ^ table[top];
}

  return reg;
```

2.4 HW implementation

The main disadvantage of the optimized SW algorithm (*fastcrc*) is the limited scalability. The exponential grow of look-up table makes 16 or 32 bit parallel implementations unrealistic and slow. The main idea of HW implementation is also in the multiple bit parallelism. Equivalent of the *slowcrc* in HW is the simple LFSR. Parallel implementation is created by "unrolling" the LFSR n times (Borrelli, 2001). It can be viewed as two registers and shift and XOR network. The size of the logic does not grow exponentially with n because of the involution property of XOR operation (A XOR A = 0). It is therefore possible to implement circuit processing more bits in parallel than the SW implementation.

For the speed of the circuit, the number of XOR operators on the critical path is important. This number is included in the following table 1. It shows, that 33 2-input XORs are on the critical path of 32-bit parallel implementation. We can see that this number corresponds to less than linear grow of the XOR elements with the number of bits processed in parallel.

Table 1. Number of XOR2 elements on the critical path

Bit Par.	1	2	4	8	16	32
XORs	2	4	6	12	17	33

The speed of the computation in HW also depends on the degree of generator polynomial. Following table shows the achievable clock frequency for 32-bit parallel implementation in XCV100-4 FPGA depending on the generator polynomial degree.

Table 2. Maximal clock frequency (XCV100-4)

Degree of g(x)		8	16	32
Clock frequency[MHz]		120	100	80

It means that for our example of CRC32, coprocessor is able to process 4B @ 80 MHz giving the maximal achievable speed of 320 MBps.

However, in reality the coprocessor is attached to the bus with a given throughput. No advantage is get from computing CRC faster than the data transfer rate on the bus. In the proposed system architecture, coprocessor would "listen" the data on the bus and compute the CRC in parallel during transfer to or from the network interface. When the whole data are transferred, coprocessor will acquire the bus and transfer the resulting CRC to its destination. In our experimental system, coprocessor is attached to 33 MHz PCI bus and we can achieve the useful speed of 128 MBps. We will denote this implementation of CRC as *pcicrc*.

Table 3. Experimental system

CPU Pentium 166 MHz
RAM 32 MB
Chipset VIA 580VPX
OS RedHat Linux 6.2

Table 4. Coprocessor

PCI based add-in board
FPGA XCV100PQ208-4
Area 293 CLB slices (24%)
Frequency 33 MHz

3. EXECUTION TIME ESTIMATION

For estimation of the execution time with external coprocessor, the full system analysis is required. The experimental system is the common PC with dedicated PCI add-on board. The main system parameters are characterized in the table 3 and 4.

In general, the time of CRC computation for the message size of n bytes can be aproximated

$$ExTime = FO + D * n$$

Where FO is the fixed overhead and D is the delay per one byte of data computation. FO and D have both HW and SW components.

$$FO = FO_{sw} + FO_{hw}$$
$$D = D_{sw} + D_{hw}$$

$FO_{s}w$ includes the delay of Linux driver call, task switch overhead and other delays that are difficult to predict. FO_{hw} covers the single read transfer of result from coprocessor. We can estimate the FO_{hw} to 7 PCI clock cycles. Software portion D_{sw} covers mainly the delay of data copying between operating system and application buffers.

For D_{hw} estimation the quantitative model of PCI based coprocessor can be used (Becvar *et.al*, 2000). In our application, CPU transfers data to coprocessor. Dhw depends mainly on the average size of burst transfers. The size of the burst is limited by the capabilities of the Host to PCI bridge and PCI bus load. From the measurements of the system seems that average burst length is $3DW = 12B$. Each burst transfer also includes address phase (1 PCI clock cycle), decoder delay (2 PCI clock cycles) additional target delay (1 PCI

Fig. 1. CRC performance

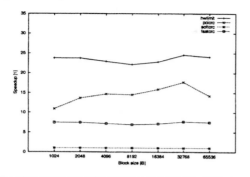

clock cycle) and turn-around cycle (1 PCI clock cycle). It means that time of single burst transfer of 12 B data is $6 + 3 = 9$ PCI clock cycles.

This means that $D_{hw} = 9/12 = 3$ PCI clock cycles per data byte. Expected throughput is therefore limited to 44 MBps.

Execution time can be finally expressed by the following equation Giving the CPU clock frequency of 166 MHz, the PCI clock cycle corresponds approx. to 5 CPU clock cycles.

$$Extime = (35 + FO_{sw}) + (3.75 + D_{sw}) * n$$
$$[CPU\,clockcycles]$$

4. MEASURED VALUES

Table 5. CPU cycles

Size[B]	pcicrc	hwlimit	slowcrc	fastcrc
1024	8460	3875	92127	12368
2048	13440	7715	183190	24642
4096	24074	15395	353235	49219
8192	46971	30755	681311	98755
16384	88528	61475	1403737	197151
32768	170513	122915	3018141	393655
65536	415429	245795	5907504	788359

The values shown in the table 5 describe elapsed CPU clock cycles. Measure was made by reading the value of the CPU's internal 64-bit counter, which is incremented each clock cycle, at start of the computing the CRC and at the end, when the result was ready to use. The results for *pcicrc*, *slowcrc and fastcrc* present an average of 10 measurements with excluding fastest and slowest time. This approach is used to eliminate the transient disturbing effect of other tasks in the system. *HWlimit* column presents an expected optimal time for the current system without SW overhead. It was computed using the ExTime equation from the previous section. The comparison of the CRC algorithms is shown in figure 1.

5. DISCUSSION OF RESULTS

The missing components in the ExTime equation can be computed from measured values. Fixed

Overhead FO is equal approx. 2794 clock cycles, D is approx. 5.53 clock cycles. It means that $FO_{sw} = 2759$ and $D_{sw} = 1.78$ clock cycles. Software overhead is significant portion in the Execution Time. More significant seems to be Fixed Software Overhead FO_{sw}, which covers the overhead of calling the standard Linux driver. On the other hand, effect of data copying between OS and application buffer is not significant comparing to PCI bus overhead. The average data throughput was limited to 27 MBps.

Even with very high software and PCI bus overheads, FPGA implementation proves to be superior over the software algorithms. Speedup of approx. 2 over the *fastcrc* and 14 over the *slowcrc* algorithm was achieved. For the system without SW overhead (*hwlimit*) the speedup of approx. 25 can be achieved over the *slowcrc* algorithm. From the figure 2 can be seen the amortization of the fixed overhead for larger data transfers. The speedup is growing for larger data block transfers. The exception for 65536 B transfer can be explained by the effect of interruption of the device driver itself during the long data transfer. This will lead to aditional SW overhead for larger data blocks.

SW overheads can be minimized either by allowing the direct access of application to coprocessor or by allowing the coprocessor to access data directly in the main memory. The second approach can also use longer burst transfers and furthter improve the performance.

Table 6. CRC Speed Up (slowcrc = 1)

Size[B]	pcicrc	hwlimit	fastcrc
1024	10.8897	23.7747	7.44882
2048	13.6302	23.7447	7.43406
4096	14.6729	22.9448	7.1768
8192	14.5049	22.1529	6.899
16384	15.8564	22.8343	7.12011
32768	17.7004	24.5547	7.66697
65536	14.2202	24.0343	7.49342

Fig. 2. CRC Speed Up (slowcrc = 1)

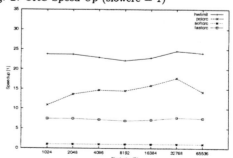

6. CONCLUSION

Effect of CPU acceleration using FPGA implemented computing kernel was demonstrated on the example of CRC computation. FPGA implementation was compared to trivial and optimized software algorithms. The effect of acceleration was measured using PCI based add-in board on the common Personal Computer. FPGA implementation proved to be faster, even when the speedup was limited by large software and hardware overheads. Identification of the critical portions of overhead can help to build systems where the capabilities of FPGA are fully exploited.

REFERENCES:

Becvar M., Schmidt J. (2001). *Reconfigurable acceleration of Intel PC: a quantitative analysis*, Proceedings of DDECS Workshop 2001, Gyor

Borrelli C. (2001), *IEEE 802.3 Cyclic Redundancy Check*, XAPP209, Application Note, Xilinx

Boudreau, Steen (1971), *Cyclic Redundancy Checking by Program*, AFIPS Proceedings, Vol. 39.

Hutchings B.,Villasenor J. (1998), *The Flexibility of Configurable Computing*, IEEE Signal Processing Magazine, Vol. 9., pages 67-84,

IEEE Std. 802.3 -2000, *IEEE Standard for CSMA/CD Access Method and Physical Layer interface (Ethernet)*, pp. 81-82

IEEE Std. 1596-1992, *IEEE Standard for Scalable Coherent Interface (SCI)*, pp.62-64

USB SIG (1995), *Cyclic Redundancy Checks in USB* , White paper, 1995

Wicker,S.B. (1995),*Error Control Systems for Digital Communications and Storage*. Englewood Cliffs: Prentice-Hall.

Wecker, S., (1974) *A Table-Lookup Algorithm for Software Computation of Cyclic Redundancy Check (CRC)*, Digital Equipment Corporation memorandum.

IFAC
Publications
www.elsevier.com/locate/ifac

ADAPTIVE CURRENT FILTERS USING FPAAs

Adam Błaszkowski, Władysław Ciążyński, Marek Garlicki, Józef Kulisz

Institute of Electronics
Silesian University of Technology
Gliwice, Poland

Abstract: The paper presents a proposal of structure and frequency description method of field programmable analogue array (FPAA) for adaptive filtering using a new structure switched-current (SI) cell. The presented new FPAA core contains analogue multipliers used as the basic functional block to change the filter parameters. More detailed description of the four-quadrant S^3I switched-current multiplier designed for use in FPAA core is to be found in this paper. A generalisation of frequency-domain description of SI circuits with periodically time-variant topology designed for analysis of output signal spectra of SC/SI circuits by means of various sampling functions has also been given. *Copyright © 2001 IFAC*

Keywords: adaptive filters, computer-aided circuit design, frequency spectrum, integrated circuits, multipliers, sampled-data signals, solid state cells, switched networks.

1. INTRODUCTION

A typical field programmable analogue array (FPAA) consists of configurable analogue cells that can perform many different functions over the applied input signals, and programmable interconnection network to provide required connections of the configured cells. Cell programming data is held in the local cell register and comprises cell component values and cell configuration. The cell interconnection information is held in the special array configuration register.

General structure of the SI FPAA designed in the Institute of Electronics, SUT Gliwice, using the TANNER RESEARCH software package for the IC design is presented in Fig. 1, see also (Błaszkowski and Ciążyński, 2000). The circuit contains eight programmable SI cells, each of them having two analogue inputs and single output.

The goal of this work is to enhance flexibility of the designed cell by inserting of the additional multiplier block. For this purpose the new structure of SI cell using S^3I switched-current multiplier was proposed. The modified structure of SI cell using multiplier block consisting of six S^3I multipliers is shown on Fig. 2.

To increase the accuracy of sub-circuit simulation authors introduce the concept of sampling function, which allows description of signal shapes in the more realistic way, without spending the big computational effort on numerical evaluating of the transient states. The simplified relations based on the description in the operator domain and applying an appropriate sampling function at the output lead to significant change in the calculated spectra, especially for higher signal frequencies. Some measurements were also conducted to verify how selection of sampling functions affects accuracy of the spectrum calculation.

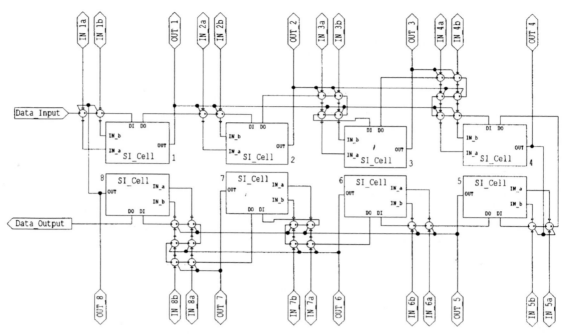

Fig. 1. Core circuit of the SI FPAA at the cell and switch level

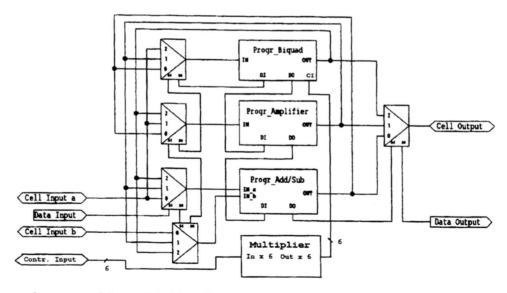

Fig. 2. Internal structure of the modified SI cell

2. ADAPTIVE FILTERS

The term "adaptive filter" is commonly used to refer to a linear system that reshapes the frequency components of the input signal to generate an output signal with some desirable features in such a way, that the output is a good estimate of the given desired signal. The process of selecting the filter parameters and/or coefficients so as to achieve the best match between the desired signal and the filter output is often done by optimising a performance function defined in a statistical or deterministic framework (Farhang-Boroujeny, 1998).

There are two general types of analogue circuits used in the adaptive filtering technique: filters with controlled coefficients and those with controlled parameters. Current-type filters can use analogue multipliers for parameter control via separate inputs. To achieve desired response one can vary filter coefficients or use coefficient input as second signal input and calculate the cross correlation function of input signals.

To implement the adaptive filters, blocks such as delays, adders, integrators, variable gain amplifiers and multipliers are needed. Analogue - both continuous and discrete - and also digital implementations are found in the literature.

3. CONTINUOUS AND SWITCHED CURRENT TECHNIQUE

In recent years the role of analogue integrated circuits has changed in such a way that they are mainly interfacing the complex digital processing systems to their inherently analogue external environment. Quest for ever smaller and cheaper electronic systems has led manufacturers to integrate entire systems onto a single chip. The single mixed-mode IC contains now both a digital signal processor and the entire analogue interface circuits required for interacting with the external analogue transducers and sensors. As a rule the digital part of such a mixed-mode IC takes much bigger part of the silicon surface. It is therefore natural that the technology be tailored to optimise digital part performance. This means in practice that only the analogue interface circuits that are fully compatible with digital CMOS technology can be manufactured economically.

State-of-the-art analogue circuits are contributing much of the current-mode approach. The current-mode designs allow for gain independent of bandwidth, give lower noise, wider dynamic range, higher speeds – although these performances are very technology and application specific.

Traditionally in the analogue interface portion of the mixed-mode designs the switched-capacitor (SC) technique has been employed, which require high quality linear capacitors. Those are usually implemented using two layers of polysilicon. The second layer of polysilicon is not needed by the purely digital circuits and become unavailable as processes shrink to deep submicron range. The trend towards submicron processes is also leading to a reduction in supply voltages, directly reducing maximum voltage swing available in SC circuits (so also their dynamic range). The realisation of the high-speed high gain op-amps is becoming more difficult.

The difficulties faced by the SC technique, and other "voltage-mode" analogue interface circuits, in coping with the advance of digital CMOS processing technology has revived interest in "current-mode (SI)" techniques, in which signals are represented by current samples. SI circuits do not require linear floating capacitors or op-amps.

That technique fully deserved for its name of "analogue technique for digital technology" (Toumazou, et al., 1993).

4. CURRENT-TYPE ANALOGUE MUTLIPLIERS

Analogue multipliers are basic functional blocks in many non-linear electronic circuits (e.g. modulators, mixers etc.). They are also used to change the filter parameters in adaptive filter circuits. This paper presents four-quadrant S^3I switched-current multiplier designed for use in adaptive filter circuit (Leenaerts, et al., 1996; Manganaro, et al., 1998).

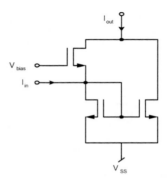

Fig. 3. Current squarer.

Multiplication of the input currents I_x and I_y is realised by evaluating their quadratic terms:

$$(x + y)^2 - x^2 - y^2 = 2xy \qquad (1)$$

The squarer circuit (Bulk, and Wallinga, 1987) is shown in Fig.3. For this circuit the relationship between the input and output current gives:

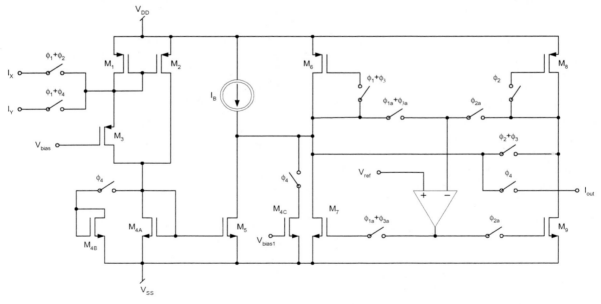

Fig. 4. Circuit schematic of S^3I multiplier

211

$$i_o = I_b + e_1 + \frac{(i_i + e_2)^2}{4I_b} \qquad (2)$$

where:

i_i and i_o are input and output currents respectively,

e_1 is output offset error,

e_2 is input offset error,

I_b is the bias current, which depends on V_{bias} and equals:

$$I_b = \frac{1}{2} I_{DSS} \left(\frac{U_{bias}}{U_{th}} \right)^2 \qquad (3)$$

The approach presented in (Leenaerts, et al., 1996; Manganaro, et al., 1998) is based on using the same squarer circuit many times while storing intermediate results in memory cells.

The algorithm consists of four steps, each step in another clock phase. Complete circuit schematic of the multiplier is shown in Fig. 4. As memory cells are controlled by signals without overlapping, there is only one amplifier required, switched between both cells. The signal on the output of the multiplier is present only in fourth phase (φ_4). In some applications output signal is required also during other phases, in this case another memory cell on the output can be built.

The simplified analysis doesn't take into account the attenuation of the memory cells, caused by transistors g_{ds} which doesn't equal to zero. Attenuation of the intermediate results causes the additional offset error as well as the nonlinearity error.

More precisely I_{out} is given by:

$$I_{out} = \frac{\eta^3}{2I_b} I_x I_y$$
$$+ \frac{1}{4I_b} \left[\eta^2 (\eta - 1)(I_x^2 + 2e_2 I_x) + (\eta^3 - 1)(I_y^2 + 2e_2 I_y) \right] \qquad (4)$$
$$+ \frac{e_2^2 I_y}{2I_b} (\eta^3 - \eta^2 + \eta - 1) + e_1 (\eta^3 - \eta^2 + \eta - 1)$$

where η is memory cell attenuation.

First term is the desired result, second is the non-linear error and the last one is the offset error. The attenuation η is very close to 1 so the term containing $\eta^2(\eta\text{-}1)$ can be neglected. The non-linear error connected with I_y can be cancelled by changing the current mirror ratio in the fourth phase (Manganaro, et al., 1998). Connecting the additional transistor M_{4B} in parallel with M_{4A} does this. The mirror ratio is changing from $1{:}1$ to $1{:}\beta$, $\beta < 1$, but the bias current I_b of the right hand side of the mirror remains constant so the additional transistor M_{4C} have to be added to cancel the extra current of $I_b(1\text{-}\beta)$. The output current equals then:

$$I_{out} = \frac{\eta^3}{2I_b} I_x I_y$$
$$+ \frac{1}{4I_b} \left[\eta^2 (\eta - 1)(I_x^2 + 2e_2 I_x) + (\eta^3 - \beta)(I_y^2 + 2e_2 I_y) \right] \qquad (5)$$
$$+ \frac{e_2^2 I_y}{2I_b} (\eta^3 - \eta^2 + \eta - \beta) + e_1 (\eta^3 - \eta^2 + \eta - \beta)$$

The non-linear error can be reduced when $\beta = \eta^3$. The offset error is thus significantly reduced.

This circuit fulfils the criteria for using in new type S^3I cell for designed FPAA and has been chosen for use in new type SI cell.

5. ANALYSIS AND SIMULATION FOR SI NETWORKS

SI circuits may be analysed and simulated at various modelling levels. Most generally one can distinguish two approaches to SI networks simulation:

- simulation in the time domain;
- simulation in the z operator variable domain.

Simulation in the time domain may be performed using general-purpose analogue circuit simulation programs, like SPICE. The main advantage of this approach is accuracy, which in fact depends only on accuracy of element modelling. Nonlinearities and most of the parasitic effects may be quite easily simulated this way. Also behaviour of signals around switching instants can be exactly modelled.

Using general-purpose SPICE-like programs has however many disadvantages. First of all it is very ineffective. Evaluating signal waveforms even for a few clock periods is time consuming. Small signal frequency and noise analyses are not possible at all. The only way to obtain frequency characteristics is to perform many time domain simulations and calculate the spectra from time domain waveforms using DFT. A great majority of the calculation time will be spent on exact modelling of the switching process. In most applications exact modelling of the switching process is not necessary. Based on this observation a new class of simulators, dedicated to switched networks, could be worked out. An example of such a program is SWITCAP3, developed at the Columbia University, New York (Yusim, 1999).

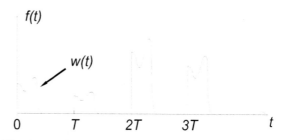

Fig. 5. Analogue signal as a sequence of sampling pulses

SWITCAP3 doesn't calculate transient states, occurring at the beginning of each phase. An assumption is made that between successive switching instants the circuit always reaches a steady state. The program instead calculates the steady state conditions by solving one set of non-linear algebraic equations per phase interval. Following SPICE, SWITCAP3 can also perform efficient small signal analysis in the frequency domain, and noise analysis.

As the transient process following switching is not modelled, in all analyses, both in time and frequency domain, it is assumed that the signals have a piecewise-constant (stepwise) form. For the same reason dynamic non-linear effects cannot be handled and model accuracy is reduced to static effects only (Yusim, 1999).

SWITCAP3 seems to be an equivalent of SPICE for switched capacitor and switched current circuits.

Methods based on analysis in the z operator variable domain assume further reduction of model accuracy. Circuit behaviour is modelled by a set of linear equations in the z domain. So generally model accuracy is reduced to linear effects. A number of such programs were developed for switched capacitor networks in the 80-s. In most of the methods the equations are represented in a matrix form and the matrices are further processed. For periodic clocks it is possible to form one matrix equation in the z-domain, valid for all phases.

De Queiroz, et al. (1993) proposed an efficient algorithm for SI circuit analysis. The key point of the method was replacing capacitors by equivalent circuits consisting of a resistance and a controlled voltage source. This way analysis of an SI network can be performed by solving one linear DC circuit per phase interval. Alternatively a matrix equation can be built, in a similar manner as for SC circuits. In fact only signal steady state conditions in every phase interval are calculated this way.

The method described above was implemented in a computer program called ASIZ (de Queiroz, et al., 1993). ASIZ solves the z-domain equations using the efficient FFT interpolation method. The main results of the analysis are partial transfer functions, which are rational functions of the z variable. As the result of the FFT interpolation numerical values of the numerator and denominator polynomial coefficients are obtained. This way ASIZ provides symbolic solution for network functions. Apart from giving the partial transfer functions ASIZ can calculate time domain waveforms, and perform sensitivity analysis using the adjoint network approach.

As mentioned before, because the transient states following switching instants are not evaluated, for frequency response calculations some additional assumptions regarding output signal shape are necessary. ASIZ allows specifying the signal shape by selecting one of two output sampling methods: sampled and held, and impulse. For the first method the output signal is considered to be sampled and held constant between successive switching instants (stepwise). For the second option the output is considered as a series of Dirac pulses.

As we see, both in SWITCAP and in the methods based on analysis in the operator domain, the information about transient states following switching instants is lost, and signal shapes have to be approximated by some simple functions (usually a stepwise function). Quality of the approximation can probably be greatly improved by introducing the concept of sampling function.

Let us assume that a signal in the continuous-time domain consists of equally spaced sampling pulses $w(t)$ having the same shape, but different "heights". The "heights" are controlled by discrete signal samples (Fig. 5 and equation 6):

$$f(t) = \sum_{n=-\infty}^{\infty} w(t-nT) \cdot f(n) \qquad (6)$$

It can easily be proved that the spectrum of the analogue signal $f(t)$ defined by (6) is obtained by multiplying the spectrum of the discrete signal $f(n)$ by a sampling function $W(j\omega)$, which is just the Fourier transform of one of the sampling pulses $w(t)$:

$$\text{where} \quad \begin{array}{l} F(j\omega) = F(e^{j\omega T}) \cdot W(j\omega) \\ W(j\omega) = \int_{-\infty}^{\infty} w(t) e^{-j\omega t} dt \end{array} \qquad (7)$$

So the concept of sampling function allows describing signal shapes in a more realistic way without spending the big computational effort on numerical evaluating the transient states. We can use instead the simplified relations based on the description in the operator domain, applying an appropriate sampling function at the output.

For multiphase circuits and unequal phase intervals the formulation needs to be further generalised. Kulisz (2000) presented the full concept together with derivation of the formulas.

The concept of sampling function discussed above was implemented as a set of Matlab functions running under Windows 95/98. The software was supposed to provide an extension to the capabilities offered by ASIZ. The functions accept as input data the symbolic transfer functions generated by ASIZ, as well as some additional parameters used to describe clock patterns and sampling functions. The functions calculate output signal spectra for two standard (sine and square) excitations, as well as equivalent transfer functions. The results can be presented as plots on the screen or they can be written to text files for further processing. Several sampling functions of various complexity and shape were implemented in the software, among them the Dirac delta and square pulse, corresponding to the impulse and sampled and held sampling methods of ASIZ. The most complex sampling pulse is described by two second order oscillatory damped functions, one for the front, and one for the back slope (8).

$$w_{osc}(t) = \begin{cases} 0 & t \in (-\infty, 0 > \\ 1 - \dfrac{\sqrt{\sigma_f^2 + \omega_f^2}}{\omega_f} \cdot \exp(-\sigma_f t) \cdot \sin(\omega_f t + \varphi_f) & t \in (0, \Delta t_k > \\ w_{osc}(\Delta t_k) \cdot \dfrac{\sqrt{\sigma_b^2 + \omega_b^2}}{\omega_b} \cdot \exp\big(-\sigma_b(t - \Delta t_k)\big) \cdot \sin\big(\omega_b(t - \Delta t_k) + \varphi_b\big) & t \in (\Delta t_k, \infty) \end{cases} \tag{8}$$

where $\varphi_f = \operatorname{arctg}(\omega_f / \sigma_f)$, $\varphi_b = \operatorname{arctg}(\omega_b / \sigma_b)$ and Δt_k stands for phase interval length

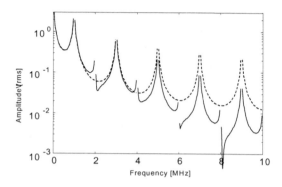

Fig. 6. Comparison of spectra calculated for the square (dotted line) and oscillatory (solid line) sampling functions.

Figure 6 shows a comparison of example spectra calculated using the old square sampling function and the new oscillatory function defined by (8).

The example shows that applying different sampling functions can lead to a significant change in the calculated spectra, especially for higher frequencies.

At the moment some measurements are made for SC circuits, to verify how selection of sampling function type affects accuracy of spectrum calculation. The results seem to be promising (Kulisz, 2001). Similar experiments on SI circuits are planned in the near future.

6. CONLUSIONS

The field programmable analogue array presented in this paper is a new type SI FPAA suitable for filters with controlled coefficients and parameters. These filters use analogue multipliers for parameter control via separate inputs. So one can vary filter coefficients to achieve desired response for adaptive filtering purposes or use coefficient input as second signal input and calculate the cross correlation function of input signals.

The presented concept of simplified frequency analysis of SI circuits using sampling function allows to describe signal shapes in a more realistic way, without spending the computational effort on numerical evaluating of the transient states. This method can be successfully used for wide range of time-discrete electronic circuits, both current and voltage type.

REFERENCES

Błaszkowski, A. and W. Ciążyński (2000). Switched-current FPAAs can use S^2I and S^3I memory cells. In: *Proc. from the IFAC Workshop on Programmable Devices and Systems PDS 2000*, pp. 43 – 48. Elsevier Science, Kidlington, Oxford.

Bulk, K. and H. Wallinga, (1987). A class of analog CMOS circuits based on the square – law characteristic of an MOS transistor in saturation. *IEEE J.Solid – State Circuits*, **vol 22**, pp. 357 – 365.

De Queiroz A. C. M., P. R. M. Pinheiro and L. P. Calôba (1993). Nodal Analysis of Switched-Current Filters. *IEEE Trans. on Circuits and Systems. – II*, **40**, 10-18.

Farhang-Boroujeny, B. (1998). *Adaptive filters. Theory and Applications*. John Wiley & Sons, Chichester, New York, Weinheim, Brisbane, Singapore, Toronto.

Kulisz, J. (2000). A Generalization of Frequency Description of SC/SI Circuits with Periodically Time-Variant Topology. *Proc. of the Int. Conf. on Signals and Electronic Systems ICSES'2000*. Ustroń, pp. 349-354.

Kulisz, J. (2001). Describing Output Signal Spectra of SC Circuits by Means of Various Sampling Functions. *Proc. of the Int. Conf. on Signals and Electronic Systems ICSES'2001*, Łódź, pp. 321-326.

Leenaerts D.M.W., G.H.M. Joordens and J.A. Hegt (1996). A 3.3V 625kHz switched – current multiplier. *IEEE J.Solid – State Circuits*, **vol 31**, pp. 1340 – 1343.

Manganaro G. and J. Pineda de Gyvez (1998). A Four – Quadrant S^2I Switched – Current Multiplier. *IEEE Trans. Circuits and Systems-II: Analog and Digital Signal Processing*, **vol. 45**, pp. 791-799.

Yusim, I. (1999). Simulation of Switched-Capacitor and Switched-Current Networks under Instant Settling Approximation. *Ph. D. Thesis*. Columbia University, New York.

Toumazou C., J. B. Hughes and N. C. Battersby (Editors) (1993). *"Switched-Currents, an analogue technique for digital technology"*. Published by Peter Peregrinus Ltd., on behalf of the IEE, London, United Kingdom.

IFAC

Publications

www.elsevier.com/locate/ifac

A UNIFIED APPROACH TO BOOLEAN FUNCTION DECOMPOSITION
FOR MULTILEVEL FPGA-BASED DESIGN

Eugene BOOLE, Katrin BOULE, Victor CHAPENKO

Institute of Electronics and Computer Science of the Latvian University,
Riga (Latvia)

Abstract: The paper is focussed on a method for constructing decomposition for Boolean function multilevel realizations based on field programmable gate arrays (FPGA). The method is based on formulating the criterion of compatibility for subsets of atomic requirements, which are defined by an initial function truth table. This criterion enables one to obtain simultaneously both input variables distribution and truth tables for all subsystems, which will be mapped to FPGA's cells of a single level of the multilevel realization. Particular restrictions for the subsets of compatibility allow to construct realization from input to output, from output to input, and to get disjunctive and conjunctive decompositions for the case of non-universal cells. The method is illustrated by examples. *Copyright © 2001 IFAC*

Keywords: combinational circuits, truth tables, decomposition method, design VLSI, multilevel structures

1. INTRODUCTION

The present range of integrated circuits used for the design of combinational circuits includes a host of programmable logical devices. The field programmable gate arrays (FPGA) have become the most popular of them. FPGA has a regular field programmable structure (Brown, *et al.*, 1992). The basic element of these types of structures (logical block) is a universal logic module, which is often implemented as a K-input lookup table (LUT) with programmable flip-flop and expander at output. Such field programmable devices are available from several VLSI chip manufacturers, such as Altera (1995), Xilinx (1994), AT&T (1995) and so on.

The complexity of a combinational circuit, realizable by an individual logical block is limited. This limitation is defined by the number of input K and number of output M variables (respectively, inputs and outputs) of the given logical block. Simple

division of all specified functions into groups of M functions one may overcome the constraint M on the number of outputs. It is sometimes possible to reduce the number of arguments of the functions down to the specified constraint M taking into account the existence of nonessential arguments (Boole and Chapenko, 1994). In the general case, in which the functions depend essentially on more than K arguments, methods based on the functional decomposition are the most efficient. These methods allow getting the circuits with a minimal number of logical blocks and a minimal number of levels, but, however, proving to be overly lengthy when there is a large number of input variables. There are many techniques to construct decomposition for a given argument partitioning, but an open problem is how to choose the proper variable partitioning (Cong and Ding, 1996). As an example of the first, a method of constructing the multiple decomposition based on a logic equation was proposed in (Boole and Chapenko, 1998). The simple removing of parentheses in the equation allows one to obtain all possible decompositions. Several approaches have been suggested for the variable partitioning into the *bound set* and the *free set* for the one-output functional decomposition. The majority of the approaches are presented in (Cong and Ding, 1996). In the present paper, a method of simultaneous constructing the multiple decomposition together with variable partitioning is considered.

2. MULTIPLE DECOMPOSITION PROBLEM

Let an initial specification of a system of partial functions $F(X)$ be given in the form of the truth table. The columns of the table are labeled by variables and functions from the respective sets $X = \{x_1, x_2, ..., x_n\}$ and $F = \{f_1, f_2, ..., f_m\}$. For partial functions the number l of rows in the table is much lesser than 2^n. The input set (ordered set of values of the input variables) in the row numbered g of the table is a cube written as a vector in ternary notation (Villa *et al.*, 1997) $X^g = \langle x_1^g, x_2^g, ..., x_i^g, ..., x_n^g \rangle$, $x_i^g \in \{0,1,-\}$, $0 \leq g < l$. Such vector represents the set of all binary vectors (minterms) that may be obtained from X^g by replacing the symbol "–" (undefined values) by zeros and ones. Hereafter this set will be denoted with X^g, too. The output set (ordered set of values of the functions) in the row numbered g is written as the ternary vector $F^g = \langle f_1^g, f_2^g ..., f_j^g, ..., f_m^g \rangle$, $f_j^g \in \{0,1,-\}$. The result of the multiple functional decomposition for the function $F(X)$ can be written in the form:

$$F(X) >> \Phi(Y) = \Phi(y_1(\chi_1),...,y_r(\chi_r),...,y_v(\chi_v)), \quad (1)$$

where a sign $>>$ denotes that $\Phi(Y)$ implements the original system $F(X)$; $\Phi(Y) = \{\varphi_1(Y), ..., \varphi_j(Y), ..., \varphi_m(Y)\}$ is a system of the *base function*; $Y = \{y_1(\chi_1), ..., y_r(\chi_r), ..., y_k(\chi_v)\}$ are *encoding functions*, and $\chi_1, ..., \chi_r, ..., \chi_v \subset X$ are the subsets of the initial set of the input variables X (Boole and Chapenko, 1996).

The presentation (1) always exists, but for the most part of functions it is a problem to find the nontrivial form (1) if there is one of the following constraints:
(a) for all $r = 1, 2, ..., v$ the cardinality of χ_r cannot be greater than the number K of LUT inputs, i.e. $|\chi_r| \leq K$;
(b) the number of encoding functions Y cannot be greater than K, i.e. $v \leq K$ or $|Y| \leq K$;

In case when both of these constraints are satisfied the decision is ready – the obtained decomposition can be mapped to a two-level network. First level comprises v LUTs programmed for v encoding functions realization and second level comprises m LUTs programmed for m base functions realization.

In all other cases recursive decomposition needs to be made. Depending on the circuit synthesis from inputs to outputs or vice versa the constraint (a) or constraint (b), respectively, becomes essential. To guarantee convergence of the recursion the decomposition (1) should be nontrivial. This means additional requirements for sought-for decomposition according to the constraint (a) and (b):

(a) the number of encoding functions should be less than initial number of input variables, i.e. $v < n$;
(b) for all $r = 1, 2, ..., v$ the cardinality of χ_r should be less than n, i.e. $|\chi_r| < n$.

There is some additional reasoning for the minimization of LUT number in the network, obtained by the recursive decomposition. First of all, it is possible in the current step of recursion to obtain the encoding function implementing one of the initial function exists, that is $y_r << f_j$. In such case the r-th LUT output is the j-th output of the synthesized circuit and the function f_j should be excluded from the given system to simplify the next decomposition step. The second reasoning is a possibility to obtain $|\chi_r| = 1$. In such case the encoding function y_r is equivalent to the initial variable (or its negotiation) and will not require the additional LUT. This reasoning means to avoid entering of new encoding function while all facilities of the entered functions are exhausted (or when $|\chi_r| < K$).

3. THE METHOD OF DECOMPOSITION

In (Boole and Chapenko, 1998), the orthogonality sets $O^X(g,h)$ and $O^F(g,h)$ for truth table rows with numbers g and h were discussed. The orthogonality set $O^X(g,h)$ ($O^F(g,h)$) includes all variables (functions), which are opposite in vectors X^g and X^h (F^g and F^h). More formal:

the variable x_i is included in the orthogonality set $O^X(g,h)$ if and only if $x_i^g x_i^h \in \{0,1\}$ and $x_i^g \neq x_i^h$, and

the function f_j is included in the orthogonality set $O^F(g,h)$ if and only if $f_i^g f_i^h \in \{0,1\}$ and $f_i^g \neq f_i^h$.

For the consistent system $F(X)$ any pair (g, h) for which the orthogonality set $O^F(g,h) \neq \emptyset$ predetermines the *dichotomy* (X^g, X^h), because $O^X(g,h) \neq \emptyset$ and so the sets X^g and X^h are disjoint (Unger, 1969). Let O be the set of all such *unordered* pairs (g,h) and set $U = \{(g_r, h_s)| r = 1, 2, ..., R, s = 1, 2, ..., S\}$ is its subset, for which the sets $G = \{g_r | r = 1, 2, ..., R\}$ and $H = \{h_s | s = 1, 2, ..., S\}$ are disjoint.

Definition. The set $U \subseteq O$, is the *encoding set* for O if and only if for any pair $(g, h) \in U$ the orthogonality set $O^X(g, h)$ is not empty, i.e. $O^X(g, h) \neq \emptyset$.

For the truth table, including the minterms only, any

set U with disjoint sets G and H is the encoding set. But in case of k-cubes with $k \geq 1$ this is not true and it is necessary to verify the *compatibility* for all pairs (g,h) included in the set U. If X^g and X^h are considered as sets of minterms the above definition is the same as the compatibility definition given in (Unger, 1969) for corresponding dichotomies (X^g, X^h). More over, all the rules for dichotomies, formulated thereat, can be applied for the initial set O reducing before the set O will be partitioned.

The encoding set U defines two subsets of the initial set of input vectors, which are not intersected, and as result the function defined in accordance with:

$$y(X^k) = \begin{cases} 0, & if \quad k \in G; \\ 1, & if \quad k \in H; \\ -, & in\ other\ case. \end{cases}$$

will be the encoding function for the system $F(X)$.

The orthogonality sets $O^X(g,h)$ for all pairs (g, h) with $g \in G$ and $h \in H$ may be presented as rows of the covering table (Villa *et al*., 1997). The column covering for this table allows to get the *covering subset* $\chi_r \subseteq X$, which will distinct all pairs $(g, h) \in U$ and allow the consistent definition of the encoding function $y_r(\chi_r)$. Such covering subsets will be used to reduce the number of variables for encoding functions.

Proposition 1. Any partition of the set O into encoding subsets U_1, U_2, ..., U_r ..., U_v defines the decomposition (1).

Really, the sets U_1, ..., U_r ..., U_v allow to define consistent encoding functions $y_1(\chi_1)$, ..., $y_r(\chi_r)$, ..., $y_k(\chi_v)$. Simultaneously any pair (g,h) for which $O^F(g,h) \neq \varnothing$ is included in some set U_r. Hence one can construct the encoding function $y_r(\chi_r)$ which will differ the reactions $Y^g = \langle y_1^g,...,y_r^g,...,y_v^g \rangle$ and $Y^h = \langle y_1^h,...,y_r^h,...,y_v^h \rangle$ to any input set from the original sets X^g and X^h. So the system of base function $\Phi(Y) = \{\varphi_1(Y), \varphi_i(Y), ..., \varphi_m(Y)\}$, defined by new truth table with vector $Y^g = \langle y_1^g,...,y_r^g,...,y_v^g \rangle$ replacing vector X^g in accordance with the sets U_1, U_2,...,U_k and the old vector $F^g = \langle f_1^g,...,f_m^g \rangle$ ($g = 1, 2, ...,l$) will implement the original system $F(X)$.

So, the method of decomposition calls for the partition of set O to the encoding sets. The set O is defined by the system $F(X)$ given in the current step. In the first step it is the original system, in all others it is a resulting base system from the preceding step if the circuit is synthesized from inputs or resulting encoding functions if the circuit is synthesized from outputs. In these two cases the additional conditions to a searched partition are distinct, too.

In case of the *synthesis from inputs to outputs* the cardinality of set χ_r, which covers the set U_r, is constrained by the number K of LUT inputs, i.e. $|\chi_r| \leq K$, ($r = 1, 2, ..., v$). For the decomposition nontriviality the number v of encoding functions should be less than n, i.e. $v < n$. The optimization criterion is the number v decreasing, forwarding the recursion convergence. To minimize the number of LUTs in the current step one can take into account the possibility to get the covering set for some U_r with one variable.

In case of the *synthesis from outputs to inputs* the number v of encoding functions is constrained by the number K, i.e. $v \leq K$. For nontriviality the cardinality of the set χ_r should be less than number n of variables in the original system $F(X)$, that is $|\chi_r| < n$, $r = 1$, $2, ..., v$. The goal of optimization is to get as many covering sets χ_r with K or one variables as possible. The first forwards the recursion convergence and the second decreases the number of LUT in the current step.

The end of the recursion sets in when all resulting functions (base functions in first case or encoding functions in the second) are dependant on K or lesser variables. Examples of synthesis for both cases will be considered in part 4.

Next two propositions are formulated for one Boolean function and for logic blocks representing the simple K-AND and K-OR gates.

Proposition 2. To get the decomposition

$$f(X) = y_1(\chi_1) \cdot y_2(\chi_2) \cdot ... \cdot y_K(\chi_K)$$

it is necessary and sufficient to find the partition of the set O into encoding subsets U_1, U_2, ..., U_K with

$$G_1 = G_2 = ... = G_K = G_F,$$

where G_F is the ON-set of $f(X)$.

Proposition 3. To get the decomposition

$$f(X) = y_1(\chi_1) \vee y_2(\chi_2) \vee ... \vee y_K(\chi_K)$$

it is necessary and sufficient to find the partition of the set O into encoding subsets U_1, U_2, ..., U_K with

$$H_1 = H_2 = ... = H_K = H_F,$$

where H_F is the OFF-set of $f(X)$.

4. EXAMPLE

The system $F(X)$ given in table 1 is now used to illustrate the above concepts and synthesis procedure. The 3-LUT will be considered as the logic block of the given FPGA. It is necessary to find the network from such 3-LUTs to realize the system $F(X)$. The network will be synthesized both from input to output and vice versa.

Table 1 The initial specification of system $F(X)$

g				X						F	
h	x_1	x_2	x_3	x_4	x_5	x_6	x_7	x_8	x_9	f_1	f_2
1	0	0	0	0	1	1	0	0	0	1	0
2	0	0	0	0	1	1	0	0	1	0	1
3	0	0	0	0	1	1	1	0	0	0	1
4	0	0	0	0	1	1	1	1	0	1	1
5	0	0	1	0	0	0	0	0	0	1	1
6	0	0	1	0	0	0	0	1	1	0	0
7	0	1	0	1	1	1	0	0	0	1	0
8	0	1	1	0	0	0	0	0	0	1	1
9	0	1	1	1	0	0	0	0	0	1	1
10	0	1	1	1	0	0	0	1	1	1	1
11	0	1	1	1	0	1	0	0	0	0	0
12	0	1	1	1	1	0	0	0	0	0	1
13	1	0	0	1	1	1	0	0	0	0	1

The set O consists of 60 pairs and is not showed for the lack of area. However, the covering table for a set U of all pairs with $O^F(g,h)$ including f_2 is considered. Such set has sets $G = \{1, 4, 5, 7, 8, 9, 10\}$ and $H = \{2, 3, 6, 11, 12, 13\}$ which are equivalent to the ON-set and OFF-set of the function f_1. The reduced covering table after removing covered rows is shown in table 2.

Table 2 The covering table for the encoding set

	$O^X(g,h)$								
g,h	1	2	3	4	5	6	7	8	9
1,2									1
1,3						1			
1,13	1			1					
3,4							1		
6,10		1		1					
7,13	1	1							
9,11							1		
9,12					1				

The minimal column coverings for it are: $\{x_1,x_2,x_5,x_6, x_7,x_8,x_9\}$, $\{x_1,x_4,x_5,x_6,x_7,x_8,x_9\}$ and $\{x_2,x_4,x_5,x_6,x_7,x_8, x_9\}$. Thus, the function f_1 can not be realized in one 3-LUT. One can verify that the second function has five essential variables (for example, $\{x_1,x_2,x_6,x_7, x_9\}$). So, it is necessary to look for the decomposition (1) for the given system $F(X)$.

As mentioned above the set O consists of 60 pairs. The straight partition of it taking into account, that each encoding function must have no more than 3 variables, will give the next encoding sets and covering sets (see table 3).

Table 3 The encoding sets requiring 5 LUTs

	G	H	χ
1	1,7,13	2,3,4,5,6,8,9,10,11,12	x_3, x_7, x_9
2	1,2,3,7,12	4,5,6,8,9,10,11,13	x_1, x_5, x_8
3	4,5,8,9,10	6,11	x_2, x_6, x_8

The base function has 3 variables and can be realized in one 3-LUT, but because of the given LUT has one output the circuit will have 5 LUTs. More economic circuit can be got if to take into account that the second function depends on 5 variables. This implies that one may find an encoding set U_1, covered by three variables, and two sets, covered by one variable (LUTs are not required). For the second function the set U_1 will be used and own encoding sets will be formed. The residue of set O' is partitioned to two sets, one of which is covered by three variables, but second by one variable. The resulting encoding sets and their covering sets are shown in table 4.

Table 4 The encoding sets requiring 4 LUTs

	G	H	χ
1	1,4,6,7,11	2,5,8,9,10,12	x_2, x_6, x_9
2	1,6,7,11	13	x_1
3	1,6,7,11	3,4	x_7
4	1,4,7	3,13	x_1, x_7, x_8
5	1,2,3,4,7,12,13	5,6,8,9,10,11	x_5

The truth table for each encoding function is obtained from the initial table by replacing the original functions with encoding function. The sets G and H are the ON-set and OFF-set, respectively (or vice versa). In the variable part of the table all columns corresponding to the variables not included in the encoding set should be removed. The truth table for base function is formed from the initial truth table by replacing the original variables with encoding functions, formed before. In truth tables for encoding functions $y_1(x_2, x_6, x_9)$, $y_2(x_1, x_7, x_8)$ and base functions $\varphi_1(x_5, y_1, y_2)$, $\varphi_2(x_1, x_7, y_1)$ shown in tables 5 and 6 the identical rows are removed and rows are rearranged. Both base functions have three variables and can be realized in 3-LUTs. Because the base functions implement the original functions the circuit is ready.

Table 5 The truth tables for encoding functions y_1 and y_2

x_2	x_6	x_9	y_1		x_1	x_7	x_8	y_2
0	0	0	0		0	0	0	1
0	0	1	1		0	1	0	0
0	1	0	1		0	1	1	1
0	1	1	0		1	0	0	0
1	0	0	0					
1	0	1	0					
1	1	0	1					

Table 6 The truth table for base functions φ_1 and φ_2

x_5	y_1	y_2	φ_1
0	0	-	1
0	0	1	1
0	1	-	0
0	1	1	0
1	0	1	0
1	1	0	0
1	1	1	1

x_1	x_7	y_1	φ_2
0	0	0	1
0	0	1	0
0	1	1	1
0	1	1	1
1	0	1	1

The resulting circuit from four LUTs is shown in figure 1.

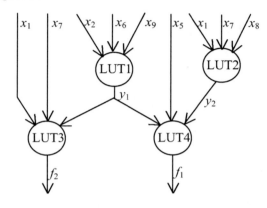

Fig. 1. Circuit synthesized from inputs to outputs.

Now the circuit will be synthesized from outputs to inputs. As was shown above the function f_1 depends on seven variables, but f_2 depends on five variables from the same set. It means that function f_2 can be used as a variable for f_1 realization. The partition for the first step of decomposition is shown in table 7. As result the encoding function $y_1(x_1,x_2,x_6,x_7,x_9)$ implements the original function f_2. This implies that for next step the initial system consists of one function f_1. But the decision for it is ready - first three encoding sets from table 4. The resulting circuit is shown in figure 2.

Table 7 The encoding sets for synthesis from outputs

	G	H	χ
1	1,6,7,11	2,3,4,5,8,9,10,12,13	f_2
2	1,2,3,7,12,13	5,6,8,9,10,11	x_5
3	2,3,12,13	4	x_8

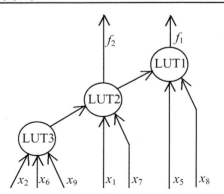

Fig. 2. Circuit synthesized from outputs to inputs.

5. CONCLUSION

The proposed method helps to find the multiple decomposition through the partition of the original set into the compatible encoding sets. Therefore, it is possible to find all the variant decompositions. One can use the method for circuit synthesis both from inputs to outputs and from outputs to inputs. Moreover, it becomes possible to solve the problem through the use of methods of search for compatible sets and methods of finding coverings well developed under practical conditions. Note that though the proposed method is designed for retrieving the decompositions (1) that describe a two-level scheme, it has been applied in recursion retrieve of a decomposition describing a multi-level scheme in an FPGA basis. For this purpose it is necessary to repeatedly apply the method of retrieving the decomposition given in (1), while all base function and encoding function become realizable on the given element basis.

REFERENCES

ALTERA (1995). *Data Book*. Altera Corp., San Jose, CA.

AT&T Microelectronics (1995). *AT&T Field-programmable Gate Arrays Data Book*. AT&T Corp., Brekley Heights, NJ.

Boole E. S. and V. P. Chapenko (1994). Partitioning a system of Boolean functions under specified constraints. *Automatic Control and Computer Sciences*, **1**, 3-11.

Boole E. and V. Chapenko (1998). Boolean function decomposition for PLD-based realizations by logic equation solution. In: *Proc. Intern. Conf. "Programmable Devices and Systems" PDS'98*. Gliwice, Poland, 115-122.

Brayton R. K., G. D. Hachtel and A. L. Sangiovanni-Vincentelli (1990). Multilevel logic synthesis. In: *Proc. IEEE*, **78**, 264-300.

Brown S. D., R. J. Francis, J. Rose, and Z. G. Vranevic (1992). *Field-Programmable Gate Arrays*. Kluwer Academic Publishers, Boston-Dordrecht-London.

Cong J. and Y. Ding (1996). Combinatorial Logic synthesis for LUT Based Field Programmable Gate Arrays. *ACM Trans. On Design Automation of Electronic System*, **1**, 145-204.

Unger S. H. (1969). *Asynchronous Sequential Switching Circuits*. WILEY-INTERSCIENCE, New York-London-Sydney-Toronto.

Villa T., T. Kam, R. K. Brayton, A. Sangiovanni-Vincentelli (1997) *Synthesis of finite state machines: logic optimization*. Kluwer Academic Publishers, Boston-Dordrecht-London.

XILINX (1994). *The programmable Logic Data Book*. Xilinx Inc., San Jose, CA.

IFAC
Publications
www.elsevier.com/locate/ifac

THE VARIABLE ORDERING FOR ROBDD/ROMDD-BASED FUNCTIONAL DECOMPOSITION

P. Buciak, H. Niewiadomski, M. Pleban, H. Selvaraj *, P. Sapiecha

Institute of Telecommunications,
Warsaw University of Technology,
Nowowiejska 15/19, 00-665 Warsaw, Poland,
e-mail: sapiecha@tele.pw.edu.pl
*University of Nevada, Las Vegas,
4505, Maryland Parkway, Las Vegas, NV 89154-4026, USA

Abstract: A *ROBDD/ROMDD*-based decomposition of the Boolean/multi-valued functions is presented in this paper. A modification of the *sifting* variable ordering algorithm has been proposed. This new form of sifting heuristic takes into account the cardinality of *the Cut-node-set* during *ROBDD/ROMDD* construction. The computer results show that this novel version of sifting algorithm makes the decomposition process of *ROBDD/ROMDD* much easier. *Copyright © 2001 IFAC*

Keywords: *BDD*, functional decomposition, sifting heuristic.

1. INTRODUCTION

The functional decomposition means dividing a complex function into several relatively easier and independent sub-functions. The reason for using such a strategy is a reduction of the input task complexity, according to the well-known "divide and conquer" paradigm. Functions with fewer variables can be implemented independently, and are relatively easier to design. Recently, the functional decomposition is being applied in computer-aided design of *VLSI* (*PLA*'s and *FPGA*'s) (Sasao, 1999; Wan and Perkowski, 1992), as well as in data mining of unmanageably large decision tables (Łuba, *et al.*, 2001; Zupan, *et. al.*, 1997).

In the previous research, the authors focused on the generalization of the Ashenhurst-Curtis decomposition of the decision tables (Łuba, *et al.*, 2001; Sapiecha, *et al.*, 2000). This decomposition method is based on the minimization of the column multiplicity in a decision chart. To solve this problem, various graph coloring algorithms were applied by the authors. The computer tests of the decomposition strategy were conducted particularly in the two computer science fields: a digital circuits design and also in a learning of the weights for the artificial neural networks.

* *Digital circuits*. The initial function is an incompletely specified Boolean or multi-valued function, which is usually given in the truth table form. In such a situation, decomposition involves breaking large functions, which are difficult to implement, into several smaller ones, which can be implemented more easily. Direct application of decomposition methods was a mapping of Boolean and *MVL* function into Field Programmable Gate Arrays (*FPGAs*) and Programmable Logic Arrays (*PLAs*).

* *Neural networks*. Initially, a multi-valued function is composed of training patterns for the considered network, i.e. sets of values for input variables and corresponding expected output values. For large input space, performance of training algorithm decreases significantly, and this makes it often very difficult to obtain reasonable results. When the initial function is decomposed into several smaller ones, the networks can be more easily trained on generated sub-functions.

Unfortunately, the variable partitioning for the functional decomposition appeared to be a crucial problem in the authors' study. This problem seems to be very difficult and complex from a computational point of view. Therefore, authors have been looking for a tool, which can help to solve this problem. There is a branch in logic synthesis, which gives a solid mathematical background helpful in this case, namely the synthesis based on binary decision diagrams. This is the reason for the application of the binary decision diagrams-based decomposition to the considered problem.

A *ROBDD*-based decomposition of Boolean functions is presented in this paper (Yang, *et. al.*, 1999; Lai, *et. al.*, 1996; Chang, *et. al.*, 1996; Sasao, 1999). The classical approach to *BDD*-based decomposition consists of two independent steps: construction of *ROBDD*, and then decomposition of *ROBDD*. In the first step, a variable ordering heuristic is more oriented towards the minimisation of the *ROBDD* size, than towards the minimisation of the cardinality of *the cut-node set* of *ROBDD* – which is a very important parameter in the next step of decomposition. This problem was the reason for our research. Therefore, a modification of the well-known *sifting* variable ordering algorithm (Rudell, 1993) is proposed in this article. This new form of sifting heuristic takes into account the cardinality of *the cut-node set* during *ROBDD* construction. Additionally, the computational complexity of *the k-cut-node set problem* is presented. The new variable ordering strategy was implemented as a software package. The results of our research show that the minimisation of the cardinality of the cut-node set of *ROBDD* makes the decomposition process of *ROBDD* much easier.

2. FUNDAMENTAL CONCEPTS

At the beginning, it is necessary to introduce some basic definitions. The Ordered Binary Decision Diagram is a graphical representation of a logic function, which was introduced by Randy Bryant (Bryant, 1986). More formally speaking:

Definition 1 *The Binary (Multi-Valued) Decision Diagram - BDD (MDD) is a directed acyclic graph (DAG) representing a Boolean (Multi-Valued) function $f(x_0,..,x_{n-1}):\{0,1\}^n \rightarrow \{0,1\}$ ($f(x_0,..,x_{n-1}):\{0,..,m-1\}^n \rightarrow \{0,..,m-1\}$).* It can be defined as a quadruple: $BDD=(V, E, r, \{0,1\})$ (or $MDD=(V, E, r, T)$), where: *V* is a set of internal nodes, *E* is a set of edges, *r* is the root and *T* is a set of terminal nodes (*0,1* are terminal nodes in Boolean case). A *BDD (MDD)* is called *ordered* (*OBDD* and *OMDD*, respectively) if the sequence of variables is the same on each path from root to the sink.

Definition 2 *The Reduced OBDD (ROBDD/ROMDD)* is a diagram obtained from complete *OBDD/OMDD* by a process of merging the isomorphic sub-graphs and deleting nodes whose all edges point at the same node.

Example 1 The presented definitions will be explained in an example. Consider the multi-valued function:

$$f(x1,x2,x3,x4):\{0,1,2\}^3 \rightarrow \{0,1\}$$

which can be represented by means of the following decision diagram such as in the Figure 1. Such a *ROMDD* is *DAG*, which has the root and two sinks – terminal nodes labelled by the constants *0* or *1*. Each non-terminal node is labelled with a variable and has at most three out-going edges, which can be labelled with *0, 1* or *2*.

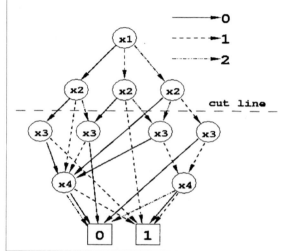

Figure 1. Example ROMDD.

According to the previous research results, the technique of cutting the diagrams for *BDD*-based decomposition is already known (Lai, *et. al.*, 1996; Chang, *et. al.*, 1996).

Definition 3 Given an *ROBDD/ROMDD* node *v* representing $f(x_0,..,x_{n-1})$ and a vector $<b_0,..,b_{i-1}> \in \{0,..,m-1\}^i$, the function *eval* is defined as: $eval(v, <>) = v$, $eval(v, <b_0,..,b_{i-1}>) = v'$, where *v'* is the *ROBDD/ROMDD* representing function $f(b_0,.., b_{i-1}, x_i,.., x_{n-1})$.

3. BDD-BASED DECOMPOSITION

The functional decomposition theory was studied by Ashenhurst (Ashenhurst, 1959), Curtis (Curtis, 1962), Roth and Karp (Roth and Karp, 1962) and Łuba (Łuba, 1995), Wan and Perkowski (Wan and Perkowski, 1992), Sasao (Sasao, 1999), Bratko (Zupan, *et. al.*, 1997) and many other researchers (Yang, *et. al.*, 1999; Lai, *et. al.*, 1996; Chang, *et. al.*, 1996). In this section, we present a method of finding good decompositions by *ROBDD* application. The definitions of decomposition and *Cut_set* in *ROBDD/ROMDD* representation will be given first. Then, basing on the concept of *Cut_set*, a well-known *MDD*-based decomposition algorithm will be described.

Definition 4 A multi-valued function $f(x_0,..,x_{n-1})$ is said to be *decomposed* under *the bound set* $\{x_0,..,x_{i-1}\}$ and *the free set* $\{x_i,..,x_{n-1}\}$, where $0<i<n-1$, if f can be transformed to following form:
$$f(x_0,..,x_{n-1}) = h(g_0(x_0,..,x_{i-1}),.., g_{j-1}(x_0,..,x_{i-1}), x_i,..,x_n).$$

Definition 5 Given an *ROBDD/ROMDD* representing $f(x_0,..,x_{n-1})$ with a variable ordering: $x_0 < ..< x_{n-1}$ and a bound set $B=\{x_0,..,x_{i-1}\}$ we define:
$Cut\_set(f, B)=\{u \mid u=eval(v, p),\ p\in\{0,..,m-1\}^i\}.$

According to the results described in papers (Lai, *et. al.*, 1996; Chang, *et. al.*, 1996), a *ROMDD*-based decomposition algorithm will be presented.

Algorithm 1. Decomposition
Input: Function $f(V)$ represented in an *ROMDD* and a bound set B;
Output: Decomposition with respect to B;
1. Compute *the Cut_set* with respect to B. Let $Cut\_set(f, B)=\{u_0,..,u_{k-1}\}$; Encode each node in *the Cut_set* by $\lceil \log_m k \rceil = j$.
2. Let the encoding of u_q be q;
3. Construct v_h to represent function h by replacing the top part of v_f by a new set of variables: $g_0,..,g_{j-1}$ such that $eval(v_h, q)=u_q$ for $0 \leq q < k-1$, $eval(v_h, q)=u_k$ for $k-1 \leq q < m^j$;
4. Construct v_{gi}' s to represent g_p' s , for $0 \leq p < j$ by replacing each node u with encoding: $b_0,..,b_{j-1}$ in *the Cut_set* by terminal node b_p.

Here, the Theorem 1 (Lai, *et. al.*, 1994) can be considered.

Theorem 1 Given an *ROBDD* with a variable ordering: $x_0<..<x_{n-1}$ representing $f(x_0,..,x_{n-1})$ and a bound set $B=\{x_0,..,x_{i-1}\}$ the $Cut\_set(f, B)=\{u_0,..,u_{k-1}\}$ and the decomposition algorithm returning *OBDD's*: $v_h, v_{g0}.., u_{g\,j-1}$, then:
$f(x_0,..,x_n)=h(g_0(x_0,..,x_{i-1}),..,g_{j-1}(x_0,..,x_{i-1}), x_i,.., x_{n-1})$, where: $h, g_0,..,g_{j-1}$ are the functions denoted by: $v_h, v_{g0}.., u_{g\,j-1}$.

In this moment it is important to point out, that each node in the *Cut_set* corresponds to: *a distinct column* in the Ashenhurst-Curtis method, *a compatible class* in the Roth-Karp decomposition algorithm, *a colour* in the Perkowski graph-based decomposition, and also *a block in intermediate partition* in the Łuba partition-based decomposition method.

Example 2 Once again consider the function: $f(x_1,x_2,x_3,x_4):\{0,1,2\}^3 \rightarrow \{0,1\}$, which was described in the Example 1. Assume that *MDD* for function f with variable ordering: $x_1 < x_2 < x_3 < x_4$ is given. In such a case, it is easy to observe that set $Cut\_set(f, \{x_1,x_2,x_3\})$ consists of: two vertices for

variable $x4$ and two sinks: 0, 1. Therefore, the following decomposition exists:
$$f(x_1,..,x_4)=h(g_0(x_1,x_2,x_3), g_1(x_1,x_2,x_3),x_4).$$

Now consider the following decision problem.

The k-Cut_set problem:
Input: A multi-valued function f and two natural numbers: s, k;
Output: Is there a bound set B of cardinality k, such that $Cut\_set(f, B)$ consists of s vertices ?

This fact expresses the computational complexity of above problem.

Theorem 2 *The k-Cut_set problem* belongs to *NP*-complete class.

Proof (Sketch): It is easy to see that *the k-Cut_set problem* belongs to *NP*-class. To prove that this problem is hard for all problems from *NP*-class, *the t-test collection problem* can be reduced in polynomial time to considered problem (Aussiello, *et. al.*, 1999).

4. VARIABLE REORDERING - A NEW VERSION OF SIFTING ALGORITHM

It is known that *the variable ordering* is the most important problem in the *BDD* construction process (Bollig and Wegener, 1996; Bollig, *et. al.*, 1995; Meinel and Theobald, 1998; Rudell, 1993). According to the Theorem 1, the quality of the decomposition strongly depends on the variable order of the diagrams. Unfortunately, the problem of reducing nodes number for a given *BDD* is *NP*-hard, so the exact algorithms to solve this problem have been unsuitable, up to now (Wegener, 1994).

To reduce the number of nodes of the diagram, a well-known *sifting algorithm* was used, which gives sub-optimal solutions (Rudell, 1993). The main idea of this algorithm can be expressed in the following way. First of all, take one variable and move it up and down the diagram, while other variables remain in constant positions and leave the variable in the position, where number of nodes in the diagram is the least. Then consequently, take another variable, that has not been moved yet, and do the same. Then another one, until every variable has been moved. This procedure is described exactly as the Algorithm 2 in *pseudo-C* code.

In the main loop of this algorithm, every variable is moved up and down in the diagram. The variable is sifted down unless it is the bottom variable. After that, it is sifted up from its beginning position unless it is the top variable. During that time, if the diagram has less nodes than the best already found it is saved as the new best one. The diagram with the best order found is returned at the end. The time complexity is $O(n^2 k \log k)$, where n - is the number of the function variables and k - is the number of edges in the diagram.

Algorithm 2. Sifting
Input: $X - MDD$ to be sifted;
Output: X – sub-optimal MDD;
{ **variables:**
  $MDD\ Y$; // begin for variable
  $MDD\ Z$; // working
  for(every variable a of the diagram)
  { $Y = X$;
    $Z = Y$;
    while(the variable a is not the bottom one)
    { in Z swap variable a with adjacent variable below;
      if(the new diagram Z has fewer nodes than X)
      { $X = Z$; }
    }
    $Z = Y$;
    while(the variable a is not the top one)
    { in Z swap variable a with adjacent variable above;
      if(the new diagram Z has fewer nodes than X)
      { $X = Z$; }
    }
  }
  return X;
}

The sifting algorithm gives some improvements to the variable order and thus to the decomposition. But the goal of this heuristic is *to reduce number of nodes* in the whole diagram, which is not always suitable for the decomposition. To improve the quality of decomposition, a modification of the concept - a "better variable order" was defined. In the decomposition process, *the minimisation of cardinality of the Cut_set(f, B)* seems to be much more important. In Algorithm 3, the variable ordering is computed in a little different manner than it was computed in the previous case. In the case of modified sifting, the aim of the algorithm is to decrease the cardinality of the *Cut_set(f, B')* where B' - is a bound set consisting of x variables,

and x - is an input parameter. In the case, when these two diagrams have exactly the same cardinality of their cut sets, the one with fewer nodes is more optimal.

Algorithm 3. Modified sifting
Input: $X - MDD$ to be sifted, B - bound set variables number;
Output: X – sub-optimal MDD;
{ **variables:**
  $MDD\ Y$; // begin for variable
  $MDD\ Z$; // working
  for(every variable a of the diagram)
  { $Y = X$; $Z = Y$;
    while(the variable a is not the bottom one)
    { in Z swap variable a with adjacent variable below;
      if(the cardinality of $Cut\_set(Z,B)$ is lower than
        cardinality of $Cut\_set(X,B)$ or if they are equal,
        Z consist of fewer nodes than X)
      { $X = Z$; }
    }
    $Z = Y$;
    while(the variable a is not the top one)
    { in Z swap variable a with adjacent variable above;
      if(the cardinality of $Cut\_set(Z,B)$ is lower than
        cardinality of $Cut\_set(X,B)$ or if they are equal,
        Z consist of fewer nodes than X)
      { $X = Z$; }
    }
  }
  return X;
}

---

5. COMPUTER TESTS

The new variable ordering strategy was implemented as a software package at Warsaw University of Technology, in the Institute of Telecommunications. The *gcc* compiler was applied. The computer experiments were conducted on a 2×*Pentium III* 800 MHz machine running *GNU/Linux* operating system. The tests were performed for a set of standard binary benchmarks from Collaborative Benchmarking Laboratory (*http://www.cbl.ncsu.edu/*), having from 6 to over 30 inputs.

Table 1 - Results for binary benchmarks

Name	O_{nr}	I_{cnt}	C_{opt}	C_{heur}	C_{avg}	Q_{heur}	Q_{avg}	Name	O_{nr}	I_{cnt}	C_{opt}	C_{heur}	C_{avg}	Q_{heur}	Q_{avg}	Name	O_{nr}	I_{cnt}	C_{opt}	C_{heur}	C_{avg}	Q_{heur}	Q_{avg}
5xp1	3	6	2	2	3.67	1.00	1.83	5xp1	0	7	3	4	5.14	1.33	1.71	Sse	7	8	2	3	2.98	1.50	1.49
Adr4	2	6	2	2	3.33	1.00	1.67	5xp1	1	7	5	5	6.33	1.00	1.27	Vg2	5	8	3	5	7.43	1.67	2.48
Beect	1	6	2	3	3.17	1.50	1.58	5xp1	2	7	3	3	6.71	1.00	2.24	Vg2	7	8	3	5	7.43	1.67	2.48
Beect	2	6	2	4	3.17	2.00	1.58	Bulm	1	7	2	3	3.29	1.50	1.64	Clip	0	9	5	6	10.31	1.20	2.06
Beect	3	6	3	4	3.17	1.33	1.06	Dk16	0	7	10	10	12.71	1.00	1.27	Clip	1	9	6	6	13.35	1.00	2.23
Beect	4	6	2	4	3.67	2.00	1.83	Dk16	1	7	9	10	10.86	1.11	1.21	Clip	2	9	4	4	17.34	1.00	4.34
Beect	5	6	3	4	3.33	1.33	1.11	Dk16	2	7	7	7	9.62	1.00	1.37	Clip	3	9	4	4	18.60	1.00	4.65
Beect	6	6	2	4	3.67	2.00	1.83	Dk16	3	7	7	7	9.29	1.00	1.33	Cse	1	9	8	9	11.67	1.13	1.46
Bulm	0	6	3	3	3.17	1.00	1.06	Dk16	4	7	7	7	9.86	1.00	1.41	Cse	4	9	4	5	7.23	1.25	1.81
Con1	0	6	3	3	3.50	1.00	1.17	Dk16	5	7	5	5	6.71	1.00	1.34	Cse	5	9	2	2	3.43	1.00	1.71

Name	G	Icnt	Copt	Cavg	Cheur	Qavg	Qheur
Conf	0	6	3	3	3.50	1.00	1.17
Dk14	1	6	3	3	3.83	1.00	1.28
Dk14	2	6	3	3	3.33	1.00	1.11
Dk14	4	6	3	4	3.67	1.33	1.22
Dk14	5	6	3	4	3.83	1.33	1.28
Dk14	6	6	3	3	3.17	1.00	1.06
Dk14	7	6	3	3	3.67	1.00	1.22
Ex7	0	6	3	3	3.67	1.00	1.22
Ex7	1	6	3	4	3.67	1.33	1.22
Ex7	2	6	2	2	2.33	1.00	1.17
Ex7	3	6	3	3	3.50	1.00	1.17
Ex7	4	6	2	2	3.00	1.00	1.50
Ex7	5	6	2	2	3.00	1.00	1.50
F51m	2	6	2	2	3.67	1.00	1.83
I20	0	6	3	3	3.67	1.00	1.22
Misex1	1	6	3	3	3.50	1.00	1.17
Misex1	5	6	3	4	3.67	1.33	1.22
Misex1	6	6	3	4	3.50	1.33	1.17
Opus	7	6	2	2	2.50	1.00	1.25
Opus	9	6	2	2	2.67	1.00	1.33
Sqr	2	6	3	3	3.83	1.00	1.28
Sqrt8	1	6	3	3	3.67	1.00	1.22
Sse	8	6	2	3	2.33	1.50	1.17
Tav	2	6	2	3	2.83	1.50	1.42
Tav	4	6	2	3	2.67	1.50	1.33
Tra11	0	6	3	3	3.50	1.00	1.17
Tra11	1	6	3	3	3.67	1.00	1.22
Tra11	2	6	3	3	3.83	1.00	1.28
Tra11	3	6	2	3	3.33	1.50	1.67
Tra11	4	6	3	3	3.83	1.00	1.28

Name	G	Icnt	Copt	Cavg	Cheur	Qavg	Qheur
Dk16	6	7	6	7	8.38	1.17	1.40
Donfl	0	7	7	7	9.33	1.00	1.33
Donfl	1	7	5	8	8.05	1.60	1.61
Donfl	2	7	7	10	9.38	1.43	1.34
Donfl	3	7	7	7	8.52	1.00	1.22
Donfl	4	7	8	8	9.05	1.00	1.13
Duke2	3	7	4	4	4.90	1.00	1.23
Duke2	25	7	2	2	2.71	1.00	1.36
F51m	1	7	3	3	6.71	1.00	2.24
Misex1	2	7	4	4	5.05	1.00	1.26
Misex1	3	7	3	4	4.10	1.33	1.37
Misex2	7	7	2	3	2.95	1.50	1.48
S8	0	7	4	4	4.67	1.00	1.17
S8	1	7	3	4	4.52	1.33	1.51
S8	2	7	2	3	2.86	1.50	1.43
S8	3	7	3	3	4.14	1.00	1.38
Sqn	2	7	4	6	6.57	1.50	1.64
Z4	1	7	2	2	5.33	1.00	2.67
Adr4	0	8	3	3	7.07	1.00	2.36
Adr4	1	8	3	3	7.96	1.00	2.66
Duke2	7	8	3	3	4.14	1.00	1.38
Duke2	10	8	4	4	5.96	1.00	1.49
Duke2	21	8	3	3	4.36	1.00	1.45
Duke2	22	8	3	3	4.30	1.00	1.44
f51m	0	8	5	5	10.20	1.00	2.04
Root	2	8	4	4	8.46	1.00	2.12
Root	3	8	6	6	13.12	1.00	2.19
Root	4	8	11	11	17.77	1.00	1.62
Sqrt8	0	8	5	5	12.16	1.00	2.43
Sse	5	8	2	3	3.21	1.50	1.61

Name	G	Icnt	Copt	Cavg	Cheur	Qavg	Qheur
Cse	6	9	2	2	2.83	1.00	1.42
Cse	7	9	4	6	5.67	1.50	1.42
Misex2	12	9	4	4	5.20	1.00	1.30
Opus	1	9	4	7	6.29	1.75	1.57
Opus	3	9	3	4	4.50	1.33	1.50
Opus	6	9	3	4	3.99	1.33	1.33
Sse	1	9	4	5	6.45	1.25	1.61
Alu2	0	10	6	8	14.71	1.33	2.45
Alu2	1	10	14	28	25.53	2.00	1.82
Alu2	4	10	14	16	21.53	1.14	1.54
Cse	0	10	7	9	12.83	1.29	1.83
Cse	2	10	5	8	10.67	1.60	2.14
Cse	3	10	6	8	10.71	1.33	1.79
Cse	9	10	4	4	6.42	1.00	1.61
Cse	10	10	3	4	4.01	1.33	1.34
Misex2	3	10	3	3	3.08	1.00	1.03
Sao2	0	10	6	7	8.52	1.17	1.42
Sao2	1	10	5	5	8.89	1.00	1.78
Sao2	2	10	5	6	9.52	1.20	1.91
Sao2	3	10	6	7	8.96	1.17	1.49
Sse	2	10	5	6	6.94	1.20	1.39
Sse	3	10	6	7	7.54	1.17	1.26
Sse	4	10	4	5	5.72	1.25	1.43
Duke2	0	11	3	4	4.73	1.33	1.58
Duke2	24	14	4	6	7.20	1.50	1.80
Misex2	9	14	2	3	3.26	1.50	1.63
Misex2	10	14	2	3	3.09	1.50	1.54
Misex2	11	14	2	3	3.16	1.50	1.58
Vg2	0	14	2	4	6.06	2.00	3.03
Vg2	2	14	2	3	6.06	1.50	3.03

Table 2 - Results for multi-valued benchmarks

Name	G	I_{cnt}	C_{opt}	C_{avg}	C_{heur}	Q_{avg}	Q_{heur}
Mm3	3	5	5.00	7.80	5.00	1.56	1.00
Mm4	3	5	7.00	16.20	7.00	2.31	1.00
Mm5	3	5	9.00	30.60	9.00	3.40	1.00
Monks1te	4	6	2.00	2.93	2.00	1.46	1.00
Monks1tr	4	6	2.00	4.00	2.00	2.00	1.00
Monks2te	4	6	2.00	2.93	2.00	1.46	1.00
Monks2tr	4	6	4.00	4.13	5.00	1.04	1.25
Monks3te	4	6	2.00	2.67	3.00	1.34	1.50
Monks3tr	4	6	4.00	5.07	4.00	1.26	1.00
Pal2	4	6	2.00	4.40	2.00	2.20	1.00
Pal3	4	6	2.00	8.40	2.00	4.20	1.00
Tic-tac-toe	5	9	5.00	7.41	7.00	1.48	1.40
Nursery	4	8	3.00	23.24	9.00	7.75	3.00

Two software tools were used. Namely, the *MADD* - a tool to create decision diagrams from decision tables to find sub-optimal variable ordering for decomposition (as described in this article), and the *HOSEA* - a tool that applies exact serial and parallel decomposition to decision tables using given variable ordering (Łuba, *et. al.*, 2001; Sapiecha, *et. al.*, 2000).

The main goal of our experiments was to test the quality of functional decomposition based on variable ordering, which was the result od *sifting heuristic* application. A natural measure of a decision diagrams-based decomposition quality is the cardinality of the *Cut_set - Card(Cut_set)* for short.

The tests were performed as follows. First, a parallel decomposition was applied to each function in order to divide it into 1-output sub-functions. the argument reduction algorithm was applied to each single output, in order to find only these inputs, this particular output depends on. After that, an exhaustive search of all possible two-level serial decompositions with 5 bound variables was performed on each sub-function, to find the optimal one (the one with smallest *Card(Cut_set)*) and to find the average *Card(Cut_set)* used for all decompositions. Then, a decision diagram for this sub-function was constructed, and *sifting heuristic* was applied to it. After finding a sub-optimal variable ordering with this heuristic, the *Card(Cut_set)* used for two-level decomposition with this variable ordering was counted.

Table 1 shows the results of these tests in a following way. *Name* - is the name of the test benchmark. O_{nr} - is the number of output of this function (outputs are numbered starting from 0). I_{cnt} - is the number of inputs to this sub-function (which can be smaller than the total number of function inputs, since not all outputs are dependent on values of all inputs). Consequently, C_{opt} - is the *Card(Cut_set)* in the optimal decomposition, C_{heur} - is the *Card(Cut_set)* in the decomposition performed using variable ordering found with *sifting heuristic*, and C_{avg} - is the average *Card(Cut_set)* of all possible decompositions. Q_{heur} and Q_{avg} are quality measures calculated as follows: $Q_{heur}=C_{heur}/C_{opt}$, and $Q_{avg}=C_{avg}/C_{opt}$. This table shows only nontrivial examples, where decompositions with different *Card(Cut_set)* number exist ($Q_{avg}>1$).

Table 2 shows the results for a set of well-known multi-valued benchmarks. Results are shown in a similar way, except that (since all functions had only one output) there is no need to give the number of output and the number of inputs. The one additional parameter *G* is given, which is the number of bound variables (for all binary benchmarks *G=5*).

It can be observed that in more than 50% cases a *sifting heuristic* allowed to find optimal decomposition ($Q_{heur}=1$). Otherwise, to find an optimal one, it would be necessary to perform exhaustive (or nearly exhaustive) search, which requires exponential time. Additionally, in over 80% cases it gave better results that average ($Q_{heur}<Q_{avg}$), which is the expected *Card(Cut_set)* when, instead of exhaustive search, one tries to find and apply a random variable ordering. These results allow to say that application of decision diagrams and *sifting heuristic* can be a valuable tool for solving variable ordering problems of decomposition.

6. CONCLUSIONS

In this paper, the *ROBDD/ROMDD*-based decomposition of Boolean and multi-valued functions was presented. The modification of the well-known *sifting* variable ordering algorithm was proposed by the authors. This new form of sifting heuristic takes into account very important parameter for *ROBDD*-based decomposition, namely the cardinality of *the cut-node set*. According to the authors' knowledge, this novel sifting method is the first method directly oriented to solve the problem

of decomposition. The new variable ordering strategy was implemented as a software package. The experimental results showed the benefits of new sifting heuristic application in comparison with the results achieved using classical methods.

REFERENCES

Ashenhurst R. (1959). The decomposition of switching functions. Ann Comp. Lab. Harvard Univ., vol. 29.

Ausiello G. and many others authors (1999). Complexity and Approximation. Springer.

Bollig B., M. Löbbing, I. Wegener (1995). Simulated Annealing to improve variable ordering for OBDDs. Proc. Int'l Workshop Logic Synthesis.

Bollig B., I. Wegener (1996). Improving the variable ordering of OBDDs is NP-complete. IEEE Transactions on Computers, vol. 45, No. 9.

Boros E., V. Gurvich, P. Hammer, T. Ibaraki, A. Kogan (1994). Decomposition of partially defined Boolean functions. DIMACS TR 94-9.

Bryant R. E. (1986). Graph-based algorithms for Boolean function manipulation. IEEE Transactions on Computers, vol. C-35, No. 8.

Chang S., M. Marek-Sadowska, T. Hwang (1996). Technology mapping for TLU FPGA's based on decomposition of Binary Decision Diagrams. IEEE Transaction on Computers, vol. 15, No. 10.

Curtis A. (1969). New Approach to the design of switching circuits. Van Nostrand, Princeton, NJ.

Lai Y., M. Pedram, S. Vrudhula (1994). EVOBDD-based algorithms for integer linear programming, spectral transformation, and functional decomposition. IEEE Transaction on Computers, vol. 13, No. 8.

Lai Y., K. Pan, M. Pedram (1996). OBDD-based functional decomposition: algorithms and implementation. IEEE Transaction on Computers, vol. 15, No. 8.

Łuba T. (1995). Decomposition on multiple-valued functions. Proc. IEEE-ISMVL, USA.

Łuba T., H. Niewiadomski, H. Pleban, H. Selvaraj, P. Sapiecha (2001). Functional decomposition and its application in design of digital circuits and machine learning", IASTED Inter. Conf. On Applied Informatics.

Meinel C., T. Theobald (1998). Ordered Binary Decision Diagrams and their significance in computer-aided design of VLSI circuits – a survey. Electronic Colloquium on Computational Complexity, Report No. 39.

Perkowski M., R. Malvi, S. Grygiel, M. Burns, A. Mishchenko (1999). Graph coloring algorithms for fast evaluation of Curtis decompositions. DAC.

Roth J., R. Karp (1962). Minimization over boolean graph. IBM J. Res. And Develop.

Rudell R. (1993). Dynamic variable ordering for ordered binary decision diagrams. DAC.

Sapiecha P., H. Selvaraj, M. Pleban (2000). Decomposition of Boolean relation and functions in logic synthesis and data analysis. Inter. Conf. Rough Sets and Current Trends in Computing, Canada.

Sasao T. (1993). FPGA design by generalized functional decomposition. Kluwer Academic Publishers.

Sasao T. (1999). Switching theory for logic synthesis. Kluwer Academic Publishers.

Wan W., M. Perkowski (1992). A new approach to the decomposition of incompletely specified multi-output functions based on graph coloring and local transformations and its applications to FPGA mapping. DAC.

Wegener I. (1994). The size of reduced OBDD's and optimal read-once branching programs for almost all Boolean functions. IEEE Transactions on Computers, vol. 43, No. 11.

Yang C., V. Singhal, M. Ciesielski (1999). BDD decomposition for efficient logic synthesis. International Conference on Computer Design.

Yang C., M. Ciesielski, V. Singhal (2000). BDS: a BDD-based logic optimization System. DAC.

Zupan B., M. Bohanec, J. Demšar, I. Bratko (1997). Feature transformation by function decomposition. IEEE Expert, Special Issue on Feature Transformation.

IFAC
Publications
www.elsevier.com/locate/ifac

EXTENSIBLE DEVELOPMENT FRAMEWORK FOR IMPLEMENTATION OF GRAPHICALLY EXPRESSED DESIGN TOOLS

Jindřich Černohorský [*,1] Gustav Hrudka [*]
Petr Kovář [*] Eliška Ochodková [*]

* VŠB – Technical University of Ostrava, Czech Republic

Abstract: The paper describes the principles of construction of a general development extensible framework oriented to the implementation of software tools applicable in design of various computer-based design and modelling systems. The application framework is based on exploitation of Black Box Component Builder development framework and extensions can be oriented both to hardware and software system development. Copyright © 2001 IFAC

Keywords: CASE, Design systems, Graph theory, Modelling, Programming environments, Real-time, Software engineering

1. INTRODUCTION

Every CASE tool usually uses an environment, which enables manipulation with block–oriented diagrams. There may be found two main problems in such type of application. The first requirement is to develop an open system with all possibilities they are necessary to realize block scheme oriented tool. The second part of the problem is related to the analysis of scheme situation. Methods of the graph theory can be used to analyze the scheme situation because the scheme can be represented by sets of nodes and edges. Edges can be oriented or not and nodes and edges can have some attributes (name, value etc.) and they can represent different entities in the scheme (like platforms, class packages etc.). The graph theory offers a lot of very powerful methods to analyze such type of data. The goal is to develop an open development environment providing all facilities that are necessary to create and analyze the block diagrams. This environment is desired to be easily modifiable with respect to specific application goal (to

[1] Supported by the Czech Grant Agency, Grant no. 102/01/0803

Fig. 1. Architecture of "framework" for system development and architecture design

solve specific software architecture using visual modelling methods). The situation is illustrated in figure 1.

Now there exist a lot of methods in software engineering, which use visual modelling to design system architecture. They are usually object–oriented or component–oriented. These methods are very popular because they are very illustrative and they use objects as elementary structural units in all phases (analysis, design, implementation).

The object–oriented design is based on objects, classes, their relationships and behavior. Visual modelling methods use specific graphic notation for system representation. The UML is one of the methods that use visual modelling for software design and it is very often discussed. Some of such methods may be too complex and difficult to understand or they may be incomplete to solve specific software requirements.

However control–engineering applications are not purely software oriented. There are tasks also dealing with hardware elements of various types that are subjects of design and which could be also somehow related to software parts of the system. They can or need not use graphs for modelling of their features. In all design phases the system design data created by one method should be accessible by any other method preferably. Therefore it is important so as to all tasks to have an access to the common data and structures and so as to easily mutually communicate. The common CASE systems usually does not cover these extra–software requirements.

The idea is to create a flexible graphical framework for realization of schemes representing not only software system architecture but also its hardware aspects. The framework must also provide an interface with functions they are necessary to analyze the scheme situation using methods of the graph theory. This way should framework adopt new elements, which are necessary for design of any specific software architecture. For example some methods like DARTS or CODARTS (which are very well defined to design RT system architectures) are not implemented in CASE based systems yet.

Some software–based systems must guarantee specific level of their reliability. Very effective way how to establish these requirements is to use particular design patterns related to system reliability aspects. Many safety critical systems use design patterns, which duplicate parts of their channels. The homogeneous or heterogeneous redundancy or watchdog components are the most frequently used solutions in safety critical systems. Their design can be predefined in the framework environment and a designer can choose the most suitable pattern for his project. Some of these design patterns are listed below:

- Homogeneous redundancy pattern (Critical channels are exact clones)
- Diverse redundancy pattern (Critical channels are duplicated with light redundancy channel or different but equivalent processing channel)
- Monitor–Actuator pattern
- Watchdog pattern
- Safety executive pattern

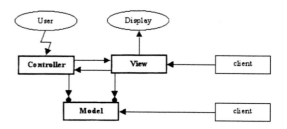

Fig. 2. MVC paradigm

2. FRAMEWORK FOR REALIZATION OF DEVELOPMENT ENVIRONMENT

The realization of application providing possibilities for manipulation with block diagrams is quite complex especially if we want to keep large flexibility of application. That's why the framework is used to implement graphical development environment because it usually offers an existing solution in some specific problem domain. Some frameworks are oriented to CAD's, visualization or simulation systems etc. Some of them use component technology in their own architecture. They are useful especially if they can provide or they can use components providing an OLE container for document embedding. On the other hand these frameworks can be also too complex and difficult to understand in a short time. That is why they usually collect a set of applicable design patterns to make their use and development easier. Component frameworks, that are based on MVC (Model–View–Controller) paradigm are very useful because they enable easy and sometimes an intuitive decomposition into three main problem parts: Model, View, Controller, see (Szyperski 1998), (Douglass 1998).

3. BLACKBOX COMPONENT FRAMEWORK

BlackBox component framework (BFC) is used to implement the kernel of the graphical environment. BFC is the component framework based on MVC architecture implemented in Component Pascal. BFC was designed and implemented by Oberon microsystems, Inc. Technopark Zrich, Switzerland.

The most fundamental advantage of the BCF lies in its design. In engineering, design is the art of reconciling different, and often conflicting, goals into one elegant and relevant whole. The design of the BCF and its component framework leads to a unique combination of advantages:

- **Rapid development** combined with **evolvability**
- **Safety** combined with **efficiency**
- **Power** combined with **simplicity**
- **Integration** combined with **platform independence**

230

The term *framework* is used because BCF offers class library, which predefines the structure of a certain category of applications. The most important feature is that BCF defines a general abstraction for containers. On Windows, container views look and feel like ActiveX containers; on Mac OS, container views look and feel like Open-Doc containers. Abstract interfaces can be implemented by the user to concretize and reach his particular application goals. BCF also enables that user modules are dynamically linked to the framework kernel and they can be added to the framework without recompiling the whole project again. These aspects make the BCF very useful to implement our application goal, see (Hrudka 2001).

4. GRAPH THEORY AS A BASE FOR ANALYZING METHODS

By developing some CASE tool a block oriented scheme is made. Such a scheme (see figure 5) can be studied as a mathematical structure – a graph. In the next paragraph a suitable graph structure and its implementation in the BFC will be presented. Also some examples will be shown what kind of important information can be gained from the scheme using graph algorithms.

4.1 Graph implementations

Usually a graph G consists of a nonempty vertex set $V(G)$ and an edge set $E(G)$ where each edge consist of two (possibly equal) vertices called the endpoints.

$$G = (V, E),$$
$$\text{where } V \neq \emptyset \wedge E \subseteq P_2(V).$$

But the diagram in figure 5 has a more complex structure than a simple graph. There are not only vertices connected by edges. There are two basic relationships between the vertices (boxes):

- two vertices are connected by an edge
- a couple of vertices have some same property: they are inside of another vertex

A common extension of a graph is a hypergraph, where an *edge* is formed not by two but by a couple of vertices.

$$H = (V, E),$$
$$\text{where } V \neq \emptyset \wedge E \subseteq P(V).$$

Such a structure is closer to the diagram mentioned above, but there is only one kind of relationship can be described by the structure H.

There has to be a new structure introduced to describe all the relationships between vertices. Since a vertex should consist of a couple of other vertices, a set of basic elements U will be defined. A vertex set V is a nonempty set of all subsets of U. Edges are pairs of vertices from the set V.

$$F = (V, E),$$
$$\text{where } U \neq \emptyset \wedge \emptyset \neq V \subseteq P(U) \wedge E \subseteq P_2(V).$$

The elements of V are a kind of hypervertices but can be called simply vertices since the graph F has the same high–level structure as G. The low–level information given by the subsets $P(U)$ gives a more detailed view on the matter. On the other hand a great profit is achieved by setting $E(F)$ as $P_2(V)$ because then F resembles to a simple graph.

If $v_1, v_2 \in V$ and $v_1 \subseteq v_2$ let us say $v_1 \in v_2$. So it is correct to speak about vertices that are *inside* of another vertex.

The implementation is:

```
HYPERVERTEX
{ NAME
  list of pointers to EDGES
  list of pointers to HYPERVERTICES
  CAPACITY
  ... other parameters
}

HRANA
{ NAME
  boolean IS_ORIENTED
  pointer to HEAD HYPERVERTEX
  pointer to TAIL HYPERVERTEX
  CAPACITY
  ... other parameters
}
```

The advantages of this structure are:

- It describes the structure of picture 3 well. Vertices can have some property in common (to be inside of another vertex), the vertices are connected with edges.
- F can be treated as a simple graph. Using transparent methods all common algorithms will run on F!
- F can be both oriented or not oriented.
- Multiple edges or loops can be easily implemented as well using the list of pointers.
- Even more information can be described by the structure of F, for example edges between vertices of different level (see figure 6).

On the other hand there might appear some complications while programming transparent methods.

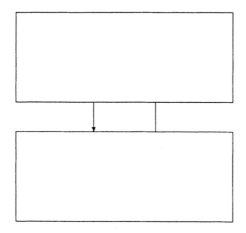

Fig. 3. A raw scheme

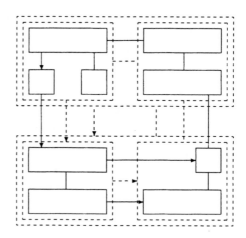

Fig. 5. The complete model

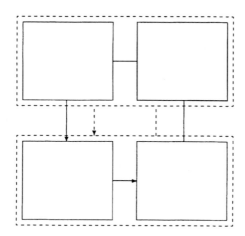

Fig. 4. Some more details added

4.2 Hypervertices, edges and their interpretation

It is natural to construct a model in such a way, that a raw scheme is done (see figure 3). Later more some details are added (see figure 4). Finally the complete model is done (see figure 5).

During the construction of the model several kinds of edges are added, but not all of them have the same importance. Some of them are useful only during the drawing of the model, some of them appear as parallel edges on different levels (see figure 6). The editor and the low–level methods concerning incidence must work properly. It is obvious, that it might need interactive approach.

4.3 Applications – gaining information from the graph

Graph theory gives a great amount of information studying a graph structure, see (E. Fuchs 2000), (Plesník 1983). Some basic algorithms are con-

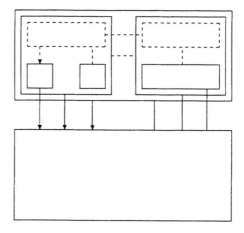

Fig. 6. Edges between different levels

cerning connectivity problems, minimum cuts or spanning tree problems or maximal flow problems.

Graph connectivity. In not oriented graphs it can be necessary to learn whether a connection between the vertices exists or not. In oriented graph there can be a connection that goes only in one direction. In electricity or electronic schemes it gives an information about possible shortcuts. In a communication network such algorithms shows whether an information can run from one point to other point or gives a list of all vertices where an information can arrive. In general the influence of a local information on the whole structure can be obtained.

Minimal cuts. Edge cuts and vertex cuts give the information about stability of a system. What happens if one or more edges (vertices) are missing? What is the weakest part of a model? Algorithms looking for the minimal cuts can help in studying the stability especially in information networks. By running the graph algorithms it is easy to find the channel or node whose malfunction would cause a collapse of the whole system (in production schemes, planning diagrams, data flow, etc.).

Minimum spanning tree. In special situations it is useful to find the minimal configuration that keeps the system running.

Maximum flow problem. It might be useful to learn about the upper limits of a system if all capacities of all edges are given. What is the capacity of an information network? What amount of information, goods or energy can go through the system?

5. CONCLUSION

Framework providing this graphical capabilities and interfaces can be used to many particular application goals. In this case, we assume the development environment for design and implementation of control systems or distributed systems but there are a lot of applicable domains in common engineering systems. Methods of scheme analysis are general and can be applied to many specific applications. Applications that use framework technology in their architecture are also more reliable because the main part of their architecture is shared and tested for the longest time. Frameworks then can be used to solve very large and complex applications.

REFERENCES

Douglass, B.P. (1998). *Real-Time UML, Developping Efficient Objects For Embedded Systems.* Addison–Wesley.

E. Fuchs, P. Kovář (2000). *Graph algorithms.* Masaryk University. Brno.

Hrudka, G. (2001). Komponentní framework jako ruchlá cesta ke spolehlivým a výkonným aplikacím. *Autos2001.*

Plesník, J. (1983). *Graph algorithms.* Slovak Science Academy. Bratislava.

Szyperski, C. (1998). *Component Software – Beyond Object Oriented Programming.* Addison–Wesley. ACM Press New York.

IFAC
Publications
www.elsevier.com/locate/ifac

TIMING-DRIVEN ADAPTIVE MAPPER FOR LUT-BASED FPGAS

Martin Daněk [1]

*Department of Computer Science and Engineering, Czech
Technical University, Faculty of Electrical Engineering,
Karlovo nám. 13, Praha 2, 121 35*

Abstract: The timing-driven adaptive mapper is a rule-based mapper that opti-
mizes its performance for a user-specified FPGA architecture. It uses the XCS
classifier system with a feedback function that considers area requirements and
signal delays. Current results are presented and a modification of the classifier
system is presented that improves its performance. The most important advantage
is that the mapper discovers on its own the optimal mapping policy for the target
FPGA device. The mapper has been tested on Xilinx XC4000 devices, but it
can use any device given an architecture specification and a proper feedback are
provided. *Copyright © 2001 IFAC*

Keywords: computer-aided circuit design, classifiers, rule-based systems, adaptive
algorithms, self-adapting algorithms, machine learning

1. INTRODUCTION

Mapping should transform an abstract Boolean
network to a network of technology dependent
units. In the case of Xilinx XC4000 devices the
technology dependent units are up to four-input
look-up tables and edge-triggered D-type flip-
flops. The task of technology mapping in the case
of LUT-based FPGAs is to find a set of clusters of technology dependent units that cover the
whole Boolean network. A good cluster selection
mechanism should take into account both the
internal structure of the configurable logic block
(CLB) of the target device (for an example, see
Xilinx (1994, 2001)) and routing delays in order
to achieve good area and performance results.

The mapping task is an NP-complete problem
(Murgai *et al.* (1995)). The prevailing way to
tackle it is to implement a heuristic algorithm

based on a previously gained human knowledge
of the target architecture. This approach may
be limiting for more complex devices, because
they may contain features not recognizable at first
sight. In addition, there is the question of considering signal delays in the mapping phase. Up-
to-date mappers approximate delays with a unit
delay per one level of logic, or they use signal delay
values that were generated in previous iterations
(Cong *et al.* (1994)). The unit-delay approach is
very fast and is optimal in situations where the
design topology is a sort of a planar graph. The
iterative approach is time-consuming, because it
requires several mapping runs interleaved with
running placement and routing algorithms, and
it does not guarantee the exactness of delay es-
timations, because they are reliable only in cases
when small modifications take place.

The adaptive mapper presented here uses techniques from the field of artificial intelligence (AI)
to overcome the dependence of performance on
the human factor involved in the design of the

[1] This work has been supported by the Grant Agency of
the Czech Republic under Grant No. 102/99/1017.

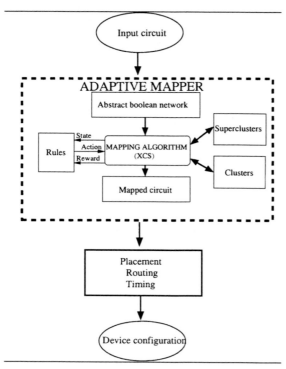

Fig. 1. Adaptive mapper - program flow.

mapping algorithm. It is shown that it is capable of learning an optimal mapping policy for the Xilinx XC4000 FPGAs. Further, it is demonstrated that the gained knowledge can be successfully transferred to different designs. Finally, a modification of the AI algorithm is described that improves its performance.

2. ADAPTIVE MAPPER

The adaptive mapper (Fig. 1) is a rule-based single-pass mapper. It is based on Wilson's XCS classifier system (Wilson (1995)). The mapper operates in two modes: in the first, training mode it evolves a set of mapping rules that perform well for specific designs; in the second, mapping mode it uses the pre-evolved set of rules to map a user-specified design in a single pass. The training mode consists of repeated mapping trials, when the mapper builds an effective population of rules. The training mode is executed only once to adapt the mapper to a specific FPGA architecture, the intended usual way of using it is the mapping mode.

2.1 Classifier system

The classifier system (Holland (1975, 1992, 1995)) is a rule-based machine that operates in the problem environment; it works in two phases: first, it senses the current state, chooses a suitable rule and performs a corresponding action, second, it

receives a feedback from the environment for the action taken and rewards the rule. Rules with poor performance are replaced with rules generated by a genetic algorithm. This evoutionary factor is what makes this approach interesting when compared to classical algorithms: the classifier system is capable of adapting to individual problem environments.

The classifier system calls elementary mapping actions and it evolves sequences of rules that lead to a minimal CLB usage and small critical signal path delays. The feedback provided to the system is a reward based on the reduction of the number of CLBs used in the design and on the estimated critical signal path delay, the reward is provided to the classifier system during the mapping process.

The adaptive nature of the classifier system ensures that the final rules take into account global properties of the FPGA chip. Initially the mapper is aware only of local properties of the chip - the internal architecture of CLBs. At the end of the run it knows how to assemble efficient sequences of actions that maximize the feedback function. The quality of the final rules depends on the quality of the feedback function and on the features of the training set.

2.2 Mapping actions

There are three groups of mapping actions: actions that construct clusters that stand for lookup tables (LUTs), actions that assign these clusters and flip-flops to superclusters (a supercluster is equivalent to a CLB), and actions that modify the current position in the netlist.

The first group contains these actions:

- assign a block to a cluster,
- add a block to a cluster.

The actions in the second group reflect the architecture of the configurable logic block:

- assign a cluster to FLUT, GLUT, or HLUT
- assign a flip-flop to DXFF, or DYFF
- generate a CLB.

The third group consists of actions that implement different *move backward/forward* commands.

The total number of actions available to the mapper is 15, the actions are binary coded using four bits (the only objective here is to use the most efficient encoding, because each rule contains only one action).

2.3 Rewards

The processing of rewards is based on reinforcement learning techniques (Barto and Sutton (1998)). To formulate the reward function correctly, it is necessary to declare the goals of the task correctly. The final goal is to generate efficiently a high-performance mapping of a design, which means

- to use as few CLBs as possible,
- to minimize the critical path delay,
- to accomplish it in as few steps as possible.

This suggests that the reward should reflect the decrease in the number of CLBs used in the design and the decrease in the critical path delay, and that each (unsuccessful) execution of an action should be given a negative reward. A closer analysis of the interaction of the actions suggests to introduce another goal to reduce the sparsity of receiving rewards from the environment:

- to use CLBs to their capacity (see Xilinx (1994)).

The reward received on a successful completion of an action then is:

$$reward = \quad \varphi * (\Delta numCLB) +$$
$$\psi * (\Delta \sum pathdelay) +$$
$$\omega * CLBusage,$$

where $\Delta numCLB$ is the change in the number of CLBs used by the design at a time step t, $\Delta \sum pathdelay$ is the change in the sum of all (two-point) signal path delays at a time step t, and $CLBusage$ is the utilisation of the possibly generated CLB or zero. Actual values of φ, ψ, ω are stated in Section 3.

2.4 Perception of the environment

In addition to the reward function, the performance of the mapper is determined by the quality of its perceptions of the mapping environment, i.e. by the structure of the sense vector. The sense vector must reflect all possibly important facts encountered during the mapping process, because the mapper cannot reconstruct them on its own.

The sense vector used in the adaptive mapper consists of equal groups of flags, the groups represent the current block to be processed and blocks connected to its inputs and outputs (at present, only the maximum of four input and four output blocks are considered), plus an extra group of flags that validate the previous groups and that reflect the utilisation of LUTs and D-FFs of the CLB currently under construction (the CLB that will

be generated by the next *generate CLB* action). Each group contains this information:

- addable to FLUT, GLUT, HLUT,
- addable to DXFF, DYFF,
- connected to the current FLUT, GLUT, HLUT
- connected to the current DXFF, DYFF.

All flags are coded by one bit, the length of the resulting sense vector is 102 bits. Although this represents a large state space, the advantage of such a coding is the clear meaning of genetic operations performed by the classifier system.

3. EXPERIMENTS

The adaptive mapper has been tested to prove that it can adapt without human intervention to a user-specified architecture of a logic block and that a rule-set evolved for one design can be successfully used for different designs, the results are presented below. Several test runs have been performed, the target architecture for all runs was the XC4000 CLB, the delay estimation used the unit delay model, and the evaluation was done for MCNC91 benchmark circuits.

Parameter values used in experiments were:

- Reward function: $\varphi = 7, \psi = 12, \omega = 4$ - these parameters define the importance of CLB packing vs. critical signal path delay reduction.
- Reinforcement learning: $\beta = 0.2, \gamma = 0.8$ - these parameters define the learning rate and the coupling between two consecutive actions.
- Exploration policy: Constant exploration strategy (Wilson (1996)) was used, the exploration probability was 0.2. Each explore run (a complete mapping) was followed by one exploit run of the mapper.

All plots shown contain values only for the exploit runs.

3.1 Adapting to a specific architecture

The principal results of this experiment are shown in Figures 2 and 3 (two figures are shown to demonstrate the performance for examples of combinatorial and sequential circuits), and a statistical overview of the learning task is shown in Figures 4 and 5.

At the beginning, the mapper uses randomly generated mapping rules, this corresponds to the initial region with low performance. Then, as more mapping trials are performed, the quality of the

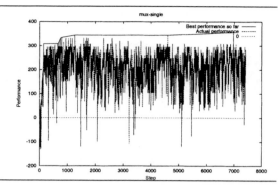

Fig. 2. Improvement in performance - combinatorial circuit (MUX), training mode, single run.

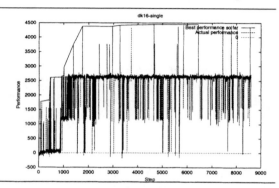

Fig. 3. Improvement in performance - sequential circuit (DK16), training mode, single run.

mapped circuit in terms of the overall reward improves.

It is good to notice the difference between Figures 2 and 3. The initial flat region lasts for more mapping steps in the case of the sequential circuit than for the combinatorial circuit. This is due to the fact that to learn an optimal mapping policy for sequential circuits is more difficult than for combinatorial circuits. Sequential circuits employ an additional logic element not found in combinatorial circuits (a flip-flop), and the critical path there is formed by stages of combinatorial circuits connected by flip-flops, which means that changes caused by mapping actions have a more local effect.

The difference between Figures 4 and 5 shows that simple circuits can be mastered more quickly and efficiently than more complex circuits. This is in accordance with results stated in (Lanzi (1997)), because a smaller circuit has a smaller number of blocks that must be visited by the mapper to build an efficient mapping policy.

3.2 Exploiting the pre-evolved rule-set

The major assumption that makes the adaptive approach presented here reasonable is that rule-sets need to be evolved only once and that they can be shared among different designs.

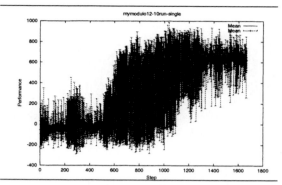

Fig. 4. Statistical analysis - simple sequential circuit (MODULO12), training mode, 10 independent runs, mean and standard deviation.

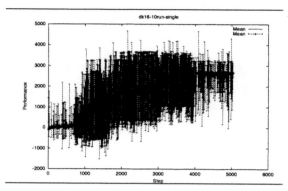

Fig. 5. Statistical analysis - sequential circuit (DK16), training mode, 10 independent runs, mean and standard deviation.

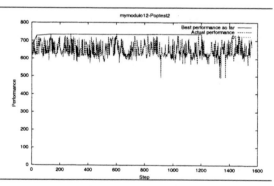

Fig. 6. Rule-set sharing - simple sequential circuit (MODULO12), pre-evolved rules taken from MODULO12, Step 8000 (figure not shown), training mode, single run.

The results shown in this section show first that it is possible to store a well-performing rule-set and to successfully use it for the same design (Figure 6), and second that rules evolved for a simpler design can be used to improve the performance of the mapper for a much more complicated design (Figures 7, 8). This induction phenomenon is an interesting feature of the mapper (and of the classifier system) which has not been mentioned in the literature to this date.

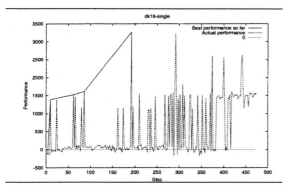

Fig. 7. Rule-set sharing - sequential circuit (DK16), no pre-evolved rules, training mode, single run, enlargement of the initial region of Figure 3.

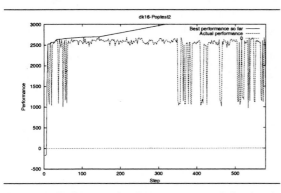

Fig. 8. Rule-set sharing - sequential circuit (DK16), pre-evolved rules taken from MODULO12, Step 8000 (figure not shown), training mode, single run.

4. DISCUSSION

The experiments done and the study of the available resources indicate that the mapping task is more complex than any tasks solved by classifier systems so far:

(1) The task itself is highly sequential, which means that the classifier system has to evolve long chains of rules.

(2) The task consists of several episodes, each episode modifies the mapping environment.

(3) The mapping environment is non-Markov, which may indicate the necessity to use an internal memory in the classifier system.

(4) The states of the search space (or gates and flip-flops in the circuit) are not visited uniformly in the learning process, to the contrary of the assumptions of the Q-learning method (Barto and Sutton (1998)) used in the XCS classifier system.

(5) The actions used by the classifier system are not rewarded equally. The reward is most often generated by the *generate CLB* action, but the effect of this action depends on all other actions that are not rewarded most of the time. The Q-learning mechanism ensures

that the reward is distributed equally among all actions to reflect their effects, but only in the limit case. Under usual circumstances, the performance of all actions that do not receive a (direct) reward from the environment will be underestimated (or overestimated in the case of *generate CLB*). This suggests that the actions should be viewed as a hierarchy according to the probability of receiving a reward from the environment:

- bottom: positioning actions - *moveBk, moveFw,*
- middle: cluster generation - *assign LUT, assign DFF, add to LUT*
- top: supercluster generation - *generate CLB*

To speed up the adaptive mapper, it is necessary to improve the processing of information done by the classifier system. This can be done either by reducing the state space by omitting some information from the sense vector, or by 'hardwiring' some a priori knowledge in the classifier system structure. The first approach would rather be avoided, because it would only increase the amount of hidden information of the non-Markov environment, and thus either decrease the performance, or increase the learning complexity. The second approach is viable.

One modification of the structure of the classifier system is described below.

4.1 Multi-population classifier system

The idea of the modification of the classifier system shown here is based on the action hierarchy discussed above and on ideas presented in (Wiering and Schmidhuber (1996)).

In classical classifier systems there is only one population of rules (or a set of actions) and the system considers all rules (actions) at every time step. This approach is not practical for the mapping task, because the classifier system has to learn an efficient ordering of actions that is known (or can be deduced) beforehand.

An optimal sequence of actions would look like *move somewhere, do something, move, do, move, move, do, move, do, generate CLB*. Each optimal sequence contains *move to a neighbour block* and *assign/add to a cluster* actions and is terminated with a *generate CLB* action. It is unlikely that performing two non-positioning actions at one position will make sense, because a block can be part of only one cluster. On the other hand it makes sense not to perform any non-positioning action at a position for the same reason.

This describes an information that the classifier system has to learn to perform optimally and that

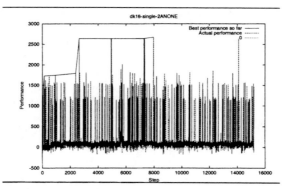

Fig. 9. Single-population mapper - sequential circuit (DK16), no pre-evolved rules, training mode, single run.

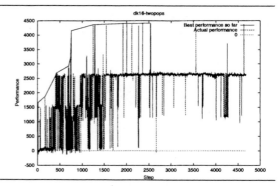

Fig. 10. Two-population mapper - sequential circuit (DK16), no pre-evolved rules, training mode, single run.

is known beforehand. It can be easily implemented by dividing the actions the classifier system works with to several sets and by supplying a finite state machine (FSM) that uses the sense vector and a history of past actions to switch among the sets. This results in reducing the information that the classifier system has to discover, because it is contained in the FSM, and also in reducing the search space that needs to be sampled by the genetic algorithm, because the action subsets are usually smaller than the original action set and the sense vector can be modified (reduced) according to the character of the actions in each set. The reward is calculated as if there was only one set of actions, so that the action sets influence each other.

In the case of the adaptive mapper, it makes sense to have a population of positioning actions and a population of all other actions supplemented with a *do nothing* action. The FSM has only two states that correspond to the two populations and alternates between them when

- the last action was a non-positioning action, or
- the last action was a positioning action and the current block has not been included in any cluster yet.

The performance of equivalent single- and multi-population mappers is shown in Figures 9, 10. It is evident that the two-population mapper performs better than the single-population one.

5. CONCLUSION

This paper examines the possibility of constructing a classifer system-based adaptive mapper for LUT-based FPGAs. An implementation of the mapper that is based on the XCS system (Wilson (1995)) has been described. It has been demonstrated that the adaptive mapper can adapt to a specific architecture of a configurable logic block,

and that pre-evolved rule-sets can be used for different, more complex designs. In addition, a modification of the adaptive mapper that uses a multi-population XCS system has been demonstrated that outperforms the single-population mapper.

REFERENCES

Barto, A. G. and R. S. Sutton (1998). *Reinforcement learning*. The MIT Press.

Cong, J., Y. Ding, T. Gao and K. Chen (1994). LUT-based FPGA technology mapping under arbitrary net-delay models. *Computers and Graphics*.

Holland, J. H. (1975, 1992, 1995). *Adaptation in natural and artificial systems*. The MIT Press.

Lanzi, P. L. (1997). A model of the environment to avoid local learning with XCS in animat problems. *Technical Report N. 97.46, Dip. Di Elettronica e Informazione, Politecnico di Milano*.

Murgai, R., R. K. Brayton and A. Sangiovanni-Vincentelli (1995). *Logic synthesis for field-programmable gate arrays*. Kluwer.

Wiering, M. and J. Schmidhuber (1996). HQ-learning: Discovering Markovian subgoals for non-Markovian reinforcement learning. *Technical Report IDSIA-95-96*.

Wilson, S. W. (1995). Classifier fitness based on accuracy. *Evolutionary Computation*.

Wilson, S. W. (1996). Explore/exploit strategies in autonomy. In: *From Animals to Animats 4: Proceedings of the Fourth International Conference on Simulation of Adaptive Behaviour* (P. Maes, M. Mataric, J. Pollack, J.-A. Meyer and S. W. Wilson, Eds.). The MIT Press/Bradford Books.

Xilinx (1994). XAPP043 - improving XC4000 design performance. *Xilinx Application Notes*. Xilinx.

Xilinx (1994, 2001). *Programmable logic data book*. Xilinx.

IFAC

Publications
www.elsevier.com/locate/ifac

APPLICATION-ORIENTED FAST FAULT SIMULATOR FOR FPGAS

Stanisław Deniziak and Krzysztof Sapiecha

Cracow University of Technology
Warszawska 24, 31-155 Cracow, Poland

Abstract: In this paper a new efficient approach to bit-parallel fault simulation for
FPGA-based sequential systems is introduced and evaluated with the help of ISCAS89
benchmarks. Reduced Ordered Ternary Decision Diagrams (ROTDD) are used to
describe functions of LUTs. This lead to substantial reduction of both, the number of
simulated faults and calculations needed for simulation. Moreover, an approach
presented in this paper is able to handle internal CLB faults in addition to
interconnection faults. *Copyright © 2001 IFAC*

Keywords: fault diagnosis, digital systems, simulators, testability, ternary logic.

1. INTRODUCTION

Field Programmable Gate Arrays (FPGAs) are
widely used for rapid prototyping and manufacturing
of complex digital systems. There are many different
FPGA architectures. The most popular are SRAM-
based FPGAs that implement functions via look-up
tables (LUTs). As the use of FPGAs in commercial
products has steadily increased, the problem of
FPGA testing and fault diagnosis is getting more
important.

FPGA testing can be applied to either unprogrammed
(Manufacturing-Oriented Testing) or programmed
(Application-Oriented Testing) FPGAs. Recent
works deal mainly with manufacturing-oriented
testing (Huang, *et al.*, 1998; Inoue, *et al.*, 1998;
Doumar, *et al.*, 1999; Harris and Tessier, 2000).
Because of the complexity of FPGA testing, each
method targets a specific FPGA part e.g.
interconnect, Logic Cell (LC), etc. FPGA
reprogramming is slow. Hence, the main problem of
manufacturing-oriented FPGA testing is to reduce the
number of configurations required for testing. On the
contrary the goal of the application-oriented testing is
to test a given FPGA configuration. In this case, only

such part of the FPGA is tested which is used for the
given application. It has been shown that in this
approach time required for automatic test pattern
generation can be reduced over 90% (Renovell, *et al.*,
2000). Moreover, no reprogramming is required, so
time needed for testing is significantly reduced, too.

In recently published application-oriented test
generation method (Renovell, *et al.*, 2000) a classical
ATPG system with the stuck-at fault model was
applied. Since FPGA-based systems are composed of
logic cells implementing various Boolean functions,
stuck-at fault model could not be satisfactory. ATPG
system for FPGAs should concern internal faults of
logic cells as well as interconnection faults. In the
case of LUTs, memory oriented fault models or
functional fault models may be used for modelling
internal faults.

One of the most essential tasks in ATPG system is
the fault simulation (Levendel and Menon, 1986). It
is used to evaluate the fault coverage for a given set
of test vectors, to build up dictionaries of faults and
to analyse testability of digital systems being
designed. Fault simulators are also a core of ATPG
systems based on genetic or random test pattern

generation (Guo, *et al.*, 1999). Recent research has shown that algorithms based on bit-parallel fault simulation technique are the most efficient ones. PROOFS (Nierman, *et al.*, 1992), PARIS (Gouders and Kaibel, 1991) and HOPE (Lee and Ha, 1996) are the examples of fast bit-parallel fault simulators for synchronous sequential circuits. All of these simulators require gate-level circuit representation and can not be directly applied for FPGA-based systems. For this purpose more convenient seems to be architectural fault simulators (Hsiao and Patel, 1995; Deniziak, 2001). However, they are less efficient (in terms of module evaluations per second) and time required for simulation of systems consisting of many thousands of logic cells may be not acceptable. Moreover, architectural fault simulators usually consider only stuck-at faults of interconnections.

In this paper a new efficient approach to bit-parallel fault simulation for FPGA-based systems is introduced and evaluated with the help of ISCAS89 benchmarks. Reduced Ordered Ternary Decision Diagrams ROTDDs are used to describe functions of logic cells. This lead to substantial reduction of both, the number of simulated faults and calculations needed for simulation. Experimental results have shown that the fault simulator is much faster than HOPE fault simulator (one of the fastest gate-level fault simulators). Moreover, fault simulator presented in the paper is able to handle functional fault models in addition to stuck-at fault model. According to our best knowledge it is the first fast fault simulator dedicated for FPGA-based systems.

In the next section concepts of ROTDDs and HTDDs are introduced. Section 3 describes our bit-parallel fault simulator based on HTDD. The paper ends with experimental results and conclusions.

2. FUNCTIONAL REPRESENTATION OF CONFIGURED FPGAS

Configured FPGA may be modelled as a network of connected modules. Each module corresponds to one logic cell or D-type flip-flop. Each LC implements the Boolean function of its inputs. Since the number of LC inputs is very limited (for LUTs usually equals 4), functions of LCs can be efficiently represented with Binary Decision Diagrams (BDDs). A BDD is a rooted directed acyclic graph as defined by Akers, (1978). An *Ordered BDD* (OBDD) is a BDD whose labels along every directed path occur in the same order and no label appears more than once in each path. A *Reduced OBDD* (ROBDD) is an OBDD where each node represents a distinct logic function (Bryant, 1986). Two edges labelled with 0 (left) and 1 (right) come out from each nonterminal node of a ROBDD. Edges corresponding to current values of

variables will be called *active* ones. Marking off active edges allows to obtain *active paths* that are characterized by the following features:
- there is only one active edge coming out from a node (belonging to one active path) and one inactive edge (not belonging to any active path),
- each active path ends with a leaf (labelled with 0 or 1),
- there is exactly one active path coming out from the root,
- a variable appears on a given path no more than once,
- every inactive edge connects two active paths or an active path and a leaf.

Let p_j^k denotes an active path, where j is serial number of the path, and k is its label:

$$k = \begin{cases} 0, \text{if the path ends with 0 - leaf} \\ 1, \text{if the path ends with 1 - leaf} \end{cases}$$

p_0^k denotes a path reaching the root of a BDD representing Boolean function f (i.e. the path describing actual behaviour of f under the given test pattern).

As far as fault simulation of sequential circuits is concerned BDDs show the following disadvantages:
1. only binary logic is taken into account (while three-valued logic is used in sequential circuit simulation);
2. representation of internal lines of a circuit is not possible.

Ternary Decision Diagrams (TDDs) (Jennings, 1995) and Hierarchical TDDs (HTDDs) (Sapiecha, *et al.*, 1998a) eliminate BDDs' disadvantages mentioned above.

In a TDD three ordered edges, related to *0*, *x* and *1* respectively, come out from each node. Marking off active edges in a TDD we select active paths which are similar to those in a BDD but with the following exceptions:
1) two inactive and one active edge come out from each node,
2) *0*, *1* and *x* may label a leaf.

In Fig. 2 active edges are darkened.
The following rules are used to create HTDDs:
- each module function is represented by its corresponding ROTDD,
- ROTDDs are grouped according to logic levels,
- there are links between nodes and its corresponding ROTDDs.

The example FPGA-based circuit and its HTDD representation are presented in Fig. 1 and Fig. 2 respectively.

HTDDs have the following attributes:
- HTDD size is linearly dependent on the size of the circuit,

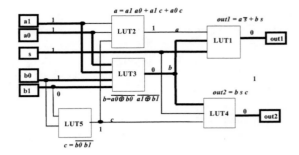

Fig 1: Example FPGA-based circuit

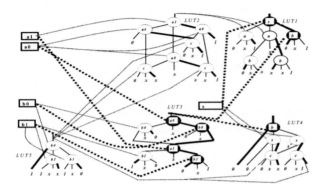

Fig. 2: HTDD representation of the circuit from Fig.1

- all internal lines (also branches) and primary inputs of the circuit have corresponding nodes.

Moreover, a HTDD can be simply derived from a structure of the circuit because compound ordering and reduction strategies are used only for creation of generic ROTDDs.

Synchronous sequential circuits are turned into combinational ones by cutting off lines and removing all flip-flops. The outputs of the flip-flops are called pseudo-primary inputs(PPIs) and the inputs of the flip-flops are called pseudo-primary outputs(PPOs). In HTDD representation of such circuits PPIs are considered like primary inputs(PIs) and PPOs are considered like primary outputs(POs).

The so-called *global path* P_0^k is calculated according to the following rules:

- if the line is an output(PO or PPO) then the path p_0^k of ROTDD associated with this line belongs to the path P_0^k.
- if a HTDD node belongs to the path P_0^k then the path p_0^k of ROTDD, which is associated with the node, belongs to the path P_0^k.

In Fig.1 and Fig.2 circuit lines and HTDD nodes belonging to P_0^k are darkened.

3. HTDD-BASED FAULT SIMULATION

One of the most efficient algorithms of fault simulation for gate-level is the bit-parallel one. It is fast and can be easily implemented with small memory requirements. But it is difficult to apply this algorithm for circuits composed of High Level Primitives, like LCs. For this purpose the deductive fault simulation is much more convenient.

3.1. Algorithm of fault simulation

Let v be a node of ROTDD labelled with an input line i of LC. From the definition of ROTDD it results what follows:

Corollary 1:
1. If a node v does not belong to a path p_0^k ($k=0,1$ or x) then variable labelling v does not influence logical function of o so none of faults of line i propagates to the output o.
2. If the node belongs to p_0^x path then i/x fault propagates to the output o.
3. If the node belongs to p_0^k path (where $k = 0$ or 1), then :

- if there is an inactive edge labelled with $l=0$ or $l=1$, coming out from the node v and reaching a leaf $\bar{k}$ or a path $p_j^{\bar{k}}$ then fault i/l propagates to o output,
- if there is an inactive edge labelled with $l=0$ or $l=1$, coming out from the node v and reaching a leaf x or a path p_j^x then fault i/x propagates to o output,
- if an inactive edge labelled with $l=0$ or $l=1$, coming out from the node v reaches a leaf k or a path p_j^k then the state of o in fault-free and faulty circuits are the same so fault i/l does not propagate to o output,

Collorary 1 can be used immediately to determine fault propagation inside modules.

In HTDD a node can appear many times on a path. The number of times a node appears on a path in HTDD equals the number of different propagation paths to the output of the circuit from the line corresponding to this node. Since each appearance of the node in the HTDD corresponds to one propagation path thus applying Corollary 1 to each of the nodes requires an execution of propagation procedure of a one fault along a one path. However, it may happened that an inactive edge of a node which belongs to a path p_0^k reaches a path p_j^l ($l\neq k$) and there is an identical node on this path that has inactive edge reaching a path p_i^k .That situation corresponds to fault self-masking. Therefore, any appearance of a fault on a path p_0^k is necessary but not sufficient condition for the fault to be propagated to the output of the circuit.

In sequential circuits faults may propagate to PPOs, what is visible on PPIs in the next time frame. The following cases are possible:
- faulty value originates only at the faulty site(such faults are called single-event faults),
- faulty value originates also at some PPIs(such faults are called multiple-event faults).

Decreasing the number of faults simulated in parallel (so called *sensitive faults*) can be done immediately after a global path P_0^k for a given test pattern has been determined. This is as follows:
- For single-event faults if a node corresponding to the faulty line does not belong to the path P_0^k then the fault is marked as insensitive (with no extra calculations).
- For multiple-event faults if a fault or at least one faulty effect associated with a given fault (only nodes corresponding to PPIs are taking into account) appears on the path P_0^k then fault is marked as sensitive, otherwise it is insensitive.

Hence, sensitivity determination of a fault with the help of HTDD does not depend on lengths of fault propagation paths, so it is more efficient then the calculation of fault propagation using Single Fault Propagation (SFP) method. Moreover, a global path P_0^k is created during fault-free simulation of a circuit, which is performed only once for a particular test pattern.

A draft of HTDD-based fault simulation algorithm is as follows (Sapiecha, *et al.*, 1998b):

begin
  Read the circuit structure
  for *every input pattern* **do**
  begin
    Fault-free simulation of a circuit and finding P_0^k
    for *each of faults* **do**
    if *it is single-event fault*
      if *the fault does not belong to P_0^k mark the fault as insensitive*
      else *candidacy test of sing.-event faults*
    else if *the fault and all faulty effects of this fault do not belong to P_0^k mark the fault as insensitive*
    Deductive bit-parallel simulation of all sensitive faults
  end
end

Candidacy test of single-event faults is done according to the following rules:
a) for a stem faults:
- if the stem has a dominator (Lee and Ha, 1996): if the dominator is sensitive and the stem fault propagates to the dominator (check-out is done according to Corollary 1), fault is mapped into the fault on the dominator, otherwise

- if fault of any fanout of the stem propagates to nearest FFR output then fault is marked as sensitive;
b) for non-stem faults: let fault be in FFR(s), if stem s is sensitive and fault propagates to s (checking is done according to Corollary 1), fault is mapped into the fault on stem s.

Rules a) and b) are similar as the ones used in HOPE, but in this simulator propagation is checked using SFP method. Moreover, propagation of the faults of stem without dominator is tested only to the next logic level.

3.2. Fault propagation through LC modules

A new deductive bit-parallel fault simulation technique is applied for calculation fault propagation through HLPs. The technique is bit-parallel because in every pass n faults (where n is equal to the CPU word) are simulated in parallel. The technique is also deductive because fault propagation through modules is determined using deductive fault list propagation. A draft of the method of calculation of fault lists propagated from inputs to the output of the LC is the following:

Propagate(O)
  {
    L^0 (O)= [0..0];
    L^x (O)= [0..0];
    L^1 (O)= [0..0];
    EvaluateLists(I_0,[1...1]);
  }

where:
  I_0 is a first node in TDD (a root),
  L^0 (l) – fault list for which state of the line l is equal 0
  L^x (l) – fault list for which state of the line l is equal x
  L^1 (l) – fault list for which state of the line l is equal 1.

Fault lists are represented with an n-bit word where each bit corresponds to one fault. The value of 1 means that fault is on the list, 0 otherwise. Function *EvaluateLists* is the following:

EvaluateLists(I, L)
{
  for(i in {0,x,1}) {
    if(Val(I)==i){
      if(i==0) Mask = !(L^x | L^1) ;
      else if(i==x) Mask = !(L^0 | L^1);
      else Mask = !(L^x | L^0);
    L' = L & Mask;
    }
    else L'=L & L^i(I);
    if(L'!=[0..0])

$$if \; (Node(e^i(I\,))==leaf \; \textbf{and} \; Val(Node(e^i(I\,)))$$
$$!= Val(O)) \; L^{Val(Node(ei(I\,)))} \, (O)+=L';$$
$$\textbf{else} \; EvaluateLists(Node(e^i(I\,)), L');$$
```
    }
}
```

where: $Val(I)$ – state of the line I in the fault free
circuit,

$Node(e^i(I\,))$ - returns node ending edge
corresponding to the value i and
starting from the node I (*leaf*
means that edge ends with leaf).

Function *EvaluateLists* recursively traverses TDD paths. In each pass of simulation L' contains 1's on positions corresponding to faults for which path just traversed is the active one. So, if the value of a leaf ending the path is not equal to the value of the output in a fault-free module, than 1's in the L' correspond to the faults which propagate to the module output. Further traversing of the path is dropped when L' is empty (the path is not active for all simulated faults).

3.3. Fault model

Two fault models were adopted in HTDD fault simulator: stuck-at-0 and stuck-at-1 faults of all lines in the circuit, and stuck-at-0 and stuck-a-1 faults of all leafs in TDDs (but only paths composed of $\{0,1\}$ edges are taking into consideration).

The first fault model reflects interconnect faults, while the second one corresponds to internal LC faults. In the case of LUTs the TDD-based fault model corresponds to stuck-at-0 and stuck-at-1 faults of memory cells. It should be pointed out that TDD faults imply most of faults of LC pins. Efficient fault collapsing significantly reduces the number of faults needed for simulation.

4. EXPERIMENTAL RESULTS

HTDD based fault simulator was run on 450 MHz Pentium III PC. Most of ISCAS89 sequential benchmarks (Brglez, *et al.*, 1989) have been simulated. Test patterns generated for gate level implementation, by a test generator GENTEST of AT&T were applied to each circuit. Synopsys FPGA Compiler (Synopsys, 2000) has been used for the FPGA synthesis of benchmark circuits. Next, for each circuit TDD library has been generated.

Experimental results are shown in Table 1. In the first four columns circuit names, circuit sizes for gate and FPGA circuit implementations and the number of test patterns applied are given. Then, the number of faults, fault coverage and time of fault simulation (given in CPU seconds) obtained for gate-level

simulation, FPGA simulation with interconnect stuck-at fault model and FPGA simulation with functional TDD fault model are presented. For FPGA fault simulation no fault collapsing was applied.

From Table 1 it results that the CPU time for HTDD fault simulation of FPGA-based systems is significantly reduced in comparison with the CPU time for gate-level fault simulation.

5. CONCLUSIONS

In the paper a new TDD-based fault simulation technique for FPGA-based systems was presented. It combines advantages of both bit-parallel and deductive fault simulations. HTDD representation of the FPGA-based systems enables accelerating of fault simulation using techniques which effectiveness was proved in gate-level fault simulators. Presented fault simulator is capable of handling functional faults in addition to stuck-at faults. TDD based fault model seems to be very promising and will be deeply investigated in further research.

Experimental results show high efficiency of the HTDD fault simulator. Moreover, TDD representation is flexible, allowing simulation of Term-based or Mux-based as well as LUT-based FPGAs.

REFERENCES

Akers, S.B. (1978). Binary Decision Diagrams, *IEEE Trans. on Computers*, **vol.C-27**, No.6, 509-516.

Brglez, F., D.Bryan, K.Kozminski (1989). Combinational profiles of sequential benchmark circuits, *Proc. of Int. Symp. on Circuits And Systems*, 1929-1934.

Bryant, R.E. (1986). Graph-Based Algorithms for Boolean Function Manipulation, *IEEE Trans. on Computers*, **Vol. C-35**, No. 8, 677-691.

Doumar, A., T.Ohmameuda, H.Ito (1999). Design of an Automatic Testing for FPGAs, *Proc. of the European Test Workshop*, 152-157.

Gouders, N. and R.Kaibel (1991). PARIS: A Parallel Pattern Fault Simulator for Sunchronous Sequential Circuits, *Proc. of the Int. Conf. on CAD*, 542-545.

Guo, R., I. Pomeranz, S.M. Reddy (1999). A Fault Simulation Based Test Pattern Generator for Synchronous Sequential Circuits, *VLSI Test Symposium*, 260-267.

Harris, I. G. and R. Tessier (2000). Interconnect Testing in Cluster-Based FPGA Architectures, *Proc of the Design Automation Conference*, 49-54.

Hsiao, M.S. and J.H.Patel (1995). A new architectural-level fault simulation using propagation prediction of grouped fault-effects,

Proc. of the Int. Conf. on Computer Design, 628-635.

Huang, W.K., X.-T. Chen, F.Lombardi (1998). Testing Configurable LUT-Based FPGA's, *IEEE Transactions on VLSI*, **Vol.6**, No.2, 276-283.

Inoue, T., S.Miyazaki, H.Fuijwara(1998). Universal Fault Diagnosis for Lookup Table FPGAs, *IEEE Design & Test of Computers*, **Vol.15**, No.1, 39-44.

Jennings, G. (1995). Accurate Ternary-Valued Compiled Logic Simulation of Complex Logic Networks by OTDD Composition, *Proc. of the 28-th Annual Simulation Symp.*, 303-310.

Lee, H. K. and D. S. Ha (1996). HOPE: An Efficient Parallel Fault Simulator for Synchronous Sequential Circuits, *IEEE Transactions on Computer-Aided Design of Integrated Circuits and Systems*, **Vol. 15**, No. 9, 1048-1058.

Levendel, Y., P.R.Menon(1986). Fault simulation. In: *Fault Tolerant Computing: Theory and Techniques*, (ed. D.K.Pradhan), vol.1, pp.184-264, Prentice-Hall, Englewood Cliffs.

Nierman, T.M., W.-T.Cheng, J.H.Patel (1992). PROOFS: A Fast, Memory Efficient Sequential Circuit Fault Simulator, *IEEE Trans. on CAD*, **vol. 11**, no. 2, 198-207.

Renovell, M., J.M. Portal, P.Faure, J.Figueras, Y.Zorian (2000). Analyzing the Test Generation Problem for an Application-Oriented Test of FPGAs, *Proc. of the European Test Workshop*, 75-80.

Sapiecha, J. S.Deniziak, K.Sapiecha (1998a). HTDD Based Accelerating of Parallel Fault Simulators, *Proc of the 9th IEEE Int. Conference and Workshop: Engineering of Computer-Based Systems*, 350-351.

Sapiecha, K. J.Sapiecha, S.Deniziak (1998b). HTDD Based Parallel Fault Simulator, *Proc. of the 5th IEEE Int. Conference on Electronics, Circuits and Systems*, vol.2, 217-220.

Synopsys Inc. (20000). FPGA Compiler User Guide.

Deniziak, S., K.Sapiecha (2001). Developing a High-Level Fault Simulation Standard, *IEEE Computer*, **vol. 34**, No.5, 89-90.

Table 1: Fault simulation results

Circuit	Gates	LUTs	Test Patterns	Gate level (HOPE)			FPGA:interconnect			FPGA:LUT		
				Faults	FC	Time	Faults	FC	Time	Faults	FC	Time
s27	21	5	15	55	87.3	0.0	54	87.0	0.0	24	75.0	0.0
s208	106	19	111	202	2.9	0.05	170	4.1	0.02	105	0.9	0.0
s298	164	36	162	336	83.3	0.07	326	89.0	0.05	179	91.1	0.02
s344	174	43	91	394	91.6	0.05	342	92.7	0.03	199	92.9	0.02
s382	195	45	2463	419	86.9	1.20	410	91.2	1.14	224	90.6	0.60
s400	202	48	1282	443	80.1	0.85	430	83.7	0.83	236	84.3	0.42
s420	225	40	173	415	1.2	0.13	360	1.7	0.10	220	0.9	0.02
s444	199	49	1881	441	85.7	1.10	438	91.3	1.10	242	90.5	0.60
s526	263	63	754	560	75.4	0.63	564	82.8	0.63	314	81.9	0.25
s641	239	69	133	474	82.9	0.05	652	83.4	0.05	321	81.9	0.05
s713	221	65	107	453	80.8	0.05	632	81.7	0.05	308	80.2	0.03
s820	442	119	411	876	84.1	0.30	1002	87.4	0.33	612	85.1	0.13
s832	401	110	377	781	87.6	0.25	936	89.5	0.27	560	86.6	0.13
s953	547	155	16	1100	6.6	0.07	1366	6.9	0.08	780	5.8	0.02
s1238	825	204	349	1536	91.0	0.27	1692	94.9	0.25	1062	90.4	0.22
s1423	760	164	36	1724	20.4	0.15	1568	21.7	0.15	881	21.0	0.10
s1488	891	244	590	1684	91.2	0.82	1980	93.7	0.90	1268	91.5	0.40
s1494	891	247	469	1686	90.4	0.68	2014	93.3	0.75	1273	90.7	0.33
s5378	1890	457	408	4445	66.3	2.58	3994	63.5	2.52	2448	61.2	1.13
s35932	9597	2760	86	33317	87.1	27.87	24436	95.7	20.30	15222	86.8	10.20

IFAC
Publications
www.elsevier.com/locate/ifac

SOFTWARE REUSE METHODOLOGY FOR HARDWARE DESIGN

Miroslaw Forczek

Aldec-ADT, Compilers Divison, ul. Lutycka 6, 44-100 Gliwice, Poland,
mirekf@aldec.katowice.pl

Abstract: Complex systems are prototyped in software before they will be implemented into the desired hardware architecture. In many cases, the software version of the system already exists and is proven when the designer attempts to develop its hardware realization. So, why not re-use the software code for hardware implementation? This paper presents an approach for direct C code compilation into a synthesizable description in hardware language. The process consists of two stages: algorithm exploration – to detect its inherent parallelism – and hardware synthesis. This paper will focus on C algorithm paralleling and conversion from a floating-point to a fixed-point arithmetic. *Copyright © 2001 IFAC*

Keywords: Algorithmic languages, Arithmetic algorithms, Compilers, Data flow analysis, Number systems, Parallelism, Programming approaches, Programming languages, Software tools.

1. INTRODUCTION

As devices become more complex, their design processes take more time and become more expensive than before. Some of the most important improvements, which were introduced over the past several years, are RTL synthesis tools, which automate the design transformation from RTL into gate-level processes (Ciletti, 1999). Since their introduction, most of a designer's efforts stop at the RTL stage of the design specification; automated tools perform the rest of the work. A few decades ago, the algorithm's distinction into either software or hardware was introduced. The possibilities of today's highly integrated chips causing such a distinction are not so obvious now (Master and Lane, 2001; QuickSilver, 2000). The algorithms have also become more complex and change frequently. Making updates in the present hardware implementations is a costly and time-intensive process. The algorithms are now prototyped and verified in a software implementation version. Often

they exist in software form for a longer time, acquiring stability before hardware implementation is required. This is why hardware designers turn to classical programming languages such as C or C++ at the first stage of the project instead of using HDL languages, which cause new problems. One of them is making the transition from C/C++ implementation into HDL implementation. These two implementations are totally different:

- C/C++ implementation is a sequential process (in most cases) while HDL implementation is a parallel, multi-process design

- C/C++ implementation runs without any clock signals while HDL implementation must take system clock signals into consideration and must address the signals timing issues.

2. ALGORITHM EXPLORATION

Currently, the most popular design solution is a C++ library of HDL classes (Synopsys, *at al.*, 2000a, 2000b). Unfortunately, an HDL classes library is not a solution that fully automates the hardware design path. The designer still has spent a lot of time on rewriting (refining) the project code from software form into HDL form. Aldec's CHDL approach (Forczek, 2001) addresses exactly the problem of C to HDL conversion automation. Instead of enabling the C/C++ environment with HDL features and pushing the user to go through the refine steps until reaching the RTL model, CHDL enables users to synthesize the HDL code directly from the C algorithm in its natural form. CHDL is a subset of the C language (ISO/IEC, 1999). The only limitation for a C programmer when using CHDL is the reduced spectrum of available language features. There is no requirement to rewrite the C algorithm in terms of modules, processes or registers as in case of HDL classes. While manual C code refines into HDL classes, the designer explicitly specifies the algorithm's inherent parallelism (by decomposing the algorithm into processes). Also, the hardware architecture and available resources are explicitly denoted. All of this additional information (regarding an algorithm's parallelism and its preferred implementation in hardware) is not present in the CHDL description of an algorithm. Instead, it utilizes a CHDL compiler task to take all required decisions while compiling the algorithm into HDL (fig. 1).

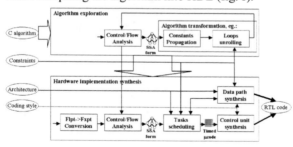

Fig. 1. C->HDL Compiler Architecture

Compiling C program into hardware realization requires two steps:
- *algorithm exploration* that will analyze data/control flow in a given piece of code and will derive back the algorithm's inherit parallelism upon sequential C description. Often some transformations are performed while this step, like constants propagation or loops unrolling.
- *hardware implementation synthesis* that will map reconstructed algorithms into user specified hardware architectures. This step includes the data path and control unit synthesis, resource allocation and task scheduling. Also, additional transformations may be required, such as a numeric-base change from a floating-point to a fixed-

point, which will be discussed later in this paper.

3. PARALLELING C ALGORITHM

It is obvious that directly translating C code into HDL code makes no sense, because this would result in a serial implementation instead of slicing it into parallel items. Specifying elementary sub-tasks that were independent of one another and that could be processed in parallel was a designer's task until now. This sort of data can now be obtained through the data and control flow analysis and this is also the first step in design path automation.

As an example, refer to a 2-D Discrete Cosine Transform algorithm (MathWorks, 1999; Intel, 1999). Assuming that the software implementation is already in place, the formula would appear as illustrated below (SIPG, 1998):

```
void fct2d(double f[], int nrows, int ncols)
{
  int u,v;
  // ...
  for (u=0; u<=nrows-1; u++) {
    for (v=0; v<=ncols-1; v++) {
      g[v] = f[u*ncols+v];
    }
    fct(g,ncols);
  }
  for (v=0; v<=ncols-1; v++) {
    for (u=0; u<=nrows-1; u++) {
      g[u] = f[u*ncols+v];
    }
    fct(g,nrows);
    for (u=0; u<=nrows-1; u++) {
      f[u*ncols+v] =
g[u]*two_over_sqrtncolsnrows;
    }
  }
}
```

This algorithm works as follows:
- 1-D DCT (*fct()* function) is performed on each row of the matrix **f**,
- the 1-D DCT is performed on each column from the result matrix after rows processing,
- the whole result matrix is scaled with a constant coefficient.

Upon the data and control flow analysis of this example, it can be found that there are few groups of elementary tasks (fig. 2):
- *fct()* on each row of **f**,
- *fct()* on each column of result from previous processing,
- scaling each element of result from previous processing.

Once elementary sub-tasks are extracted, they provide valuable information for data processing unit requirements (functionality and quantity). On the other hand, the synthesized design must meet user constraints, thus a unit may be allocated for different sub-tasks in time and tasks scheduling must be involved.

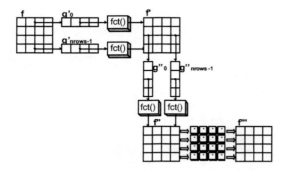

Fig. 2. The data and control flow diagram extracted from source code analysis.

4. NUMERIC BASE CONVERSION

A lot of algorithms that should be implemented in hardware for speedup, are defined in real domain. None of the popular logic synthesis tools support floating-point arithmetics (Synopsys, 1999; Synplicity, 2000); the possible solutions for this problem are:

- explicit use of floating-point units within design,
- transform the algorithm into its fixed-point arithmetic version.

The explicit use of floating-point units is a rather costly solution, but it preserves full accuracy of calculations across the whole numeric domain. On the other hand, transforming an algorithm into a fixed-point version simplifies the design complexity, gives calculation speedup and reduces its size, but is not always applicable. However, there is quite a large group of data processing algorithms that require finite accuracy in a range of numeric domain. These algorithms can benefit from transforming them into fixed-point realizations without loss of their applicability.

The most popular fixed-point representation is an extension of binary integral values encoding. The register is divided into two parts: integral part and fractional part with a specified sizes: m and n, respectively. Since a real domain includes negative values, a fixed-point register must also include a sign bit. We will denote $R[s,m,n]$ as a fixed-point register R with a single sign bit, m bits wide integral part and n bits wide fractional part. In fact, the real value r stored in a fixed point register $R[s,m,n]$ can be seen as an integer value $i = r*2^n$ stored in a classic signed register $I[s,m+n]$. Fixed-point representations also re-use negative values encoding U2 and most of the arithmetic operators. Thus the $flpt->fxpt$ conversion is an easy task, especially if an HDL language that support signed registers is targeted, such as VHDL (IEEE, 1993) or Verilog 2001 (IEEE, 2000).

The automated $flpt->fxpt$ conversion algorithm also requires a strategy for operands size control. For this purpose, the fixed-point operands will be classified into:

- *registers* – corresponding to the variables from original algorithm source,
- *intermediates* – temporary results from arithmetic operations.

The *registers* are synthesized to physical units while *intermediates* are just wired in most cases. For most of the operators, the size of the *intermediates* must be extended in comparison to the operands in order to avoid loss of operation results. The example of sizing rules is given in table 1.

Table 1. Fixed-point operations results sizing rules

Operation	Operands/Result
	$a[s,m,n]$, $b[s,m,n]$
$c = a+b$	$c[s,m+1,n]$
$c = a*b$	$c[s,2m+1,2n]$

In case of complex expressions, the *intermediates* from one operation become operands to another operation. This introduces another problems of *operands adjusting*, *results storage* and *size expansion*. Both problems concern case when sizes of two fixed-point values don't match one another. Before the operands can be processed with an arithmetic operator unit, their sizes must be equalized as well as the size of the operator unit must be determined. The *operands adjusting* method simply chooses the maximum widths for integral and fractional parts of the operands and apply *expand* operation on them, accordingly to the given equations:

$$a[s,m_1,n_1] => a'[s,max(m_1,m_2),max(n_1,n_2)]$$
$$b[s,m_2,n_2] => b'[s,max(m_1,m_2),max(n_1,n_2)] \quad (1)$$

Similarly, when the final result must be stored in a *register*, its size once again needs to be adjusted. First of all, register size can be adjusted to match the size of result. But if such changes are not possible (due to sizing strategy rules), the result must be *expand* or *shrink* before storing it in a register. In most cases, implementations of expand and *shrink* can be wired without resources allocation.

Our algorithm for $flpt->fxpt$ *conversion* is controlled by three constraints values:

- *initial integral/factional* parts sizes, which are applied to the first in chain operands – mostly these are *registers* that hold values used initially within algorithm
- *maximum integral/fractional* parts sizes for *registers*, applied at *results storage* method to prevent from unnecessary registers size expansion cascade
- *maximum integral/fractional* parts sizes for *intermediates*, applied at operands *adjusting method* to prevent from unnecessary intermediates size expansion cascade.

Also, notice that automated *flpt->fxpt conversion* can be used not only for C->HDL compilation purposes, but also for behavioral HDL code to refine the RTL subset. One of the implementations for this method was integrated with Aldec's Verilog compiler. It allow transforming codes for modules that exploit *real* type into *vectors* based on. Also it preserves modules interfaces.

5. CONCLUSIONS

This paper present a general outlook of an automated design path that re-use software code for hardware implementation. We focused on a direct C to HDL conversion. There is no need to manually rewrite the algorithm in HDL manner for precise parallelism of the algorithm. As an alternative, the elementary sub-tasks can be extracted from data and the control flow analysis. The data and control analysis is key to derive back algorithm parallelism from sequential code. It was also shown how a real domain algorithm could be transformed from a continuous state into its discrete version. This can produce efficient and low cost implementations for a variety of data processing algorithms like: FFT (Izydorczyk *at al.*, 1999), DCT (SIGP, 1998), Color Spaces Transforms (Pillai, 2001) and so on. There are also several issues that were not addressed in this paper. These, among others, are:
- selecting implementation architecture,
- synthesizing of data path and control unit,
- resources allocation and tasks scheduling.

All these issues are important elements of a fully functional automated C to HDL compilation system.

6. REFERENCES

Ciletti, M.D. (1999). *Modeling, Synthesis and Rapid Prototyping with the Verilog HDL*. Prentence Hall, New Jersey.

Forczek (2001). CHDL - An Approach For Hardware Design At The System Level. In: *The International Workshop on Discrete-Event System Design. DESDes'01*. Przytok.

IEEE (2000). *IEEE Standard Hardware Description Language Based on the Verilog Hardware Description Language", IEEE Std P1364-2000 (Draft 5)*. The Institute of Electrical and Electronics Engineers, Inc.

IEEE (1993). *IEEE Standard VHDL Language Reference Manual. IEEE Std 1076—1993*. The Institute of Electrical and Electronics Engineers, Inc.

Intel (1999). *Intel Signal Processing Library. Reference Manual*. Intel Corporation, USA.

ISO/IEC (1999). *International Standard ISO/IEC 9899: 1999(E), Programming Languages – C*.

Izydorczyk, J., G. Plonka and G. Tyma (1999). *Signals Theory. (in polish: Teoria sygnalow)*. Helion, Gliwice.

Master, P. and K. Lane (2001). Powering up 3G Handsets for MPEG-4 Video. In: *Communication Systems Design*, http://www.cdsmag.com/main/2001/01/0101feat2.htm

MathWorks (1999). *Image Processing Toolbox. Users Guide. Version 2*. The MathWorks Inc., Natick.

Pillai, L. (2001). Color Space Converter. In: *Application Note: Virtex-II Family*. Xilinx. http://www.xilinx.com/technotes/xapp283.pdf

QuickSilver (2000). QuickSilver: Technology Backgrounder. In: *QuickSilver Technology*. http://www.quicksilvertech.com

SGIP (1998). A Fast Discrete Cosine Transform. Signal and Image Processing Group, Bath.

Synopsys, CoWare and Frontier Design (2000a). *SystemC. Version 1.1 User's Guide*. Synopsys Inc., CoWare Inc., Frontier Design Inc. http://www.systemc.org

Synopsys, CoWare and Frontier Design (2000b). *Functional Specification for SystemC 2.0*. Synopsys Inc., CoWare Inc., Frontier Design Inc. http://www.systemc.org

Synopsys (1999). *FPGA Compiler II / FPGA Express. Verilog HDL Reference Manual*. Synopsys Inc.

Synplicity (2000). *"Synplify" Synplicity Synthesis Reference Manual*. Synplicity Inc.

IFAC
Publications
www.elsevier.com/locate/ifac

USING GENETIC ALGORITHMS TO DESIGN EFFICIENT BUILT-IN TEST PATTERN GENERATORS

Tomasz Garbolino, Norbert Henzel and Andrzej Hławiczka

Institute of Electronics, Silesian University of Technology, Gliwice, Poland

A method of finding a structure of the Cellular Automata (CA) based on Genetic Algorithms (GA) is presented in the paper. The mentioned CA, henceforth denoted as DTI-LFSR (Intrainverted Linear Feedback Shift Register with D and T flip-flops), is composed of D and T flip-flops with either active-high or active-low inputs. The DTI-LFSR structure found by the GA should generate vector sequence containing at least some deterministic test patterns, which covers all or the majority of stuck-at faults in a given Circuit Under Test (CUT). Remaining faults, if any, are covered by the pseudo-random patterns produced by the same DTI-LFSR. The advantages that make the DTI-LFSR the efficient Test Pattern Generator (TPG) are short testing time, low area-overhead and high operating frequency. *Copyright © 2001 IFAC*

Keywords: test generation, genetic algorithms, deterministic tests, linear feedback shift registers, T-type flip-flops.

1. INTRODUCTION

In the recent years the Built-In Self Test (BIST) has emerged as a promising alternative to using the external Automating Tests Equipment (ATE) in solving the VLSI testing problem. An efficiency of a BIST implementation is characterised by the test application time and the hardware overhead that are required to achieve complete or sufficient fault coverage. The issue of random pattern resistant faults (also called hard faults) needs to be addressed in BIST design process, if a test sequence of reasonable length, generated by a TPG, is to provide acceptable fault coverage. Several methods have been developed to solve mentioned problem. Some of them modify the CUT, while other ones alter pseudo-random test sequence applied to the CUT inputs. The first category involves test point insertion techniques, which increase detection probability of hard faults (Cheng, Lin, 1995), (Tamarapalli, Rajski, 1996). The methods belonging to the second category encompass

the set of techniques called mixed-mode (Muradali, et al., 1990; Wunderlich, 1990) and semi-deterministic (Kanaragis, Tragoudas, 1996; Kiefer, Wunderlich, 1997; Novak, 1999; Novak et al., 2001) BIST. The methods coming from the both groups have some essential drawbacks compared to the use of the classical TPG based on Internal Exclusive-OR type LFSR (IE-LFSR). They are necessity of CUT redesigning, high area overhead and, in some cases, timing penalties.

The technique of calculating an efficient LFSR seed for BIST that is free from the above disadvantages was proposed in (Fagot et al., 1999). The proposed algorithm tries to find in the vector sequence produced by the IE-LFSR implementing a given primitive polynomial, such a sub-sequence of a given length that provides very high fault coverage and simultaneously detects the majority of hard faults. The first pattern of that sub-sequence becomes the seed of IE-LFSR. The essential drawback of the above method is the fact that it only takes into account the IE-LFSR structure. Probability that given set of deterministic test vectors is produced in very short time by such a type of register is rather low. This stems from the fact that the IE-LFSR is composed of D-type flip-flops only. Strong shift dependency between the adjacent stages of the IE-LFSR leads to high correlation between consecutive vectors produced by that register (Garbolino, Hławiczka, 1999). Due to the same reason the vector sequence generated by the IE-LFSR displays poor random properties (Garbolino, Hławiczka, 1999). In addition, for a given length of the IE-LFSR the authors of (Fagot *et al.*, 1999) narrow their interest down to one or few primitive polynomials

Other solution to the considered issue was developed in (Garbolino, Hławiczka, 2001). There is shown the method of finding the TPG structure that is able to produce a vector sequence composed of several deterministic test patterns. That sequence is expected to cover the majority of faults in the CUT. The remaining faults, if any, should be covered by a random vector sequence generated by the same TPG.

Unfortunately, the demand that deterministic test patterns have to be generated by the TPG as a sequence of consecutive vectors turned out to be a limitation of the idea proposed in (Garbolino, Hławiczka, 2001). Such a sequence usually contains only a few deterministic test patterns and in consequence relatively long pseudorandom pattern sequence has to be additionally generated to provide 100% fault coverage. In some cases achieving full fault coverage is even impossible in acceptable number of clock cycles.

The above limitation has been overcome in the present work. Genetic algorithm (GA) (Goldberg, 1989; Michalewicz, 1996; Arabas, 2001) is used to find the TPG structure generating vector sequence in witch all or at least majority of deterministic test patterns is embedded. On the contrary to the method proposed in (Garbolino, Hławiczka, 2001) the deterministic test patterns do not have to appear in the consecutive clock cycles at the outputs of the TPG. The redundant vectors included in the sequence may serve as pseudorandom patterns detecting the faults which are not detected by the deterministic test patterns present in that sequence.

The rest of the paper is organised as follows. Section 2 explains authors' motivation to research into a specific type of Cellular Automata (CA) registers and their application as TPGs. In section 3 basic concepts concerning evolutionary computation techniques are discussed. Section 4 presents application of genetic algorithm (GA) to finding the structure of TPG that provides high fault coverage in relatively short testing time. Some experimental

results are included in section 5 and the paper is concluded in section 6.

2. MOTIVATION

The proposed TPG has a form of Cellular Automata (CA) with one global feedback (see Fig. 1). Each of its stages can take the form of a cell chosen from the set of four ones {D, oD, T, oT}. Symbols D, oD, T and oT denote D-type flip-flop with active high input, D-type flip-flop with active low input, T-type flip-flop with active high input and T-type flip-flop with active low input, respectively. The cells

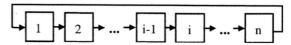

Fig. 1 The overall structure of n-bit DTI-LFSR, which is composed of cells D, oD, T or oT

implement the following CA rules: 240, 60, 15 and 195, respectively. The discussed above CA register is henceforth called DTI-LFSR (Intrainverted Linear Feedback Shift Register with D and T flip-flops).

DTI-LFSR has several essential advantages. Garbolino and Hławiczka (1999) demonstrated that CA having one global feedback and containing T-type flip-flops in addition to D-type ones, denoted there as DT-LFSR, generates vector sequence that has better random properties than that obtained at the outputs of the IE-LFSR. Novak (1999) showed that CA of the same form like one discussed above but consisting of T-type flip-flops exclusively produces weighted random patterns after being only seeded with the right, previously determined vector. Such a CA is able to generate pseudo-exhaustive test sets for large CUT cones without any reseeding. Other virtue of the DTI-LFSR is a number if its different available structures, which amounts to 4^n for n-bit register. What is particularly important is that each of those structures generates different sequence of vectors. Moreover, their state diagrams may also differ in number and length of cycles and even shape

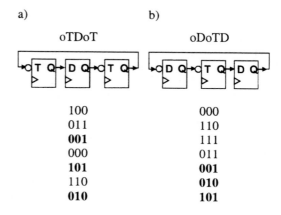

Fig. 2 Two example 3-bit DTI-LFSR registers and their output vector sequences

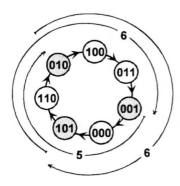

Fig. 3 State diagram of the example 3-bit DTI-LFSR having the following structure oTDoT

- i.e. they may be composed of set of cycles, may take a tree-like form or constitute a hybrid combination of the above two shapes. There are two 3-bit DTI-LFSR registers depicted in Fig. 2. Their structures may be denoted in abbreviated symbolic form as oTDoT (Fig. 2a) and oDoTD (Fig. 2b). The vector sequences generated by the above registers are also shown in the figure.

Notice three vectors: 001, 101 and 010, which are nearly uniformly distributed in the vector sequence generated by the register oTDoT and appear in three consecutive clock cycles in the sequence generated by the register oDoTD. Furthermore, the state diagram of the example DTI-LFSR which structure is oTDoT (see Fig. 2a), is presented in Fig. 3. Notice that the length of the vector sequence covering all three above-mentioned vectors may vary depending on the initial state (seed) of the DTI-LFSR. This is illustrated in the figure in the form of three angles ended with arrows, which represent three different vector sequences generated by the example register that contain the considered vectors. The beginning of each angle points out to the first vector while the arrow at its end shows where is the last vector of the sequence and indicates direction of the latter. The label associated with each angle informs about the number of vectors in the sequence. Thus, in order to reduce the number of clock cycles that are necessary to produce the given set of vectors at the outputs of the DTI-LFSR, finding the appropriate structure of the register and its seed is the crucial task.

3. EVOLUTIONARY COMPUTATION

Evolutionary computation (EC) is a class of stochastic optimisation methods based on some natural phenomena like genetic inheritance and Darwinian strife for survival. It has gained growing popularity during the last years. That class includes, among others, well known algorithms like genetic algorithms, evolutionary programming, evolution strategies, and genetic programming. There also exist many hybrid systems that incorporate various features of the above paradigms. Generally all those

```
procedure evolutionary algorithm
begin
    t ← 0
    initialise P(t)
    evaluate P(t)
    while (not termination-condition) do
    begin
        t ← t + 1
        select P(t) from P(t − 1)
        alter P(t)
        evaluate P(t)
    end
end
```

Fig. 4 The structure of an evolutionary algorithm

evolutionary techniques have the same structure shown in Fig. 4.

The evolutionary algorithm operates on the set (population) of m individuals, $P(t) = \{ x_1^t, ..., x_m^t \}$ for iteration (generation) t. Each individual, which is implemented in the form of some data structure, represents a potential solution to the considered problem. In each iteration t of evolutionary algorithm each i-th individual is evaluated, i.e. it is assigned a value - its "fitness" $F(x_i^t)$, which reflects the usefulness of the particular individual in constructing the optimal solution to the problem. Then, in iteration $t + 1$, a predefined number of individuals is selected with probability proportional to their fitness and they form a new population. Next, genetic operators alter some of the members of the new population in order to create new solutions. There are two main types of genetic operators: mutation and crossover. Mutation type operators are unary transformations that create a new offspring individual by a small change in a single parent individual. The crossover type operators, on the other hand, constitute higher order transformations that create new individuals by combining parts from two or more individuals. The evolutionary algorithm converges after some number of generations and than the best individual is expected to represent an optimum or at least near-optimum solution.

4. FINDING DTI-LFSR STRUCUTRE USING GENETIC ALGORITHM

There is a simple combinational circuit C17 among digital benchmarks that were introduced at ISCAS conference (Brglez, Fujiwara, 1985). It has five inputs x_1, x_2, x_3, x_4, x_5 and two outputs y_1, y_2. The schematic of the circuit is shown in Fig. 2. It is assumed that there is a single fault out of all possible stuck-at faults in the circuit. Exactly, the fault may be one of 17 stuck-at-1 (Sa-1) faults $\{f_i\}$, where $i \in \{1,2,...,17\}$, or it may be one of 5 stuck-at-0 faults $\{f_j\}$, where $j \in \{18, 19, 20, 21, 22\}$. The nodes where the above faults may appear are marked in Fig. 2 by cross symbol 'x' and label. The faults are detected by

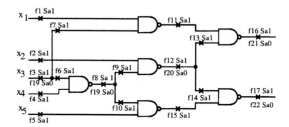

Fig.5 Gate level schematic of C17 benchmark with stuck-at faults indicated

set of four test patterns V1 = 10010, V2 = 11111, V3 = 00101 and V4 = 11010, which were calculated by Automatic Test Pattern Generation software ATALANTA (Ha, Lee, 1993).

The particular test patterns detect the following groups of faults: V1 detects faults $\{f_2, f_3, f_5, f_7, f_{16}, f_{17}, f_{20}\}$, V2 detects faults $\{f_8, f_9, f_{10}, f_{11}, f_{18}, f_{21}\}$, V3 detects faults $\{f_1, f_4, f_{15}, f_{19}, f_{22}\}$ and finally V4 detects $\{f_6, f_{12}, f_{13}, f_{14}, f_{21}, f_{22}\}$. Our aim is to design 5-bit DT-ILFSR structure that will generate all or at least majority of the above test patterns in as little number of clock cycles as possible. We try to accomplish that task by means of genetic algorithm.

Genetic algorithms (GAs) are some sub-class of evolutionary computation techniques that were devised to model adaptation process. Originally, they operated on binary strings and used a recombination operator with mutation as a background. However, they may also operate on strings of symbols belonging to some alphabet. In our case the alphabet is composed of four symbols D, oD, T and oT and each symbol is encoded as some integer value. Thus, the chromosome of each individual, which represents certain DTI-LFSR structure, is in fact an array of integers. The fitness F of each individual is calculated according the following formula:

$$F = \frac{TC + 1}{TSL} \qquad (1)$$

where:

TC - is the number of deterministic test patterns covered by a sequence that is generated by the DTI-LFSR structure represented by the individual,

TSL - denotes the test sequence length.

It reflects the main goal of the optimisation process: to find the structure of DTI-LFSR generating the vector sequence of the minimal length that simultaneously covers as many deterministic test patterns as possible. The reason a '1' is added to the TC in the numerator of the fraction in equation (1) is to enable any individual to be selected to a new population even if a sequence generated by the DTI-LFSR structure corresponding to that individual does not cover any deterministic test pattern. In the

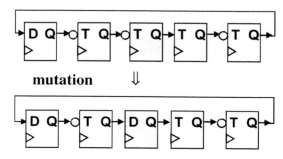

Fig. 6 Influence of the mutation operator on the DTI-LFSR structure

latter case the sequence length TSL takes its maximal value: either 2^n-1 or 10000, whichever is lower (sequence length 10000 is considered to be a maximal acceptable value of a semi-deterministic sequence).

Mutation operator substitutes a symbol at randomly chosen position in the chromosome with other one coming form the alphabet. For example, the string $x_b = DoToTToT$, which denotes the chromosome of some individual in the population, after mutation operation may take form $x_a = DoTDToT$. Symbol T in the third position of the initial string has been replaced by a symbol oD. Influence of the mutation on the DTI-LFSR structure is illustrated in Fig. 6. Fitness of the considered individual before and after mutation equals to: 0,3077 and 0,3125, respectively.

Crossover exchanges genetic material between two parents. If the parents are, say $x_{p1} = TTDDT$ and $x_{p2} = DoToDToT$, crossing the chromosomes after the second position would produce the offspring: $x_{c1} = DoToDDT$ and $x_{c2} = TTDToT$. This is illustrated in Fig. 7.

The fitness of the parent individuals equals to: 0,2632 and 0,3125, respectively, while the fitness of their

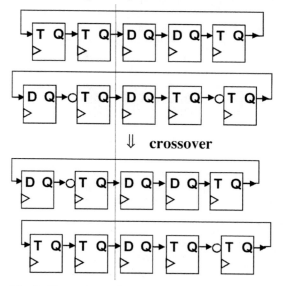

Fig.7 Illustration of crossover of two example DTI-LFSR structures

offspring is: 1,25 and 1,25, respectively. Notice that individuals x_{c1} and x_{c2} are optimal solutions of the considered problem. They generate the following sequences of deterministic test patterns: V1, V3, V2, V4 and V4, V3, V2, V1, respectively, in four consecutive clock cycles.

5. EXPERIMENTAL RESULTS

The proposed method has been used to find an effective TPG based on DTI-LFSR for other, more complicated benchmark c6288. The considered circuit has 32 primary inputs and the number of stuck-at faults for that benchmark is 7744. There are 12 deterministic test patterns that provide 100% coverage of non-redundant stuck-at faults in c6288 circuit. It has been observed that some inputs of the benchmark can be fed from the same stage of a TPG because values applied to them are identical in every test pattern. This is so called compatible input reduction (Chen, Gupta, 1998). Owing to that technique the TPG having only 11 stages is necessary to apply test sequence to c6288 benchmark. The best

structure of the 11-bit DTI-LFSR and its seed that have been found by the GA are as follows:

ToToTDoDoDoDoDDoTD, 01010111111

That DTI-LFSR structure, starting from the above seed generates test vector sequence providing 100% coverage of non-redundant faults in c6288 benchmark. The test sequence length is 69 clock cycles. Notice that the above DTI-LFSR implements

them were used to generate vector sequences beginning from the same seed like DTI-LFSR did and providing 100% fault coverage for c6288 circuit. The shortest sequence that was found is 85 clock cycles long and it is produced by the IE-LFSR implementing primitive polynomial $p_{IE}(x) = x^{11} + x^9 + x^6 + x^5 + 1$.

Curves representing percentage of uncovered faults in c6288 benchmark as a function of test sequence length for the discussed above two vector sequences generated by the DTI-LFSR and IE-LFSR, respectively, are shown in fig. 8 (Attention! Scale of vertical axis is logarithmic). Noteworthy are "jumps" in fault coverage provided by the sequence generated by the DTI-LFSR in contrast to relatively smooth fault coverage curve connected with the sequence produced by the IE-LFSR. This is a result of embedding some of deterministic test patterns, which detect large number of faults, in the former of the above sequences.

Moreover, worth to stress is the fact that the test sequence produced by the above-mentioned structure of DTI-LFSR not only is shorter than one generated by the IE-LFSR but it is also shorter than the test sequence produced by the DTI-LFSR structure found in (Garbolino, Hławiczka, 2001). The length of the latter is 74 clock cycles.

Other advantage of the TPG based on the DTI-LFSR in comparison with the plain IE-LFSR is low area overhead of the former. Cost of the mentioned above 11-bit DTI-LFSR manufactured in AMS 0,8 μm

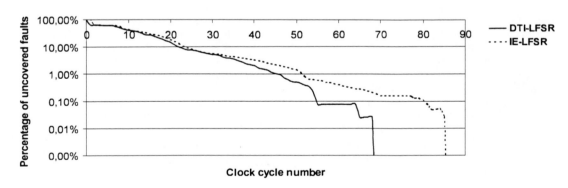

Fig. 8 Fault coverage in c6288 benchmark provided by test sequences produced by DTI-LFSR and IE-LFSR

polynomial $p_{DTI}(x) = x^{11} + x^7 + 1$, which is neither prime nor primitive.

In order to evaluate the quality of the above TPG in comparison with a plain IE-LFSR, several structures of the latter were examined. In details, only some of IE-LFSRs implementing primitive polynomials of degree 11 and containing no more than 3 XOR gates in the linear feedback were taken into account. The latter restriction was imposed to make a cost of compared DTI-LFSR and IE-LFSRs similar. All of

technology (AMS, 1996), which uses specially designed T-type flip-flop (Garbolino et al., 2000), is 83 equivalent NAND gates, while complexity of the 11-bit IE-LFSR equals to 86 NAND gates.

6. CONCLUSIONS

A new technique of designing built-in TPG for digital circuit is proposed in the paper. It is based specific linear CA structure (DTI-LFSR) composed of D- and T-type flip-flops with either active-high or active-low inputs, which has a single global feedback. The proposed idea assumes neither the primitive polynomial that is to be implemented by the register nor the initial vector of the generated test sequence. Instead, a GA is used to find DTI-LFSR structure that generates short vector sequence containing some of deterministic test patterns and providing high fault coverage in a CUT.

The advantages of proposed methodology are:
- deterministic test patterns embedded in the sequence produced by the DTI-LFSR cover some of random-resistant fault,
- randomness of the vector sequence produced by the DTI-LFSR is better than in the case of the plain IE-LFSR (Garbolino, Hławiczka, 1999)
- area overhead of the DTI-LFSR is lower and its operational speed is higher than in the case of the plain IE-LFSR.

Practical experiment presented in the paper proved that DTI-LFSR designed following suggested methodology outperforms the IE-LFSR as a TPG.

7. REFERENCES

AMS (1996). 2.0-Micron, 1.2-Micron, 1.0-Micron and 0.8-Micron Standard Cell Databook, Austria Mikro Systeme International AG, Austria.

Arabas, J. (2001). *Wykłady z algorytmów ewolucyjnych*. Wydawnictwa Naukowo-Techniczne, Warszawa, Poland.

Brglez F. and H. Fujiwara (1985). A Neutral Netlist of 10 Combinational Benchmark Circuits and Target Translator in FORTRAN. *Proc. of IEEE International Symposium for Combinational Circuits and Systems*, pp. 696-698.

Chen, C.-A., K. Gupta. (1998). Efficient BIST TPG Design and Test Set Compaction via Input Reduction. *IEEE Trans. on CAD of ICs and Systems*, **Vol. 17, no. 8**, pp. 692-705.

Cheng, K.T. and C.J. Lin (1995). Timing Driven Test Point Insertion for Full-Scan and Partial-Scan BIST. In: *Proc. of IEEE International Test Conference*, pp. 506-514. USA.

Fagot C., O. Gascuel, P. Girard and C. Landrault (1999). On Calculating Efficient LFSR Seeds for Built-In Self Test. In: *Proc. of IEEE European Test Workshop – ETW'99*, pp. 7-14. Constance, Germany.

Garbolino, T. And A. Hlawiczka, A (1999). A New LFSR with D and T Flip-Flops as an Effective Test Pattern Generator for VLSI Circuits. In: *Lecture Notes in Computer Science (Proc. of EDCC-3 – Third European Dependable Computing Conference)* (J. Hlavicka, E. Maehle, A. Pataricza, Eds.), Vol. 1667, pp. 321-338. Springer Verlag Press, Berlin.

Garbolino, T., A. Hlawiczka, A. Kristof (2000). Fast and Low-Area TPGs based on T-type Flip-Flops can be Easily Integrated to the Scan Path. In: *Proc. of European Test Workshop – ETW'00* pp. 161-166. Computer Society Press, Cascais, Portugal.

Garbolino, T. and A. Hławiczka (2001). On the Design of New Efficient Cellular Automata Structures for Built-In Test Pattern Generators. In: *Proc. of The 8th International Conference Mixed Design of Integrated Circuits and Systems – MIXDES 2001*, pp. 195-200. Zakopane, Poland.

Goldberg, D. (1989) *Genetic Algorithms in Search, Optimization, and Machine Learning*. Addison-Wesley Publishing Company, Inc., Reading, Massachusetts, USA.

Ha D. and H. Lee (1993). On the Generation of the Test Patterns for Combinatorial Circuits, *technical report 12.93*. Department of Electrical Engineering, Virginia Polytechnics Institute and State University, Virginia, USA.

Kanagaris, D. and S. Tragoudas (1996). Generating Deterministic Unordered Test Patterns with Counter, In: *Proc. of IEEE VLSI Test Symposium*, pp. 374-379. Princeton, New Jersey, USA.

Kiefer, G. and H-J. Wunderlich (1997). Using BIST Control for Pattern Generation. In: Proc. of IEEE International Test Conference, pp. 347-355. USA.

Michalewicz, Z. (1996). *Genetic Algorithms + Data Structures = Evolution Programs*. Springer-Varlag, Berlin Heidelberg, Germany.

Muradali, F., V.K. Agarwal and B. Nadeau-Dostie (1990). A New Procedure for Weighted Random Built-In Self-Test. In: *Proc. of IEEE International Test Conference*, pp. 660-669. USA.

Novak, O. (1999). Pseudorandom, Weighted Random and Pseudoexhaustive Test Patterns Generated in Universal Cellular Automata, In: *Lecture Notes in Computer Science (Proc. of EDCC-3 – Third European Dependable Computing Conference)* (J. Hlavicka, E. Maehle, A. Pataricza, Eds.), Vol. 1667, pp. 303-320. Springer Verlag Press, Berlin.

Novak, O., A. Hławiczka, T. Garbolino, K. Gucwa, J. Nosek and Z. Pliva (2001). Low Hardware Overhead Deterministic Logic BIST with Zero-Aliasing Compactor. In: *Proc. of the Fourth International Workshop on IEEE Design and Diagnostics of Electronic Circuits and Systems – IEEE DDECS 2001*, pp. 29-35. Györ, Hungary.

Tamarapali N, Rajski J, Constructive Multi-Phase Test Point Insertion for Scan-Based BIST", IEEE Int. Test Conf. 1996, pp. 649-658

Wunderlich H-J (1990). Multiple Distributions for Biased Random Test Patterns. *IEEE Trans. on CAD*, **Vol. 9, no. 6**, pp. 584-593.

IFAC

Publications
www.elsevier.com/locate/ifac

ZERO ALIASING LINEAR SPACE COMPACTORS

Krzysztof Gucwa

Institute of Electronics, Silesian University of Technology, Gliwice

Abstract: Compaction of responses appearing on the outputs of multi-output circuit under test requires using a compactor with many inputs. One of possible implementation of this compactor is using a structure which consist of space compactor; what reduces the number of outputs; and then performing time compaction. However using space compactor can cause error aliasing. There are known methods for designing zero aliasing space compactors. In this paper the new improved method for designing linear zero aliasing space compactor for modelled faults is introduced. The presented method uses the graph for compactor representation. The method gives better results than methods known before. *Copyright © 2001 IFAC*

Keywords: Test, Testability, Fault detection, Graphs, Combinational circuits

1. INTRODUCTION

In order to test a digital circuit, test patterns (stimulous) are applied to the inputs of Circuit Under Test (CUT) and responses appearing on the outputs are verified. Usually CUT has a lot of outputs so many data streams appearing on these outputs have to be validated.

Space compactor (SC) reduces the number of data streams which have to be farther verified. However compaction process can cause error aliasing. The goal is to design SC which eliminates aliasing phenomenon or significantly reduces it. Space compactor can be designed as a linear or nonlinear circuit. Nonlinear SC usually provides better compaction ratio than the linear one (Chakrabarty *et al*, 1998b) but designing nonlinear SC for CUT with a great number of outputs requires a long computation time. There are known method for designing linear SC (Chakrabarty *et al*.,1995, 1996, 1998a). A method for designing linear zero aliasing SC presented in this paper is significantly improved

in comparison to the known methods and provides a better compaction ratio.

1.1 Linear space compactor error aliasing

Linear SC consists of multi-input XOR gates.

Because SC is a linear circuit, error sequences instead of output sequences of CUT can be used. Error sequence e_{fr} is obtained as a modulo 2 addition (XOR) sequence appearing on r-th output of fault free CUT, and sequence appearing on r-th output of faulty CUT affected by f-th fault.

An error being a result of f-th fault can be masked if as an effect of space compaction we get sequences containing only zeros on all outputs of SC.

Analysing compaction process in linear SC, can be noticed that aliasing can occur in two following cases:

1. when even number of identical error sequences are provided to the same output of SC (fig.1.a)

2. when different error sequences (even or odd number, but more then two) are provided to the same output of SC and one's in this sequences are cancelled. (number of ones in error sequences at each bit is even) This case is illustrated in fig. 1.b.

Let's suppose that error sequence is m bit long. Probability of error (probability of appearing one in error sequence) at each bit position is the same, equal P_{one} and independent of any other position.
In order that two sequences was masked bits at each position must be the same (both equal one or zero) The probability of masking two chosen at random bits is equal to:

$$P_b = P_{one}^2 + (1 - P_{one})^2$$

Probability that two m-bit long sequences are masked is equal to:

$$P_{m2} = (P_{one}^2 + (1 - P_{one})^2)^m$$

Probability that i identical sequences (where i is even number) are masked:

$$P_{mi} = (P_{one}^i + (1 - P_{one})^i)^m$$

Probability of masking three errors sequences which differs from each other, supposing that they are different at two positions only is given below.

$$P_{M3} = [P_{one}^2(1-P_{one})]^2 (P_{one}^3+(1-P_{one}^3))^{m-2}$$

If error sequences differs at j positions then:

$$P_{Mj} = [P_{one}^j(1-P_{one})]^j (P_{one}^3+(1-P_{one}^3))^{m-j}$$

For $P_{one} = 0.5$ probabilities are equal respectively:
$P_{m2} = 2^{-m}$; $P_{mi} = 2^{-m(i-1)}$ $P_{M3} = 2^{-6}(2^{-2})^{-(m-2)}$
if $m >> 1$ then $P_{M3} \approx 2^{-6} 2^{-2m}$

It can can be noticed that in this case (for $P_{one} = 0.5$) P_{m2} is only an order of magnitude smaller than P_{M3}.

However this is the worst case. For real error sequences probabilities P_{one} is smaller then 0.1. In table 1 the probabilities P_{one} (calculated as a sum of ones in error sequences divided by total length of sequences) for ISCAS benchmarks are given. For great majority of checked CUTs P_{one} is smaller than 0.05.
Using the above value for calculation is clear that P_{M3} is at least 4 orders of magnitude lower than P_{m2}, and for P_{one} 0,01 is even 10 orders smaller.

Therefore only error aliasing described as a case 1 will be farther discussed. Using the notion „zero aliasing" only aliasing being result of masking an

even number of identical error sequences will be considered.

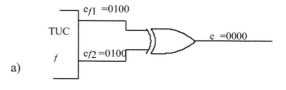

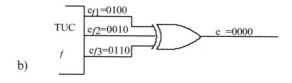

Fig 1. Two cases of aliasing error sequences in linear space compactor

2. DESIGNING OF ZERO ALIASING SPACE COMPACTOR

Definition 1.
There is a given CUT - C, having n outputs, a set of faults F and a set of test patterns T. The graph $G(V, E)$ is created. The set of vertices $V = \{ z_1, z_2, ... z_n \}$ represents the outputs of CUT. The edge $(z_i, z_j) \in E$ if and only if there exists such a fault $f \in F$, that error sequence e_{fi} being the result of fault f is the same as an error sequence e_{fj} and this sequences are not equal to zero.
Such a graph will be called $G(C, F, T)$.

Theorem 1.
If graph $G(C, F, T)$ is q colourable then exists q-output zero aliasing space compactor.
Proof:
„Graph is q-colorable" means that none of two vertices connected with edge have the same colour. So none of identical error sequences are provided to the same output of SC. As a result none of them is masked.

2.1. Reduction of graphs

The theorems which allow of elimination some edges of $G(C, F, T)$ graph will be farther presented.

Reducing graph edges requires checking if edge is implied by specific fault. For that reason further the graph $G(C, F, T)$ will be treated as a sum of all graphs $G(C, f, T)$ where $f \in F$.

Theorem 2.

If edges of $G(C, f, T)$ graph forms polygon having even number of vertices then all edges which belongs to $G(C, f, T)$ can be removed.

Proof:

Suppose that $G(C, f, T)$ graph consists of single even-vertices polygon only. Even if all of CUT outputs are assigned to the same SC output, the result of module 2 addition even number of identical sequences is not equal to 0. So vertices creating even-vertices polygon can be removed. Even if all other sequences are masked fault can not be masked. In this case no matter how the CUT outputs are assigned to inputs of CUT fault f is not masked.

(It is easy to notice that as a result of definition 1 any polygon contains all diagonals).

Chakrabarty (1998a) presented the theorem on removing triangles which is a special case of given above theorem.

Theorem 3.

If exists such an output z_i, that error sequence e_{fi} appearing on that output is not equal to 0, and for each $i \alpha j$ $e_{fi} \alpha e_{fj}$ then all edges which belongs to $G(C, f, T)$ can be removed.

Proof:

If error sequence e_{fi} is different then any other error sequence then even if all other sequences are masked fault can not be masked. In this case no matter how the CUT outputs are assigned to inputs of CUT fault f is not masked.

a_4:	0101001011o
a_3:	0110100101l
a_2:	0011010010o
a_1:	0011000011o
b_4:	1000111100l
b_3:	1011010100l
b_2:	1100110110o
b_1:	01111000011
c_0:	1100000010l

As a result of applying the above test vectors to the faulty CUT containing fault f_1 (Input a_1 stuck at 0) the following error sequences was obtained at CUT outputs:

c_4: (z_1)	0001000000o
s_4: (z_2)	0001000000o
s_3: (z_3)	0011000000o
s_2: (z_4)	0011000011o
s_1: (z_5)	0011000011o

According to definition 1 such error sequences imply edges (z_1, z_2), (z_4, z_5). In fig. 3.a graph $G(C, f_1, T)$ is presented.

Fault f_2 (Input a_1 stuck at 1) causes the following error sequences:

c_4: (z_1)	1000100000o
s_4: (z_2)	1000100000o
s_3: (z_3)	1000100000o
s_2: (z_4)	1000100000o
s_1: (z_5)	1100111100l

3. EXAMPLE

As an example of CUT 4-bit adder (7483) will be used (fig.2). The model of adder contains 119 stuck at (S-a) fault. To detect all faults 11 test vectors are used. This vectors was obtained using ATALANTA test pattern generator (TPG) (Ha *at al*. 1993).

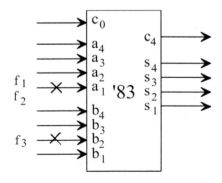

Fig.2. Four bit adder as an example of CUT

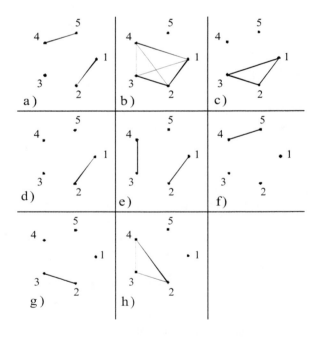

Fig. 3. All different graphs $G(C, f, T)$ for 4 bit adder

In fig. 3.b graph $G(C, f_2, T)$ is presented. CUT can contain one of 119 faults, but only 41 of them causes generation identical error

sequences on two or more outputs. These 41 faults generate eight different $G(C, f_i, T)$ graphs shown in fig.3 .

Using theorem 1 graphs shown in fig. 3c and 3h can be removed. Because for all faults which implies graph 3b (fault f_2 is an example) error sequence appearing on outputs s_4 (vertex z_5) are non zero, graph shown in fig 2b can be removed also (Theorem 3).

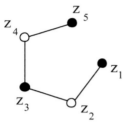

Fig. 3. Resulting graph $G(C, F, T)$

Finally graph $G(C, F, T)$ being a sum of graphs shown in fig. 3a,d,e,f,g is obtained. The resulting graph is shown in fig.4. This graph is 2 colourable. As a result of designing process 2-output SC presented in fig.4 was obtained. Final simulation confirmed that this compactor does not mask any of modelled faults.

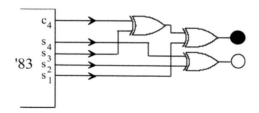

Fig. 4. Zero aliasing space compactor for 4 bit adder

If method known before was used the graph 3.b could not be removed and resulting SC would have 4 outputs (4-colourable graph).

4. EXPERIMENTAL RESULTS

The ISCAS benchmarks (Berglez, *et al.*, 1985, 1989) was used as TUCs. TUCs was stimulated by test pattern sets obtained using ATALANTA TPG, and also by much biger pseudorandom test pattern sets.

After designing SC the whole test circuit was simulated again, and none of modelled fault was masked, what confirms that masking caused by three or more different error sequences are very little probable.

As can be noticed using presented theorems in some cases, SC with less outputs then using the method known before can be designed.

Table 1.

1	2	3	4	5	6	7
Deterministic test pattern sets (ATALANTA)						
c432	7	524	50	0,0559	7	6
c499	32	758	53	0,0135	3	2
c880	26	942	54	0,0102	7	6
c1355	32	1574	85	0,0098	3	3
c1908	25	1879	113	0,0222	8	8
c2670	140	2747	108	0,0027	4	4
c3540	22	3428	154	0,0155	13	12
c5315	123	5350	120	0,0027	28	28
c6288	32	7744	33	0,0312	9	8
c7552	108	7550	192	0,0327	32	32
Deterministic test pattern sets (ATALANTA)						
s820	24	850	109	0,0122	5	5
s832	24	870	111	0,0183	6	6
s953	52	1079	91	0,0120	7	6
s1196	32	1242	141	0,0110	6	6
s1238	32	1355	152	0,0105	5	5
s1423	79	1515	69	0,0109	14	13
s5378	156	4351	259	0,0132	22	21
Pseudo random test pattern sets						
s820	24	850	9240	0,0102	5	4
s832	24	870	8148	0,0153	4	4
s953	52	1079	9600	0,0131	7	6
s1196	32	1242	9006	0,0120	5	5
s1238	32	1355	12768	0,0098	4	4
s1423	79	1515	9570	0,0106	12	11
s5378	156	4351	9440	0,0111	18	18

Column description:
1. ISCAS circuit
2. Number of CUT outputs
3. Number of faults
4. Number of test patterns
5. P_{one}
6. Number of SC outputs - previously known method
7. Number of SC outputs - method presented in this paper.

5. CONCLUSION

The presented method for designing a zero aliasing space compactor gives better results then the method known before, what was confirmed by experimental results. Using the presented theorems for reduction CUT output graph only slightly increases computation effort. For that reason presented method is useful for designing zero aliasing SC, especially for CUT having a lot of outputs where methods for designing nonlinear SC needs a long computation time.

Presented method has also two main disadvantages: designed SC provides zero aliasing only for modelled faults. Unmodelled fault can be masked. Additionally changing test pattern sets requires redesigning SC.

REFERENCES

Brglez F., Fujiwara H. (1985). A neutral Netlist of 10 Combinational Benchmark Circuits and Target Translator in Fortran. *Proceedings of IEEE International Symposium on Circuits and Systems 1985*, pp. 696-698.

Brglez F., Bryan D., Kozminski K. (1989). Combinational Profiles of Sequential benchmark circuits. *Proceedings of IEEE International Symposium on Circuits and Systems 1989*, pp. 1929-1934.

Chakrabarty K., Murray B.T., Hayes J.P. (1995). Optimal Space Compaction of Test Responses. *Proceedings of International Test Conference*, pp. 834-843.

Chakrabarty K., Hayes J.P. (1996). Test Response Compaction Using Multiplexed Parity Trees. *IEEE Transactions on Computer-Aided Design of Integrated Circuits and Systems*, **11**, Vol. 15, pp. 1399-1408.

Chakrabarty K. (1998a). Zero-Aliasing Space Compaction Using Linear Compactors with Bounded Overhead. *IEEE Transactions on Computer-Aided Design of Integrated Circuits and Systems*, **5**, Vol. 17, pp. 452-457.

Chakrabarty K., Murray B.T., Hayes J.P. (1998b). Optimal Zero-Aliasing Space Compaction of Test Responses. *IEEE Transactions on Computers*, **11**, Vol. 47, pp. 1171-1187.

Ha D., Lee H. (1993). On the generation of the test patterns for combinational circuits. *Dept. of Electrical Engineering, Virginia Polytechnic Institute and State University, Technical Rapport 12.93*,

IFAC
Publications
www.elsevier.com/locate/ifac

ATPG System and Fault Simulation
Methods for Digital Devices

Vladimir Hahanov Vitaly Pudov Iryna Sysenko
Kharkov National University of Radio Electronics, Ukraine

Abstract

Models and methods of digital circuit analysis for test generation and fault simulation are offered. The two-frame cubic algebra for compact description of sequential primitive element (here and further, primitive) in form of cubic coverings is used. Problems of digital circuit testing are formulated as linear equations. The described cubic fault simulation method allows to propagate primitive fault lists from its inputs to outputs; to generate analytical equations for deductive fault simulation of digital circuit at gate, functional and algorithmic description levels; to build compilative and interpretative fault simulators for digital circuit. The fault list cubic coverings (FLCC) allowing to create single sensitization paths are proposed. The test generation method for single stuck-at fault (SSF) detection with usage of FLCC is developed. The means of test generation for digital devices designed in Active-HDL are offered. The input description of design is based on usage of VHDL, Verilog and graphical representation of Finite State Màchine (FSM). The obtained tests are used for digital design verification in Active-HDL. For fault coverage evaluation the program implementation of cubic simulation method is used. Copyright © 2001 IFAC

1 Introduction

Field Programmable Gate Arrays (FPGA) and Complex Programmable Logic Devices (CPLD) make a deserved competition to microprocessor chips. Such success is defined by usage of Hardware-Software Co-operation Design, minimum time of digital system design (4-5 months), high-speed operation (under 500 MHz), high level of gate-array chip integration.

However, there are testing problems together with advantages of CPLD (FPGA). For solving these problems it is required to create models, methods and CAD software. The mentioned means have to support:

1) digital device testing at gate, functional and algorithmic description levels, when the digital device has a high-level integration and it is specified as FSM transition graphs, Boolean equations, multi-level hierarchical structures;

2) test generation for SSF detection with fault coverage about 100%, where the test has a form of single sensitization path cubic coverings;

3) acceptable operation speed of fault simulation algorithms;

4) design verification and diagnosis for synthesis into FPGA, CPLD;

5) possibility of concurrent execution of vector operation for test generation and fault simulation;

6) VHDL standard support for digital circuit and obtained test description;

7) opportunity of the integration into existing CAD systems of world-wide leading firms (ALDEC etc.).

Deductive fault simulation method [1,2] is more preferable because of its high-speed operation. It allows to detect all SSFs by input test-vector during one iteration of digital circuit processing. But this method is oriented to the gate level of digital circuit description. It is connected with complexity of output fault list generation for non-gate primitives. The offered cubic fault simulation method allows to process digital circuits described at gate, functional and algorithmic levels. In the other side, the solution of the mentioned problem is presented as the method of test generation for SSF detection with usage of FLCC allowing to create single sensitization path.

2 Mathematical apparatus of primitive analysis

FSM model of sequential primitive is represented below:
$$M = <X, Y, Z, f, g>,$$
where $X = (X_1, X_2, ..., X_i, ..., X_m)$, $Y = (Y_1, Y_2, ..., Y_i, ..., X_h)$, $Z = (Z_1, Z_2, ..., Z_i, ..., Z_k)$ are sets of input, internal and output State variables. The State variables are interconnected by the following characteristic equations:
$$Y(t) = f[X(t-1), X(t), Y(t-1), Z(t-1)];$$
$$Z(t) = g[X(t-1), X(t), Y(t-1), Y(t), Z(t-1)]. \quad (1)$$
Variables $Z(t)$ are external and, therefore, they are observed on output lines. Variables $Y(t)$ are internal and, therefore, they are non-observed. FSM format for sequential primitive cubic coverings, corresponding to (1), looks as follows:

$X(t-1)$	$Y(t-1)$	$Z(t-1)$
$X(t)$	$Y(t)$	$Z(t)$

.

FSM model from Fig.1 corresponds to the mentioned format. Functional sequential primitive is specified by components:
$$F^2 = <(t-1,t),(X,Z,Y),\{A^2\}>,$$
where $(t-1,t)$ are two consecutive frames in function description; (X,Z,Y) are vectors of input, internal and output variables; $\{A^2\}$ is a two-frame alphabet of State variables description [3,4]:
$$A^2 = \{Q=00, E=01, H=10, J=11, O=\{Q,H\}, I=\{E,J\}, A=\{Q,E\},$$
$$B=\{H,J\}, S=\{Q,J\}, P=\{E,H\}, C=\{E,H,J\}, F=\{Q,H,J\}, L=\{Q,E,J\},$$
$$V=\{Q,E,H\}, Y=\{Q,E,H,J\}, A^1=\{0,1,X=\{0,1\}\}, \varnothing (U)\}.$$
The primitive is described by the cubic covering
$$C = (C_1, C_2, ..., C_i, ..., C_n),$$

where $C_i = (C_{i1}, C_{i2}, ..., C_{ij}, ..., C_{iq})$ is a cube including input, internal and output coordinates $C_i = (C_i^X, C_i^Y, C_i^Z)$, q=m+h+k. For combinational circuit, the format of cubic covering description

$$F^T = <(t),(X,Z),\{A^1\}>$$

is defined by relations at (q=m+k)-dimensional vector of variables $C_i = (C_i^X, C_i^Z)$. The format specifies the multi-output combinational primitive with m inputs and k outputs. Binary Boolean function of m variables $Z=f(X_1,X_2,...,X_m)$ is defined by k=1.

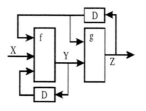

Fig.1. The primitive's state

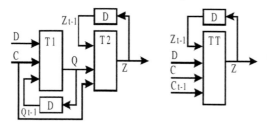

Fig.1. FSM model of flip-flop and latch

Cubic covering of two- and one-frame flip-flops:

D	C	Q_{t-1}	Z_{t-1}	Q_t	Z_t
1	1	X	1	1	1
0	1	X	1	0	1
1	1	X	0	1	0
0	1	X	0	0	0
X	0	1	X	1	1
X	0	1	X	1	1
X	0	0	X	0	0
X	0	0	X	0	0

D	C	C_{t-1}	Z_{t-1}	Z
1	0	1	X	1
0	0	1	X	0
X	X	0	1	1
X	X	0	0	0
X	1	X	1	1
X	1	X	0	0

The main feature of suggested models is compactness of truth- and transition tables for complex functional primitives and FSM descriptions; universality and completness of table models for solving problems of forward propagation and backward implication; universality and simplicity of cubic model analysis algorithms for: deterministic test generation, fault-free- and fault simulation [5].

Example 1. The obtaining of two-frame cubic covering for the counter is represented by:

V	C	A_{t-1}	B_{t-1}	A_t	B_t
0	X	X	X	0	0
1	E	0	0	0	1
1	E	0	1	0	0
1	E	1	0	1	1
1	E	1	1	1	0
1	F	0	0	0	0
1	F	0	1	0	1
1	F	1	0	1	0
1	F	1	1	1	1

=

V	C	A	B
0	X	0	0
1	E	Q	E
1	E	E	H
1	E	J	E
1	E	H	H
1	F	Q	Q
1	F	Q	J
1	F	J	Q
1	F	J	J

=

V	C	A	B
0	X	0	0
1	E	S	E
1	E	P	H
1	E	S	S

Dimensionality of tables for counting function description for n bits is defined by the relation

$$h = \frac{2^n}{n}.$$

Thus, the two-frame alphabet gives an opportunity to reduce dimensionality of transition\output tables of sequential primitives and circuits with functional complexity. The ratio of reduction depends on function type. For example, the ratio of reduction for flip-flops is 2-4 times, this one for registers is 2-3 times and this one for counters is 10-100 times.

The similar alphabet have been proposed in [4,5], but it have been used by author only for purposes of digital circuit simulation.

3 Fault simulation based on cubic algebra

Let's consider the model W=(M,L,T), where M is a primitive model represented by cubic covering $C=(C_1, C_2, ..., C_i, ..., C_n)$, L is a fault list cubic covering (FLCC), T is a test. Problems of digital circuit testing are formulated on condition that one of components (M,L,T) is not defined. Fault list cubic covering for primitive or digital circuit is specified as:

$$L=(L_1, L_2, ..., L_i, ..., L_n),$$

where $L_i = (L_{i1}, L_{i2}, ..., L_{ij}, ..., L_{iq})$ is a cube including input, internal and output coordinates: $L_i = (L_i^X, L_i^Y, L_i^Z)$, q=m+h+k; $(L_{ij}^Y, L_{ij}^Z) = \{0, 1, X\}$, "0" defines subtraction (complement) of the list L_j, "1" defines intersection L_j, X={0,1} identifies unessential fault list L_j. If L_{ir}^Z is an output observed variable, then (0)1 is an identifier of (un-) detectable faults in the cube L_i on output r, X={0,1} identifies unknown state of the output coordinate L_{ir}^Z, which may be interpreted as either 0, or 1.

FLCC L for the vector T and the primitive covering C is computed by a linear equation

$$T \oplus C = L, \qquad (2)$$

where $\oplus$ – is a binary operation XOR, which determines interaction of components T, C, L in the three-valued alphabet:

$$T_j \oplus C_{ij} = \begin{array}{|c|c|c|c|} \hline \oplus & 0 & 1 & X \\ \hline 0 & 0 & 1 & X \\ \hline 1 & 1 & 0 & X \\ \hline X & X & X & X \\ \hline \end{array}. \qquad (3)$$

The universal formula of FLCC analysis is obtained as a result of application of (3) to the test-vector T and to covering of the multi-output primitive C. The mentioned formula is used for definition of faults L_r detected at output r and it looks as follows:

$$L = \bigcup_{\forall i(T_r \oplus C_{ir}=1)} \bigcap_{j=1}^{k} L_j^{T_j \oplus C_{ij}}, \qquad (4)$$

where $L_j^{T_j \oplus C_{ij}} = \begin{cases} L_j \leftarrow T_j \oplus C_{ij} = 1; \\ \overline{L}_j \leftarrow T_j \oplus C_{ij} = 0, \end{cases}$

$\overline{L}_j$ – is considered as a line j in the fault list which should be subtracted from faults detected at non-output primitive lines;

L_j are faults which have to be intersected with non-output lists.

4 Algorithm of cubic SSF simulation for combinational circuits

The cubic deductive SSF simulation algorithm for combinational circuit is defined according to the procedure (4):

1. First, the fault-free simulation of next primitive $P_i (i = \overline{1,M})$ at the test-vector $T_t (t = \overline{1,N})$ is executed. If $t = N$, then a detected fault list $L(T)$ at the test T is formed. The end of simulation. Otherwise, if $t < N$, then move to point 2.

2. If all circuit elements are processed $(i = M)$, the comparison of two consecutive fault-free vectors is executed. If vectors are equal ($T_t^r = T_t^{r-1}$), then there is the end of simulation T_t and move to point 3. Otherwise, move to point 1.

3. Primary input fault lists are defined in the form of a complement to their fault-free state $L_j = \{j^{\overline{T_j}}\}$.

4. The fault simulation according to the procedure (4) is executed for primitive $P_i (i = \overline{1,M})$. The primitive output fault identified as $j^{\overline{T_j}}$ is added to the obtained list.

5. If $(i = M)$, then detected fault list $L(T_t)$ generation and motion to point 1 are executed. Otherwise, if $i < M$, then move to point 4.

Cubic fault simulation algorithm for digital systems at functional description level has an operating speed

$$C_k^F = b^2 \times \sum_{i=1}^{M} \{q_i \times n_i \times [(0{,}01 \times L)^2 + 3] + 2\} \approx$$

$$\approx b^2 \times (0{,}01 \times L)^2 \times \sum_{i=1}^{M} (q_i \times n_i),$$

where q_i, n_i is a number of variables and cubes in cubic covering; L is a number of lines in circuit; b is a number of faults; M is a number of functional elements; $(0{,}01 \times L)$ is an average number of active adjacent patterns of circuit lines; b is a number of non-equivalent faults in circuit.

The operation speed of digital systems simulation at the gate description level is evaluated by following expression:

$$C_k^G \Big|_{(M=G;G=q_i \times n_i)} = b^2 \times L^2 \times G.$$

In [4-7] the operation speed evaluation for concurrent algorithm is $C_p = (b^2 / W) \times G^3$ and for deductive one (CHIEFS system) is $C_d = b^2 \times Q \times G^2 \Big|_{Q=G} = b^2 G^3$, where W is a lenght of word; G is a number of equivalent gates; Q is an average number of gates sensitized by faults.

Since a number of circuit lines is less than a number of gates at least in two times, the proposed cubic simulation method has a better operaiotn speed in comparison with dedutive algo-

rithm. This gain will be bigger under processing of circuits at functional level, when a number of lines is less than a number of gates by dozens of times [6].

Example 6. To execute cubic SSF simulation for the circuit shown at Figure 2:

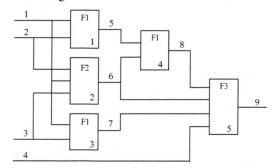

Figure 2: A digital functional module

Primitive operation is specified by the following cubic coverings:

$$
\begin{array}{ccc}
C(Fl) & C(F2) & C(F3)
\end{array}
$$

$$
\begin{array}{ccc}
\begin{array}{|ccc|}
1 & 2 & 5 \\
1 & 3 & 7 \\
5 & 6 & 8 \\
\hline
1 & X & 0 \\
0 & 1 & 1 \\
0 & 0 & 0 \\
\end{array}
;
&
\begin{array}{|cccc|}
\overline{2} & \overline{1} & 3 & 6 \\
X & 0 & 0 & 1 \\
1 & 1 & 1 & 1 \\
X & 0 & 1 & 0 \\
X & 1 & 0 & 0 \\
0 & 1 & X & 0 \\
\end{array}
;
&
\begin{array}{|ccccc|}
8 & 6 & 7 & 4 & 9 \\
X & 0 & 0 & X & 1 \\
X & 1 & 0 & 1 & 1 \\
0 & 1 & 1 & X & 1 \\
X & 0 & 1 & X & 0 \\
1 & X & 1 & X & 0 \\
X & 1 & 0 & 0 & 0 \\
\end{array}
.
\end{array}
$$

SSFs of lines identified as j^σ, where $\sigma = \{0,1\}$ is the sign of SSF, j is a number of a circuit line, are considered.

For the input test-vector T=1111 fault-free simulation is executed. It gives the vector (111101011). Then, the primary fault list $L_j = \{j^{\overline{T_j}}\}$ of a digital circuit is generated:

$$L_1 = \{1^0\}; L_2 = \{2^0\}; L_3 = \{3^0\}; L_4 = \{4^0\}.$$

The step-by-step processing of all circuit elements for output fault list obtaining is executed by (4). For the first primitive, the simulation result looks as follows:

$$L_5 = T(110) \oplus C_{Fl} \begin{array}{|ccc|} 1 & 2 & 5 \\ 1 & X & 0 \\ 0 & 1 & 1 \\ 0 & 0 & 0 \end{array} = L \begin{array}{|ccc|} 1 & 2 & 5 \\ 0 & X & 0 \\ 1 & 0 & 1 \\ 1 & 1 & 0 \end{array} = L_1 - L_2 = \{1^0\}.$$

Then, it is required to add the output line fault to the obtained list. The mentioned fault is inverse concerning the value of the test-vector coordinate. In this case, the list $L_5 = \{1^0, 5^0\}$ will be obtained.

The step-by-step processing of other primitives in the circuit gives the following results:

$$L_6 = \{1^0, 2^0, 3^0, 6^0\}; \quad L_7 = \{1^0, 7^1\};$$

$$L_8 = \{1^0, 2^0, 3^0, 5^1, 6^0, 8^0\}; \quad L_9 = \{1^0, 4^0, 7^1, 9^0\}.$$

⊕	0	1	X	Z	G	T	K	N	Q	E	H	J	O	I	A	B	S	P	C	F	L	V	Y
0/0	0	1	X	Z	G	T	K	N	Q	E	H	J	O	I	A	B	S	P	C	F	L	V	Y
0/1	1	0	X	Z	T	G	K	N	E	Q	J	H	I	O	A	B	P	S	F	C	V	L	Y
1/0	0	1	X	Z	G	T	K	N	H	J	Q	E	O	I	B	A	P	S	L	V	C	F	Y
1/1	1	0	X	Z	T	G	K	N	J	H	E	Q	I	O	B	A	S	P	V	L	F	C	Y
0/X	X	X	X	X	G	T	K	N	A	A	B	B	Y	Y	A	B	Y	Y	B	B	A	A	Y
1/X	X	X	X	X	T	G	K	N	B	B	A	A	Y	Y	B	A	Y	Y	A	A	B	B	Y
X/0	0	1	X	Z	X	X	X	X	O	I	O	I	O	I	Y	Y	Y	Y	I	O	I	O	Y
X/1	1	0	X	Z	X	X	X	X	I	O	I	O	I	O	Y	Y	Y	Y	O	I	O	I	Y
X/X	X	X	X	X	X	X	X	X	Y	Y	Y	Y	Y	Y	Y	Y	Y	Y	Y	Y	Y	Y	Y

Table 1

Table 2

0	1	X	Z	G	T	K	N	Q	E	H	J	O	I	A	B	S	P	C	F	L	V	Y
Z	1	1	Z	G	T	Z	Z	G	E	T	J	Z	1	E	J	J/X	E/X	1	J	1	E	1

The number of detected faults F at q lines generates test fault coverage:

$$Q = [F/(2 \times q)] \times 100\%.$$

For test-vector 1111, the following fault coverage is obtained:

$$Q = [4/(2 \times 9)] \times 100\% = 22\%.$$

5 Fault simulation in sequential primitive elenemts

Output fault list is a function, which is specified according to (1) by equation:

$$L_r^t = f[(T),(C),(L_X^{t-1}, L_X^t, L_Y^{t-1}, L_Z^{t-1})],$$

where $T_t \in T = (T_1,...,T_t,...,T_p)$ are two consecutive input vectors, where each coordinate is specified in the following combinations:

$$T_t = \left[\frac{0}{0}, \frac{0}{1}, \frac{1}{0}, \frac{1}{1}, \frac{0}{X}, \frac{1}{X}, \frac{X}{0}, \frac{X}{1}, \frac{X}{X}\right].$$

The two-frame format of an input vector is directed forward the sequiential FSM analysis, since, in general case, its covering is specified in the two-frame alphabet À^2. Hence, Table 1 represents $\oplus$-operation between coordinates of the test-vector and the two-frame cubic covering $T_{tj} \oplus C_{ij}$.

Each coordinate of the table is a compact form of fault lists $L_j = T_{tj} \oplus C_{ij}$. For instance, if $T_{tj} = (01), C_{ij} = P$, then, according to Table 1, at the input coordinate j is obtained:

$$T_{tj}(01) \oplus P\{(01),(10)\} = (\overline{L}_j^{t-1}\,\overline{L}_j^t) \vee (L_j^{t-1} L_j^t).$$

Table 2 determines output (observed) coordinates of FLCC at frames (t-1, t). For instance, for the coordinate $L_{ij} = V$, its value in frame t-1 is 0, and in frame t is 1, and it is specified in Table 2 by letter E. Determinations of symbols S, P are exceptions. There is an interpretation difference depending on the fact, whether an output variable is the function or the argument to the output, for which a detectable fault list is generated. In the first case, determination of the above-mentioned symbols gives (J, E), in the second one – (X, X), that indicate to the absence of FLCC for the specified output in the frame t-1.

Example 2. To execute fault simulation of CD-latch on two test-vectors. For first test-vector the following result is obtained:

$$T_1 \begin{vmatrix} 1 & 0 & X \\ 0 & 0 & 0 \end{vmatrix} \oplus C_{CD}\; \begin{array}{cc|c} C & D & Q \\ \hline H & J & I \\ L & Y & S \\ H & Q & O \\ H & P & X \end{array} = L\; \begin{array}{cc|c} C & D & Q \\ \hline Q & J & I \\ C & Y & Y \\ 0 & Q & O \\ Q & P & X \end{array}.$$

$$L(T_1) = \overline{C}_{t-1}\overline{C}_t D_{t-1} D_t \vee Q_t = D_t \vee Q_t.$$

For second vector the simulation procedure creates following cubes and fault lists:

$$T_2 \begin{vmatrix} 0 & 0 & 0 \\ 0 & 1 & 0 \end{vmatrix} \oplus C_{CD}\; \begin{array}{cc|c} C & D & Q \\ \hline H & J & I \\ L & Y & S \\ H & Q & O \\ H & P & X \end{array} = L\; \begin{array}{cc|c} C & D & Q \\ \hline H & H & I \\ L & Y & S \\ H & E & O \\ H & S & X \end{array}.$$

$$L(T_2) = C_{t-1}\overline{C}_t D_{t-1}\overline{D}_t \vee (\overline{C}_{t-1} \vee C_t)Q_{t-1} \vee Q_t = Q_{t-1} \vee Q_t.$$

6 Software implementation of method

6.1 Test generation system for finite state machine ASFTEST

ASFTEST program is intended for automatic test pattern generation (ATPG), where the initial information of informal FSM graph is represented in ASF format [7]. This program is the part of CAD Active-HDL. As a result of program's work the minimized test patterns are created in form of VHDL (Verilog) file. These test patterns are used for design validation and verification with the help of CAD VHDL and Verilog. Three strategies of test generation are possible:

1) The minimized traversal of all FSM graph nodes by solving problem of Euler circuit building;

2) The minimized traversal of all graph arcs by Hamilton circuit definition;

3) The FSM set into initial state under the condition of the next node reaching.

The choice of one of the indicated strategies depends on purposes of simulation and essentiality of test coverage improving. Also the program analyses a model by checking its correctness, generates description in VHDL and macro-instructions for Active-HDL, fixes static information in file.

ASFTEST has speed operation and functional features close to characteristics of well-known analogue STATECAD. The state of ASFTEST program is beta-version.

6.2 Test generation system for boolean equation TESTBUILDER

The program is intended for ATPG with respect to SSFs of digital designs described in language of Boolean equations.

Program operations:

1. Pseudo-random test generation in term of built-in binary code generators and decimal code generators.

2. Deterministic binary test-vector generation, where the mentioned test-vectors sensitize single logical paths in circuit.

3. Single stuck-at fault simulation with purposes of fault coverage evaluation of obtained test.

4. Test formatting in standard of VHDL - Testbench.

The program has processed:

– 10 combinational circuits from list ISCAS'85; average time of deterministic test generation is 28 minutes.

– 140 combinational and sequential circuits from PRUS; 45 sequential circuits with large complexity from PRUS;

– 22 sequential circuits from list ITC'99; average time of deterministic test generation is 2 hours.

– 216 sequential circuits; average time of deterministic test generation is 14 seconds.

– 72 combinational circuits; average time of deterministic test generation is 57 seconds.

Average complexity of design is 1000 lines. Average time of pseudo-random test generation is 5 minutes. Test coverage is more than 90 %.

6.3 Fault simulator

Fault simulator is intended for single stuck-at fault simulation of digital circuit, where the digital circuit is described at functional level in form of cubic coverings.

The problems solved by the program:

1. SSF simulation on cubic coverings of functional primitive elements.

2. Simulation of complements to states of circuit lines on cubic coverings of functional elements.

3. Algorithmic and pseudo-random test generation.

4. Length test optimization by improving its quality.

5. Optimization of number of algorithmic generators by coverage problem solving.

Initial descriptions of testing object are VHDL and representation of circuit in form of Boolean equations.

Result of program work is the test for digital design represented in VHDL (Testbench) format.

Average fault simulation time on 100 vectors for circuits with 1000 lines is 1 sec.

7 Conclusions

The cubic fault simulation method is a new technology of digital circuits processing at gate, functional and algorythmic description level. It allows to simulate all single stuck-at faults detected by test-vector during one iteration. The application condition consists in usage of digital circuits description in terms of cubic coverings of primitive elements. The proposed method effectively processes sequential digital circuits described by two-frame cubic coverings [6] as well. The last ones formalize algorithm descriptions in the form of primitives corresponding to transition graphs, FSM-charts, state tables of digital circuits.

The proposed technology of testing by the equation $T \oplus C = L$ provides the possibility of fault simulation on the basis of the cubic covering analysis. It also allows to obtain deductive formulas for any typical functional element, to design compilative simulators for processing of digital circuits at optional description level, to generate tests for digital circuits on the basis of FLCC usage, to verify results of fault simulation and test generation and to design high-speed hardware simulators.

The proposed models and methods are realized in the form of program applications. The last ones are used for test generation of digital designs based on FPGA and CPLD. The class of processed structures is FSM in the form of transition graph and Boolean equations on flip-flop circuit. Digital circuit description language is VHDL. Program applications are directed toward their use in design systems: Aldec, Xilinx.

Time of test generation depends on the total number of circuit cubic coverings:

$$W(T) = [\sum_{i=1}^{M} (n_i \times q_i)]^2.$$

References

[1] Armstrong D.B. A deductive method of simulating faults in logic circuits. IEEE Trans. on Computers. Vol. C-21. No. 5. 1972. P. 464-471.

[2] Abramovici M., Breuer M.A. and Friedman A.D., Digital System Testing and Testable Design, Computer Science Press, 1998. 652 p.

[3] Yermilov V.À. Metod otbora sushchestvennih neispravnostey dlya diagnostiki cifrovih cshem. Obshchie vyirageniya dlya neispravnostey , vozmognyih pri experimente. // Avtomatika i telemehanika, 1971, N1.-S. 159-167.

[4] Hayes J.P. A systematic approach to multivalued digital simulation// ICCD-84: Proc. IEEE Int. Conf. Comput. 1984. No. 4. P. 177-182.

[5] Birger A. G. Mnogoznachnoe deductivnoe modelirovanie cifrovih ustroystv // Avtomatika i vichislitelnaya technika.– 1982.– N4.– S.77-82.

[6] Levendel Y.H. and Menon P.R. Comparison of fault simulation methods – Treatment of unknown signal values.- Journal of digital system.- Vol.4.- 1980.- P.443-459.

[7] Hahanov V. I. Technicheskaya diagnostika elementov i uzlov personalnih computerov. K.: IZMN. 1997.308 s.

DATA STRUCTURE IN BEHAVIORAL TESTING OF LOGICAL CIRCUITS

Ewa Idzikowska

*Dpt. of Control, Robotics and Computer Science, Technical University of Poznań,
pl. M. Skłodowskiej-Curie 5, 60-965 Poznań, POLAND,
Idzikowska@sk-kari.put.poznan.pl*

Behavioral models describe the logical circuit design in a way that is independent of technology and implementation. This is an important advantage in the first stages of design. Such the model may be used in the process of test generation.
An approach to test pattern generation from the behavioral, VHDL model is presented in this paper. The data structures which are extracted from this model such as data flow graph, control flow graph and state transition graph are shown. The criteria used in the selection of test sets are based on software testing. *Copyright © 2001 IFAC*

Keywords: test generation, behaviour modelling, data flow, circuit models

1. INTRODUCTION

For production test purposes, it is feasible to have a model that accurately describes the real hardware. In the early phases of the design process such a model does not exist yet. The designer instead writes an implementation and technology independent model – high level model – in a specialized hardware description language (for example VHDL) to gather information about possible problems such as timing, resource scheduling, or concurrency, and about the best implementation. Testing based on such the model is called behavioral testing and offers a number of advantages. An automatically generated behavioral test will exercise all parts of the modeled design, and therefore it is more reliable and complete than the manually generated designer's test. The tests are usable in later refinement stages of the design. Simulation and test generation are time-consuming and both directly depend on the number of components thus behavioral test generation is expected to be much faster because the number of elements is smaller.

Behavioral testing is closely related to software testing. The main difference is, that it applies to hardware, not to computer programs. Both methods are based on a textual, formalized language to express and validate hardware on the one hand and software behavior on the other hand. Software testing is based on finding a test set which observable activates every function of the model. There are some test completeness criteria (Miller, 1980) which are based on a topological representation of the program.

These include:

- execution of all statements,
- execution of all paths,
- execution of all graph edges,
- complete execution of all subexpressions of conditionals,
- repetition of all loops n-times,
- execution of all forward paths.

These criteria have become useful in the hardware testing.

2. DATA STRUCTURES EXTRACTED FROM BEHAVIORAL MODEL

Before the process of test patterns generation begins, the data structures that are extracted from the behavioral model must be defined. These are a control flow graph, a data flow graph and a state transition graph. Two of them - control flow graph (*CF*) and data flow (*DF*) graph are described in (Idzikowska, 1998, 2001).

The control flow graph *(CF)* is a Petri net and represents a control flow in VHDL model. Places represent VHDL code statements such as assignments or wait statements. Transitions represent actions - execution of code statements from its pre-places. Time intervals or conditional statements can be associated with transitions. These conditional statements represent VHDL branch statements and can depend on control, input or data variable.

The data flow graph *(DF)* is a directed graph, not necessarily strongly connected. The nodes correspond to signals, variables or constant values; the nodes drawn as rectangles are terminals (primary input or output signals), nodes drawn as circles refer to internal variables. Directed edges of the *DF* show the flow of data from node to node and are labeled with numbers of assignment statements from VHDL that effect transformation and flow of data.

Observation of the data flow graph, for example such as it is shown at the Fig. 1 uncovers the design rule that all variables must have a source. But there is a question, whether a variable must also have a drain. The graph from Fig. 1 has not an output node and there is no I/O paths. It is the typical situation of a counter.

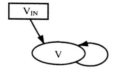

Fig. 1. A part of data graph without an I/O path

A variable *V* is called the control variable. The variable is called the control variable if there is no data path from node *V* to an output node of the data flow graph and if there is at least one reference to *V* in the conditional statement in the control flow graph.

Control variables have a drain in a reference of condition. This means that a direct observation of its current value is impossible. That's why another structure, which contains dependencies between control variables, has been defined (Veit, 1992).

The state transition graph *(ST)* is a directed graph for every control variable *V*. There is a set of states *S* being the domain of values of *V* and there are edged from a state S_k to S_l with $S_k, S_l \in S$.

State transition graph of a variable has at least one initial state and is strongly connected.

3. A STRATEGY FOR TESTING

Several information structures which describe different aspects of a model can be extracted from the high level circuits model:
- control flow graph,
- data flow graph,
- state transition graph.

The selection of test sets is based on these graphical structures.

The generic method of test selection in software testing is based on the control flow graph, but in hardware testing this approach is not useful. This is because two kinds of operations, characteristic for all tests, have to be executed:
- sensitization - transfering characteristic data to the location indicated by the operation in a selected *CF* graph node,
- propagation - propagating this data from this location to an observation output.

Data flow graph is more useful in test selection. There are assignment statements represented by edges that are members of a complete data path between an input and an output. In such a path it is possible to detect all deviations that modify the transfer function between the input and the output. Selection of all data paths replaces checking all nodes, but not all edges in control flow graph. Therefore every data path has to be repeated with different input data and all data flow graph edges have to be executed a given number of times.

4. STATE TRANSITION GRAPH IN TEST STRATEGY

Test selection based on the data flow graph is very useful in the situation of selecting complete input – output paths. As it is shown in chapter 2 these paths do not exist by definition for control variables and in this case a test would be incomplete. Not all possible functions will be activated. But as it was said the control variable make sense only if it is referred at least once for instance in a condition.

A conditional branch has always at least one depending assignment, which is executed for a specific solution of the condition. It means a specific value of the corresponding control variable. This depending assignment itself is contained in a data flow graph path, but an observation of these control variables is possible only indirectly. If the associated branch and the corresponding data assignment have been chosen, identifying a value of the control state is possible. Such an example is shown at figure 2. It is *DF* graph of 8-bit register (Armstrong, 1989). One of the processes in its VHDL model is shown at the figure 3. Output *DO* depends on the control variable *ENBLD* in this process.

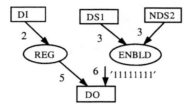

Fig. 2. Data Flow Graph of 8-bit register.

P3: process(REG, ENBLD)
begin
    if(ENBLD='1') then   *(4)*
    DO <= REG after del3;   *(5)*
    else DO <= '11111111' after del3;   *(6)*
    end if;
end process P3;

Fig. 3: VHDL Model of 8-bit register - process P3.

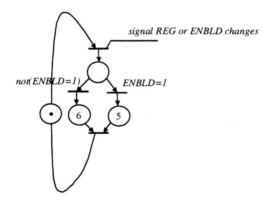

Fig. 4. Control Flow Graph of the process P3.

It is not possible to select complete input – output paths. The control variable *ENBLD* can be observed indirectly because the output variable *DO* depends on *ENBLD*. As it is shown at Fig. 4, if *ENBLD* is equal 1 – the assignment statement 5 *(DO<=REG after del3)* is executed, otherwise statement 6 *(DO <= '11111111' after del3)*. However, it is not possible to check what values the control variable had assumed between the beginning of the test and the observation. This means that the actual coding of the control variable is not relevant. It can be shown also by the following example:

A: process (RST,CNT)   *(1)*
begin
    if (RST = '1') then   *(2)*
    OUT <= '0';   *(3)*
    CNT :=0   *(4)*
    else
    if (not CNT' stable) then   *(5)*
    CNT <= (CNT+1)mod 4;   *(6)*
    end if;
    if CNT = 3 then OUT <= '1'   *(7)*
    else OUT <= '0';   *(8)*
    end if;
    end if;
end process A;

Fig. 5. VHDL model of a counter.

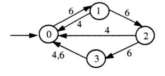

Fig. 6. Data flow graph of a counter.

Since *CNT* is a control variable which has no path to an output in the data flow graph (Fig. 6), the state transition graph for it is needed and is shown below (Fig. 7).

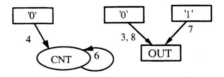

Fig. 7. State transition graph of a counter.

'1' occurs at this counter output after every 4 impulses at the input. If four *CNT* impulses are issued after a reset impulse, the output '1' can be observed, but it is not possible to say what this output value after 1, 2, or 3 impulses is. The state after a reset doesn't necessarily need to be' 0'. We can not observe the reset value. The information about the work of the counter is hidden in the length of the output sequences. Therefore state transition graph (*ST* graph) can be useful in finding tests for this control variable.

Finding tests for a control variable is possible while repeating such two steps for all edges of the *ST* graph:
-   drive the control variable into a known state,
-   execute a cycle from the known state back to this state in the *ST* graph,

The generation of test pattern for logical circuits described by their behavior is based on these three, extracted data structures – data flow graph, control flow graph and state transition graph.

5. CONCLUSION

The data structures that are extracted from the behavioural VHDL model before the process of test generation begins are shown in this paper. It is shown also, why all these structures are needed. The selection of test sets is based on these graphical structures. It is also shown why test selection based on data flow graph is better as test selection based on control flow graph and why state transition graph is needed in this process.

REFERENCES

Armstrong, J.R. (1989). *Chip-level modeling with VHDL.* Prentice Hall, Englewood Cliffs, New Jersey 1989.

Idzikowska, E. (1998). Generation of Validation Tests From Behavioral Description in VHDL, *Proceedings of The Sixth Annual Advanced Technology Workshop ATW98.* Ajaccio, France.

Idzikowska, E. (2001). Petri Net Models of VHDL Control Statements, *Proceedings of The International Workshop on Discrete-Event System Design, DEDDes'01.* Przytok, Poland.

Miller, E. (1980). Software testing, *Software Research Associates.* Seminar, San Francisco.

Veit H. H. (1992). A Contribution to the Generation of Test for Digital Circuits Described by Their Behaviour. *Doktor-Ingenieurs Dissertation*

IFAC

Publications

www.elsevier.com/locate/ifac

PCI DESIGNER APPLICATION NOTES [1]

Michal Jáchim [*,**] **Martin Jäger** [*,**]

* *Asicentrum s.r.o., Prague*
** *Faculty of Electrical Engineerig, Czech Technical*
University, Prague

Abstract: PCI interface IP core was developed as a component for research
in the area of reconfigurable acceleration. This core can be also used together
with PCI add-in card for prototyping of the PCI devices. The PCI core was
implemented using VHDL RTL description as a technologicaly independent macro
and synthetised to Xilinx Virtex FPGA family. IP Core verification plays also
very important role. For this purpose the testbench was developed. The testbench
consisting of configurable target device, a PCI bus functional model and a bus
master device model can be delivered together with the PCI core and development
board. The PCI core was also validated in the real PC environment. During
the core design several interesting issues had to be resolved. *Copyright © 2001*
IFAC

Keywords: PCI, Device, Design

1. PCI CORE DESIGN

The PCI core design should be divided into two
main parts: data path and the control logic, but
these two parts are not exactly different. For example, the parity is computed over the data and
command together. So the division of the PCI
core was made as follows: parity maintain, AD
bus control and config control. There are several
critical issues when designing the PCI core. The
first one, which affects all another, are the time
constraints. Exactly the setup time which must
be less than 7ns and the hold time of 11ns for
the 33MHz PCI . It means, that delay from the
input pads to first flip-flop must be less than
given value of 7ns, respectively delay of the path
from last FF to the output pad must be less
than 11ns. Second problem is the parity maintain block, which should provide the 36 XORs
in one clock cycle, but with the inputs directly

connected to the AD bus, which is constrainted
to 7ns (PCI Special Interest Group, 1995), or the
output connected to the PCI bus' PAR line, which
is constrainted to 11ns. As we found out, the best
solution is to take the values of the AD lines
registered, then make parity computation using
the XOR tree and the output connect directly to
the PCI bus. This can be improved by pinout and
exact PCB layout. The last, third, problem is in
the BAR's (Base Address Registers). The BAR's
together with the connected user application dra-
maticaly affects the design speed. More BAR's
makes longer combinatorial paths and the longer
interconnection distances. The maximal possible
count of the BAR's, when the PCI core is implemented without any problem is 4. When more
than 4 BAR's are needed, there must be payd
attention to the physical layout of the PCI core.
One of the main time critical issues are the fanouts
of the wiring. There is technique of signal replication widely used in the PCI core design, which
can partialy eliminate the high fanout problem.

[1] Acknowledgment: This research was in part supported
by grant 102/99/1017 of the Czech Grant Agency.

This technique is based on redundant FF which are forced into design to avoid the high fanout nets. These FF can be generated automaticaly by the Exemplar LeonardoSpectrum, but is special cases the replication can be forced by explicit part of the RTL code together with a synthesis script, where the desired FF and wires must be signed as preserved to be sure that synthesis tool will not optimize (and remove) them. The main advantage of doing the PCI core as macro block is in technological independence. Whole design is writen as VHDL RTL code, which can be synthesised for Xilinx FPGA (Shah, 1998), Altera FPGA or ASIC. The PCI core can be easily simulated for its functionality using directly RTL code. There is no need to have expensive synthesis tool for evalution or educational purposes. For the testing of the user application is available gate level VHDL code with the timing parameters for the Xilinx Virtex family devices.

1.1 Verification

The verification of the PCI core implementation can be divided into four steps (the fifth step - validation of the specification was done by the PCISIG comitee). The first step, after the RTL code is done, is the functional verification. This step checks the corectness of the RTL code and tests the basic functionality without taking care about the timing constraints. This is commonly called the RTL simulation and it should be doing continously during the writing the RTL code. When the functionality of the RTL code is just like specified, the RTL code is synthesised into the technological netlist (EDIF or Verilog) and the post-synthesis simulation model. This model is again simulated to check if the synthesis was done correctly. The simulation of the post-synthesised code can differ from the simulation of the RTL code due to resolution function. For example the the resolution function for the '0' + 'Z' gives the result of '0', when the RTL simulation is performed, but the post-synthesis simulation gives the result of 'X'. This approach is good becouse in the ideal case the 'Z' value does not affect any other logic state, but in the real world the 'Z' value is something defined not exactly. Third step of the verification is made after the Place&Route phase of the PCI core. As the result of the P&R the layout of the design and another simulation model are obtained together with the timing constraints. These constraints contain the delays of the logic blocks (technology cells), wires etc. This simulation model should be again simulated to check the time behavioral of the design. When this (post-P&R) simulation passes successfuly, there is high probability that the design will be operating correctly in the real world. Last step of the verification is to check the functionality against the real host system. For this occasion the custom PCI board for the PC can be manufactured, in and the developing the applications and software drivers can start.

2. APPLICATION DESIGN

The highest focus has been aimed at the unified interface of the application units (each controls specific address space) according to the simplicity of the application part.

If the application design consists of the several address spaces with different connection and protocol, driver of such independent components needs to solve the connection of the these components by the same number of the control state machines. The use of partly different components also leads to partly independent units, which are connected directly to the PCI core with the minimal centralization of the arbiter.

Better control over the architecture is achieved by centralized arbiter, one common application core connecting the PCI core macro and the BAR components. Time sequence of the burst read transaction with the initial latency 2 cycles is shown in the figure 2. Every signal participating on the burst transfer is registered in the scope of its logic block (synchronnous design). This automatically causes that the the whole PCI device can process first transfer on the third cycle after the start of the PCI transaction. Faster device has to have the application and the PCI core binded closer with the occurence of the combinatorial logic paths in the application part. In this case the designer can not assure the timing of the PCI core, and the whole task is on the application designer.

Principially there is no difference between the medium device with N cycles latency and slow device with the N-1 cycles latency. Each BAR component interface and communication protocol (shown in table 1) follows the description of the standart memory component synthesizable in the FPGA chips constructed by the BlockRAM cells.

Table 1. Application component interface

Signal	Direction	Size	Description
clk	IN	1	PCI bus clock
rst	IN	1	PCI bus reset
cs	IN	1	Component select
oe	IN	1	Output enable
we	IN	1	Write enable
addr	IN	8	Reduced address bus
data_in	IN	32	Input data bus
dta_rdy	OUT	1	Device ready
data_out	OUT	32	Output data bus

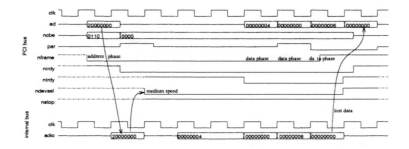

Fig. 1. Burst read transaction from prefetchable device

When performing a burst transaction there are two data items transferred at the same time due to the 2 cycle latency, first the data on the PCI bus, second the data on the path from the application to the PCI core. Every stopping/termination/abortion of the transaction is signaled with 1 cylce latency to the application and causes the second data to not to be transferred, although it is already fetch from the application. This latency problem can be omitted when using prefetchable device (memory) or must be secured when using nonprefetchable device as the application's data storage medium.

2.1 PCI communication model

The architecture and the strategy of the operating system's scheduler (in this case Linux kernel on IA32 platform) are the most important components affecting the performance of the software part of the PCI device application.

The software of the PCI device driver follows standard conventions of the writing module driver for the Linux kernel using file operations for the transferring the data between software application, kernel space driver and the hardware device.

The upper bound transfer speed of the PCI medium target only device (synchronous design, Pentium P133, 16MB RAM, Linux kernel 2.2.14, all operations were processed in the kernel level code, no system load) :

Table 2. Maximal measured transaction speed

Block write speed (target device)	51 MB/s
Block read speed (target device)	9 MB/s

Other PCI device architecture - the driver - the target application configurations lead to the same or lower transfer speeds. Also at this step some important considerations can be stated:

- Every write or read operation in the application code leads to the call of the kernel swithiching into the kernel mode and back. All commands and data should be transferred at once when possible (the buffering on the application's level). More sofisicated drivers which are intended to be working reentrant (the concurrent applications sharing the same device when running on the one CPU system or the SMP system) have to buffer data on the driver level.
- The method of buffering data suffers by increased latency of the delivery of this data (caused by the PCI bus latency and/or by the kernel switching latency).
  - (Target only - response time less than handling interrupt)
    Frequent transfers of small data amounts (e.g. handshake style communication) is the worst case, the driver has to use active waiting algorithm (repetitive reading of the device's status register to check the readiness). Every check generates the PCI bus transaction and therefore increasses the bus traffic.
  - (Target only - response time higher then handling interrupt)
    After the processing the communication the driver can shedule itself to sleep for specified period. The driver is then woken up by the interrupt handler (which reschedules the driver). The handling interrupt time depends mainly on the CPU frequency (hunderts of the CPU cycles).
  - (Master capable device)
    The driver can use active waiting algorithm of reading the memory cell known to the PCI device. The cell is initialized to known value before the start of the device's operation. Master capable device changes this value when finished and allows to exit the waiting algorithm. Reading of the memory cell value is at first cached and no subsequent memory or PCI bus transactions are performed.

For the longer response time the generating of the interrupt is the only suitable way to take CPU back.

- There is no possibility to perform burst read target transaction on some architectures, which do not allow multiple scheduling and execution of read instruction (any kind of Load/Store unit inside a superscalar CPU). When the CPU generally starts to read the data from the device by the sequence of MOV/IN instructions or by block instructions SCAS/MOVS (e.g. read of the several sequentially addressed device's registers), only the first read trensaction is executed by the Load/Store unit and others are waiting until the pervious ones are completed.

- Designer can reduce the variety of the transaction types by using optimal instructions and data types. PCI application can deny any type of the transaction (e.g. it can accept only the full width data bus transactions).

- Swappable memory pages can by swapped in/out by the kernel memory management in arbitrary time. Virtual address processed by the software application part remains unchanged, but the physical address known to the PCI device changes, therefore the PCI device in master mode has to operate only with non-swappable pages. This problem is partialy superseded in the AGP devices, where the small translate look-ahead buffer translates application addresses into the proper physical addresses, but this method requires the support from the hardware layer of the device driver.

2.2 BIOS testbench

The BIOS PCI bus configuration and the functions providing the single/burst transactions have been implemented in the behavioral VHDL description. Respected functions, which generate the desired part of the transaction, are forced to be also synchronous, although the VHDL code itself is behavioral. This feature is then utilized when analysing post-synthesis and gate-level test functionality.

The designer of the application part can write his desired transfer operations consisting of every predetermined transaction needed. At this step one should take care in a few rules of the PCI north-bridge behaviour, they can also simplify the application design.

- North-bridge can and will generate all transactions specified by PCI specification. Single or burst transactions (for memory transfers), (only) single transactions for I/O transfers.

North-bridge can also use merging, delaying, combining and other speed-up techniques. The tester has to check these transactions if the transferred data are properly generated. For this purpose conditional functions generating the data phases were implemented.

- Tester has to check all the possible events terminating the transaction (stop, retry and abort) as desired. The device can at initial developing phases stop every burst transaction and force the north-brige to start the new transaction for each data item.

- Standard flow of the BIOS PCI setup starts up with the PCI device detection and enumeration followed by the assigning the physical adresses to the BAR registers and enabling the PCI device as the target. This procedure flow is always the same with some minor differecnes among the BIOSes of the different vendors. Enabling the master capability and initialization of the applicaton is up to the driver. It is performed when starting the module (can occure more than once) or when starting the kernel (built-in drivers).

3. CONCLUSION

These PCI application's designer notes discovered the specific problems and their possible solutions during the PCI device design. Both design of the PCI device and design of the software driver according to the operating system are constrained by many host's features and limited to only several configurations, which do not significantly harm host and application performance. The notes can help the designer at first step to estimate the critical parts of his work without the need of the implementation.

REFERENCES:

Shah, N. (1998). The Challenges of Doing a PCI design in FPGAs. Online pages: www.xilinx.com. April 28, 1998.

PCI Special Interest Group (1995). PCI Local Bus Specification - production version. Revision 2.1, June 1, 1995.

IFAC
Publications
www.elsevier.com/locate/ifac

FPGA BASED SECURITY SYSTEM WITH REMOTE CONTROL FUNCTIONS

Vladimír Kašík

VSB Technical University Ostrava
Faculty of Electrical Engineering and Computer Science
Department of Measurement and Control
17.listopadu St., 708 33 Ostrava
Czech Republic
Tel.: +420-69/7321231, fax: +420-69/7323138
E-mail: vladimir.kasik@vsb.cz

Abstract: In this paper a home security system with some remote control functions is presented. The main control logic is realized with reconfigurable FPGA logic, which executes separate functions concurrently. The remote control ensures a GSM telephone access to switch heating devices and to replay a status message. In case of object violation the security system in conjunction with cameras and video recorder monitors the secured areas, runs the siren and dials predefined telephone numbers. In the FPGA is also implemented the serial communication interface for communication with PC, which performs additional user functions, e.g. event logging. *Copyright © 2001 IFAC*

Keywords: Remote control, Distributed control, Logic arrays, Logic design, Programmed control.

1. INTRODUCTION

Presented security system has been designed to ensure several security functions, such as entry and room watching, video recording of secured areas and automatic telephone dialling in case of trespass (Krček, 1997). In addition there are some additional functions designed to remote control the indoor electrical heating devices. The user is also able to find out the security system current state via voice message.

Entire the designed system is a set of autonomous subsystems with logic function distribution. The logic kernel of main security functions is used an FPGA device – the programmable logic. The co-operation of particular subsystems and the FPGA heart is described below.

2. SECURITY SYSTEM ARCHITECTURE

The main logic functions are ensured with the Main Control Unit (figure 1) based on reconfigurable FPGA logic.

Secured areas are split into two groups named entrance and room 1..n . The number of PIR detectors and cameras can be increased in each group, depending on the system configuration. The video-signal from the cameras is switched automatically and recorded on VCR in case of trespass. In this case the GSM telephone runs automatic dial up to 4 stored telephone numbers with alarm messages.

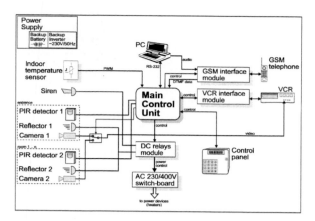

Fig. 1. General structure of the security system.

In addition the GSM with a DTMF decoder enables the remote control with common tone dialing telephone.

The remote control features include power devices (heaters) switching and the voice status messaging. Interface units, attached to VCR and GSM, adjusts the specific interface of these devices to match the main control unit connection. The voice message generator is implemented in the PC (notebook) with audio output. In addition the PC performs an event logging and the extended functions (user menu) in control panel. In this way the system has its functions distributed between main control unit and PC.

The control panel serves as a primary user interface. It consists of 4x4 matrix keyboard, 16x2 LCD display, tricolour status LED and beeper. The power electrical devices are switched by AC contactors in the switch-board. To drive these contactors the DC relays are used. A small panel with LED and 3-position switch („Manual ON" – „OFF" – „Automatic") for each power device is mounted in the front of the switch-board to enable the manual control possibility. The indoor temperature is measured with an electronic thermometer with PWM output. The measured value is used to tempering function in the winter season and to compose the voice status message. Whole the system is backed up with the Uninterruptable Power Source (UPS).

3. MAIN CONTROL UNIT

This unit ensures the most of logic functions in the design. It includes:
- FPGA – Xilinx XCS10 in this project,
- Configuration memory - serial EEPROM (ATMEL AT17C256),
- Quartz oscillator 4.000 MHz (the accurate frequency is needed for the SCI interface),

- I/O circuitry which interfaces the FPGA logic to other devices in the system, using optocouplers, bus drivers, etc.

3.1 FPGA design

The major part of the design is split into separate blocks with specific functions in accordance with figure 2. The logic design keeps general design rules to enhance reliability (Alford, 1989) and avoid possibilities of logic hazard (Bernard, et. al., 1992)

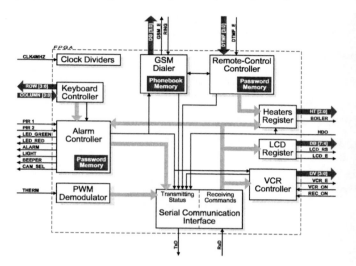

Fig. 2. Functional blocks implemented in FPGA.

The Clock Dividers block generates the appropriate clock frequencies for other function blocks.

The Keyboard Controller block generates the code of the depressed key (on the control panel). The row/column bus interface is designed to deal with a 4x4 matrix keyboard in which only one key can be depressed continuously. The encoded result is then passed to the Alarm Controller block.

The Alarm Controller block watches the key code input sequence and makes a comparison with the data stored in a password memory. The memory block is also implemented in the FPGA structure and the password could be up to 15 numbers long. If the input sequence matches the password correctly, a code-lock circuit changes its output binary value „LOCKED". The general security function is designed as a state machine (fig. 4).

The PWM Demodulator block measures the low and high state times of the input signal and sends the results through the SCI interface to the PC.

The GSM Dialer block ensures the alarm dialing using the phonebook memory. The size of this memory determines the phonebook size. In this design it is 4 phone numbers up to 15 digits long.

In advance, the GSM Dialer block watches the ring detector, notifies this event to other logic and handles the hang-up/down circuitry.

The Remote-Control Controller block becomes useful, when the user accesses the remote control functions via the GSM telephone. The control algorithm is outlined in fig. 3.

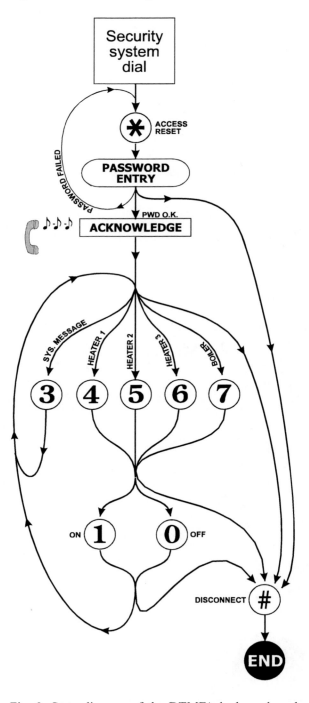

Fig. 3. State diagram of the DTMF/telephone based remote control procedures.

All the commands acceptable by the system are transmitted as a DTMF modulated tone dial sequence. After success login, a pulsing tones sounds in the phone as the acknowledge signal. Then, there is enabled to switch the heaters ON and OFF and initiate the status voice message.

The Heaters Register and LCD Register blocks latches the input data only and passes them to the corresponding devices. However, the heaters can be controlled from two data sources (with priority selection), as shown in figure 2.

The VCR Controller block makes the VCR accessible in two ways:

1) automatic control when trespass,
2) user control with menu-functions on Control Panel.

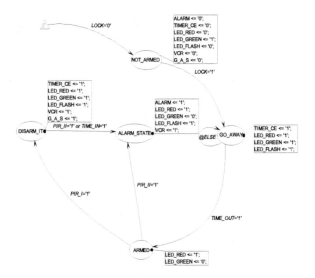

Fig. 4. The general security function state diagram.

In the SCI block is the receive/transmit controller implemented for the asynchronous 9600 Baud seial communication. The received data stands for commands in a specific binary format. The transmitted data, on the other hand, represents the status and output values of all the connected blocks. The up-to-date 8-Bytes long status sequence is transmitted periodically.

4. CONCLUSION

There has been realized one prototype of security system described above. After long stage of development and debugging the fully specified functionality has been achieved. However, the use of specific commercial electronic devices such as obsolete 2-head/mono/SP video-recorder or GSM telephone not including digital user interface, cause compatibility problems for further designs.

The only part, which could make a device-independent functions, is the digital design for FPGA device. It will be improved and used in similar designs in future.

This work has been supported by MSMT project No. J17/98.

REFERENCES

Alford, R. C. (1989) Programmable Logic Designer's Guide. Howard W. Sams & Company, Indianopolis.

Bernard, M. J., J. Hugon and R. Le Corvec (1992) From logic devices to microprocessors. SNTL, Prague.

Křeček, S. (1997) Property protection and industrial TV systems. Grada, Prague.

IFAC
Publications
www.elsevier.com/locate/ifac

TESTING FPGA DELAY FAULTS:
IT IS MUCH MORE COMPLICATED THAN IT APPEARS

Andrzej Krasniewski

Institute of Telecommunications
Warsaw University of Technology

Abstract: Testing delay faults in LUT-based FPGAs differs significantly from testing delay faults in gate networks. These differences are explained in some detail. To illustrate specific problems associated with FPGA delay fault testing, a new invalidation mechanism for non-robust tests that cannot be observed in a network of simple gates is shown. This leads to the introduction of a new class of tests – super-strong non-robust tests. Finally, key problems that must be solved to overcome difficulties associated with testing delay faults in FPGAs are identified and directions for future research in that area suggested. *Copyright © 2001 IFAC*

Keywords: digital circuits, VLSI, testability, fault detection, timing, delay analysis

1. INTRODUCTION

Most proposed techniques for testing in-system reconfigurable FPGAs are intended to check, as exhaustively as possible, all possible operation modes of FPGA components – configurable logic blocks and interconnection switching arrays. It is presumed that such an approach gives a reasonable chance that – after programming – the device will correctly operate in the field, independent of a specific function implemented by the user. Such a test strategy, normally employed by the chip manufacturer, is referred to as *application-independent testing* (or configuration-independent testing (Quddus, *et al.*, 1999) or manufacturing test (Renovell, *et al.*, 1997).

Both external testing techniques and built-in self-test (BIST) techniques can be used to implement application-independent FPGA testing. External testing techniques rely on a sequence of test patterns, derived for an assumed fault model and supplied by an external tester. Regularity of an internal structure and reconfigurability of an FPGA are usually exploited to concurrently examine its individual components (Huang, *et al.*, 1998; Inoue, *et al.*, 1998; Renovell, *et al.*, 1997; Renovell, 2000). The BIST-based techniques exercise an FPGA in a number of self-test sessions, so that during each session a selected part of the FPGA is examined using the remaining portions of the device, temporarily

configured into test pattern generators and test response compactors (Abramovici, *et al.*, 1999; Harris, *et al.*, 2001; Itazaki, *et al.*, 1997; Stroud, *et al.*, 1996).

One of the major problems in testing FPGAs is the detection of timing-related faults. Timing-related faults in FPGAs are mostly associated with interconnection delays that can account for over 70% of the clock cycle; moreover, programmable interconnections are the primary source of large variations in propagation delays. With increasing popularity of cluster-based FPGA architectures, the impact of interconnections on delay faults is becoming even more significant (Harris, *et al.*, 2001). Testing for FPGA delay faults should, therefore, focus on the interconnection structure, which means that the path delay fault model should be used. However, with the application-independent testing, it is possible to exercise only a very small fraction of interconnections patterns that can be set by programming the device. In other words, only a very small fraction of paths that exist in a particular user-defined circuit can be examined. Therefore, *the conventional approach to FPGA testing, i.e. application-independent testing, is not suitable for the detection of FPGA delay faults, even if at-speed testing based on the BIST techniques is performed.*

To deal with FPGA delay faults more effectively, the concept of application-dependent testing (also referred to as configuration-dependent testing (Quddus, *et al.*, 1999) or user test (Renovell, *et al.*, 1997) has been proposed. The idea is to thoroughly exercise only one specific configuration of the FPGA – corresponding to the user-defined function. Application-dependent testing can be based on externally provided test patterns, as proposed by Renovell, *et al.* (1997) and Quddus, *et al.* (1999), or on the BIST techniques.

The idea of BIST-based application-dependent testing of FPGAs is similar to that of application-independent self-testing – during each of several test session, a selected part of an FPGA, configured to implement a user-defined function, is examined using the remaining portions of the FPGA, temporarily reconfigured into test resources. An extension of the basic procedure for application-dependent testing of FPGAs to make it specifically oriented towards the detection of delay faults in LUT-based FPGA was proposed by Krasniewski (1999)

The development of test techniques for FPGA delay faults and the evaluation of their effectiveness cannot rely on the conventional methods for analysis of delay fault testability – testing delay faults in FPGAs differs in many respects from testing delay faults in simple gate networks. In section 2, these differences are discussed in some detail. New facts and arguments are added to the earlier observations, so that quite a complete view of the complexity of this problem is given. In the following sections, the consequences of these differences are presented.

2. TESTING FPGA DELAY FAULTS – WHAT MAKES IT SPECIFIC

Most theoretical and practical results on delay fault testing have been obtained for combinational circuits composed of NOT, AND, NAND, OR, and NOR gates, see, for example (Cheng and Chen, 1993; Fuchs, *et al.*, 1991; Lam, *et al.*, 1995; Pomeranz, *et al.*, 1998; Sparmann, *et al.*, 1995). These simple logic gates have a specific feature – they implement Boolean functions that are either positive unate or negative unate in all their variables (a single-output Boolean function is positive/negative unate in input variable x_i if changing x_i from 0 to 1 does not affect the value of the function or causes its change from 0 to 1/from 1 to 0). This is not the case with FPGAs for which LUTs – basic components of combinational logic blocks – can implement arbitrary Boolean functions. This has the following consequences.

Multiple logical paths (and delay faults) associated with each physical path. In the literature on delay fault testing, the concept of a logical path in a gate network is introduced (Sparmann, *et al.*, 1995); a logical path is also referred to as a functional path (Fuchs, *et al.*, 1991). A *logical path* is defined by a path (physical path) and the polarity of a transition (rising or falling) that is applied to the input of the

path. It is assumed that delays which contribute to the total delay of a path may depend on the polarity of transitions that occur at the inputs or outputs of the gates along the path. Thus, the propagation delay corresponds to a logical path, and so does a path delay fault. Hence, two logical paths and two path delay faults are associated with each physical path. The situation looks differently for a network of LUTs. For a given physical path, the polarity of a transition at its input does not determine, in general, the polarity of transitions at the outputs of LUTs along the path. In other words, for a given physical path, several logical paths – each defined by a pattern of transition polarities along the path – may exist. This is illustrated in Fig. 1. For path a-d-g in the gate network of Fig. 1(a), there are two logical paths; the corresponding path transition patterns are: a↑d↑g↓ and a↓d↓g↑. Each of these two logical paths may have a delay fault. For the network of LUTs in Fig. 1(b), there are eight logical paths associated with physical path a-d-g; these are defined by the following path transition patterns: a↑d↑g↑, a↑d↑g↓, a↑d↓g↑, a↑d↓g↓, a↓d↑g↑, a↓d↑g↓, a↓d↓g↑, and a↓d↓g↓. Each of these eight logical paths may have a delay fault. Clearly, for a path with n connections, not all 2^n path transition patterns may be feasible (for example, if LUT2 in Fig. 1(b) implemented an AND function, only four path transition patterns associated with path a-d-g could occur). However, in general, with an increase in the number of LUTs along the path, the number of path transition patterns and the number of delay faults increases; in the worst case – exponentially.

Non-existence of a controlling value for a basic logic component. For each basic component of a gate network – a simple logic gate, a controlling logic value is defined. A *controlling logic value*, when applied to one of the gate inputs, determines the logic value at the gate output, independent of the logic values at the remaining gate inputs. For example, a 0 is a controlling value for the AND gate. A *non-controlling logic value* is a complement of the controlling value. This simple concept does not apply for a basic logic component of an FPGA – a LUT. A LUT may not have a controlling value, i.e. the logic value at the LUT output may not be determined by the logic value at any individual input.

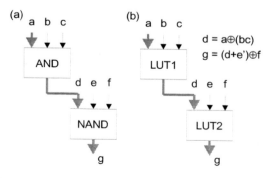

Fig. 1. A path in a network of simple gates (a) and in a network of LUTs (b)

Consider, for example, LUT1 in Fig. 1(b). It can be seen that neither 0 nor 1 at input a determines the logic value at the output d independent of the logic values at inputs b and c. Thus, LUT1 has no controlling value for input a. It can also be seen that LUT1 has no controlling value for input b and input c. Clearly, non-existence of a controlling value is not a rule. A LUT may implement a function, such that a controlling value exists for some or all inputs. For example, for function $d = a(b \oplus c)$, a controlling value exists for input a, but does not exist for input b or c.

To show that the above described features of FPGAs have really a significant impact on the testability of path delay faults, observe first that the key problem in test pattern generation, fault simulation and other procedures, is to determine necessary and sufficient conditions for a pair of input vectors to be a test for a given path delay fault. For a network of simple gates, a decision on whether or not a given pair of input vectors is a test of specific type (e.g. non-robust, robust, hazard-free, etc.) for a target path delay fault is based on checking the logic values on the side (off-path) inputs of the gates along the path; in particular, it is examined whether controlling or non-controlling values occur at the side inputs (Cheng and Chen, 1993). However, as explained above, the concept of a controlling value does not apply to LUTs. Therefore, *for LUT-based FPGAs, the conventional methods and procedures for testing delay faults, developed for networks of simple gates, are not applicable.*

A new approach, suitable for testing delay faults in LUT-based FPGAs, requires the formulation of necessary and sufficient conditions for a pair of input vectors to be a test for a given fault using the concepts that are applicable to a network built of components which implement arbitrary logic functions. Based on the results presented by Underwood *et al.* (1994), such conditions – in a form particularly suitable for networks of LUTs – have been developed (Krasniewski, 2001c). Analyses of various aspects of FPGA delay fault testability using these conditions reveal several interesting and unexpected features of this class of networks. One of such features is described in the following section.

3. A NEW INVALIDATION MECHANISM FOR NON-ROBUST TESTS

Non-robust tests, although not as effective in the detection of path delay faults as robust tests, are essential for the identification of timing-related problems simply because, for most circuits, the percentage of logical paths for which robust tests exist is quite low (Fuchs, *et al.*, 1991). There are several mechanisms that may cause an invalidation of a non-robust test, i.e. a situation in which a non-robust test fails to detect a path delay fault it targets, even though the additional delay caused by that fault makes the total path delay exceed the limit determined by the sampling time. A comprehensive survey of invalidation mechanisms for non-robust tests is given by Konuk (2000). Below, a new mechanism is shown, which is unique to circuits built of components that implement arbitrary logic functions, such as LUT-based FPGAs.

Before giving formal definitions of non-robust tests for path delay faults in a network of LUTs, the concept of a sensitization vector for an on-path input of a LUT is recalled.

Consider a LUT that implements a single-output Boolean function F of N input variables. Let $X = \{x_1, x_2, ..., x_N\}$ be the set of input variables and z be the output variable of F. Let F_1 and F_0 denote the on-set and off-set of F, respectively (the on-set/off-set of function F is the set of input vectors for which F takes the value of 1/0). Let $x \in \{0,1\}^N$ be an input vector for F, and $x^* \in \{0,1,*\}^N$ be an incompletely specified input vector for F (* represents an unspecified logic value). Let $x^*(-i)$ denote an input vector for which $x_i = *$ and all the remaining bit positions (input variables) are specified.

An input vector $x^*(-i)$ is a *(static) sensitization vector* for input x_i of a LUT if it does not determine the logic value at the LUT output, i.e. if the logic value at the LUT output depends on the logic value of x_i. Vector $x^*(-i)$ is called a non-inverting sensitization vector if $F(x^*(-i)) = x$ and is called an inverting sensitization vector if $F(x^*(-i)) = x'$.

Let $<v1,v2>$ be a pair of input vectors (an input pair) applied to the considered network of LUTs. To detect a delay fault associated with a logical path $\pi(TP)$, where TP is some path transition pattern associated with path π, an input pair $<v1,v2>$ must produce a transition on the input of the path and propagate this transition, so that at the output of each LUT along the path a transition defined by TP occurs. The logic values at the connections (output of LUTs) along path π defined by TP are called *logic values consistent with TP*. If TP defines the same polarity of a transition at the on-path input x_i and at the output of a LUT (both rising or both falling), then it is said that a *non-inverting sensitization vector for x_i is consistent with TP*; if different polarities of transitions at x_i and at the output of a LUT (one rising, the other one falling) are defined by TP, then it is said that an *inverting sensitization vector for x_i is consistent with TP*.

Below, two types of non-robust tests for delay faults in a network of LUTs, analogous to those defined for conventional gate networks, are defined. For the sake of brevity, the term "test for a logical path $\pi(TP)$" instead of the precise expression "test for a delay fault associated with a logical path $\pi(TP)$" is used.

Definition 1 [adapted from (Underwood, *et al.*, 1994)]: An input pair $<v1, v2>$ is a *weak non-robust test* for a logical path $\pi(TP)$ if

(a) the initial and final value at the input of the path are consistent with TP;

(b) for each LUT along the path, the final value at its output is consistent with TP;

(c) for each LUT along the path, v2 produces at the LUT input a sensitization vector for the on-path input of the LUT whose polarity is consistent with TP.

It can be observed that for a network of simple gates, conditions (a) and (b) in Definition 1 simplify to "a transition of appropriate polarity is produced at the input of the path" and condition (c) simplifies to "for each gate along the path, each of its off-path inputs has a non-controlling final value". Thus, a weak non-robust test for a network of LUTs is defined in the same way as a non-robust test for a network of simple gates; see, for example (Cheng and Chen, 1993).

Definition 2 [adapted from (Underwood, *et al.*, 1994)]: An input pair <v1, v2> is a *strong non-robust test* for a logical path π(TP) if <v1,v2> is a weak non-robust test for π(TP) and for each LUT along the path, the initial value at its output is consistent with TP.

Unlike weak non-robust tests, strong non-robust tests, also called restricted non-robust tests, are required to produce a transition at the output of each LUT along the considered path in response to the transition at the input of the path, regardless of the network timing parameters. This requirement holds for both conventional gate networks and, as implied by Definition 2, for networks of LUTs. The common understanding of the concept of strong non-robust tests is, therefore, that, in the case of hazard-free transitions along the path, such a test will detect its target path delay fault.

Another possible interpretation of a strong non-robust test for a network of simple gates is as follows. If a transition that propagates along the path is significantly delayed because of a large fault (large extra delay associated with the path), then – possibly except for some transitional period associated with changes on the side inputs of the gates along the path – the output of the path remains in its initial state and "waits" for the late transition to occur. This means that a large extra delay is always detected if the sampling time is set appropriately.

These interpretations, derived for conventional gate networks, are not necessarily valid for networks of LUTs, as illustrated by the following example.

Consider the network of Fig. 2(a). Several other LUTs, not shown in Fig. 2(a), can be located along the paths leading from the network inputs (a and b) to the inputs of the considered LUT (c, d, e), but to simplify the discussion, it is assumed that a transition applied to a or b propagates to the input of the LUT shown in Fig. 2(a) without changing its polarity and that the propagation delays along the paths leading

from the input of the network to the considered LUT are equal.

Assume first that the LUT shown in Fig. 2(a) implements a XOR function. Assume also that a pair of input vectors <00, 11> is applied to the input of the network as a potential test for a delay fault associated with logical path a↑...c↑z↑. Under all the assumptions taken, the input pair <00,11> produces the vector pair <000,111> at lines c, d and e, and a rising transition at line z. Thus, by Definition 2, it satisfies the requirements for strong non-robust testability for the LUT shown in Fig. 2(a). It is assumed that such requirements are also satisfied for other LUTs that are located along paths a-...-c-z, a-...-d-z and b-...-e-z. However, as shown in Fig. 2(b), a strange behavior of the network can be observed: as a results of a transition propagating from input b, regardless of the size of the delay fault associated with path a↑...c↑z↑, the output of the path assumes its expected (fault-free) final value at the expected time and keeps it unchanged. This means that, despite hazard-free transitions along the path, the fault is not detected.

This strange behavior is due to the specific function of the considered LUT. For a function which is positive unate or negative unate in all their variables (such as AND, OR, NAND, or NOR), it would not be observed. This is illustrated in Fig. 2(c) which shows the waveforms for the case when the LUT implements an AND function. It can be seen that the output z remains in its initial state until the transition applied to input a, and delayed at the beginning of the path, propagates along the entire path a-...-c-z. This means that for the case of an AND function, the fault is detected if the sampling time is set appropriately.

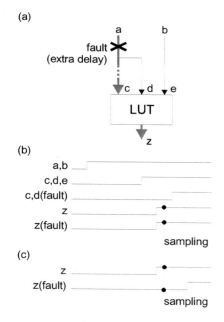

Fig. 2. Network of LUTs (a); waveforms for LUT implementing a XOR function (b) and an AND function (c)

This example shows that an input pair that satisfies the conditions given in Definition 2, derived from the general definition of a strong non-robust test for a network of components that can implement arbitrary logic functions (Underwood et al., 1994), may not have intended properties of strong non-robust tests.

To distinguish between the "traditional" strong non-robust tests, for which the output of the path does not stabilize at its final value before a delayed transition arrives, and tests which may produce behavior similar to that shown in Fig. 2(b), the concept of a super-strong non-robust test is introduced.

<u>Definition 3</u>: An input pair $<v1,v2>$ is a *super-strong non-robust test* for a logical path $\pi(TP)$ that starts at input v_i of the network if $<v1,v2>$ is a *strong non-robust test* for $\pi(TP)$ and vectors v2 and v2(i'), where v2(i') is obtained from v2 by complementing bit i, produce complementary logic values at the output of each LUT along the path.

Definition 3 can be applied to the network in Fig. 2(a) with a XOR function implemented by the LUT to check whether or not the input pair $<00, 11>$, being a strong non-robust for $a\uparrow...c\uparrow z\uparrow$, is a super-strong non-robust test for this logical path. With $v_1 = a$ and $v_2 = b$, we have v2 = 11, v2(1') = 01 and, under assumptions taken, z(v2) = z(v2(1')) = 1. Thus, $<00, 11>$ is not a super-strong non-robust test for the considered logical path. In fact, no super-strong test exists for $a\uparrow...c\uparrow z\uparrow$.

The conditions in Definition 3 cannot be translated into simple "local" requirements associated with individual LUTs, as it is the case for weak and strong non-robust tests. It does not mean that the verification whether or not a given input pair is a super-strong non-robust test for a given logical path is more difficult than in the case of a weak or strong non-robust test. It simply requires the calculation of the logic values at the outputs of the LUTs along the path for one more input vector.

4. GENERAL OBSERVATIONS

The presented discussion which shows that testing delay faults in LUT-based FPGAs is very specific and that the conventional methods for testing delay faults, developed for networks of simple gates, are not applicable to such devices, immediately leads to a practical question: How to test delay faults in LUT-based FPGAs? Below, several aspects of that problem are discussed.

Development of tools for fault simulation and test pattern generation. As stated previously, the methods for fault simulation and test pattern generation that rely on checking whether controlling or non-controlling values occur at the side inputs of the components located along the path under test are not suitable for networks of LUTs. Instead, conditions

stated by Definitions 1, 2, 3 and similar requirements for other types of tests must be applied. It must, however, be observed that fault simulation and test pattern generation (some BIST schemes may rely on deterministically generated patterns) for networks of LUTs are computationally very intensive for two major reasons:
- there exist multiple logical paths associated with each physical path;
- the testability requirements to be checked for individual LUTs are more complex than for simple gates (the requirements for robust testability, which are not shown here, are especially complex (Krasniewski, 2001b).

Evaluation of testing quality. The evaluation of test quality for delay faults, traditionally based on the concept of fault coverage calculated by means of fault simulation, is difficult, even for conventional gate networks. This is because there exist different types of testability (non-robust, robust, etc.) and therefore several different values of the fault coverage for a given test sequence. Moreover, the relationship between the path delay fault coverage by specific types of tests and "practical" measures that are meaningful for industry, such as defect level, is unknown.

For FPGAs, the problem becomes even more difficult because of the following dilemma: if there is a physical path with a very large number of logical paths, should each of these logical paths be counted when calculating the fault coverage (if so, the calculated test quality measure will strongly depend on the coverage of faults associated with that particular physical path) or some rules that reduce the number of logical paths under examination should be developed and applied.

One possible way to overcome the difficulties associated with the evaluation of quality of testing path delay faults in FPGAs is to look for some specific measures of the test quality. Some measures, based on the calculation of signal activity at LUT outputs (Krasniewski, 2000) and the calculation of delay fault activation profiles of individual LUTs (Krasniewski, 2001c), have been proposed, but definitely more work in this area is needed.

Enhancements of BIST techniques aimed at the detection of delay faults. The unique properties of in-system reconfigurable devices make it possible to adapt FPGA-oriented BIST techniques to specific requirements associated with the detection of delay faults. The first constraint is that the BIST scheme must not change the delay characteristics of the section under test (Krasniewski, 1999; Harris, et al., 2001). With this requirement satisfied, the detectability of path delay faults in LUT-based FPGAs can be improved through a modification of user-defined functions of LUTs in the section under test. There are basically two ideas for such a modification:

- to modify the network so that all delay faults associated with physical paths from a selected set of critical paths, identified using a static timing analyzer or some other method, are covered by robust tests (Harris, *et al.*, 2001) or at least are very likely to be covered (Krasniewski, 2001a); this approach is, however, effective only if a set of critical paths is relatively small which may not be the case for most speed-optimized circuits;
- to modify the network so that to increase the susceptibility of delay faults to random testing (Krasniewski, 1999, 2000).

5. CONCLUSION

It has been shown that testing delay faults in a network of LUTs differs significantly from testing delay faults in a network of simple gates. An analysis of these differences leads to the conclusion that for LUT-based FPGAs the conventional methods for testing delay faults are not applicable.

To illustrate the difficulties in testing FPGA delay faults, a strange behavior of a network of LUTs in response to a strong non-robust test is shown, which demonstrates a new invalidation mechanism for non-robust tests that cannot be observed in a network of simple gates. This leads to the introduction of a new class of tests – super-strong non-robust tests.

Finally, major problems associated with testing delay faults in FPGAs are identified and directions for future research in that area are suggested. As FPGAs become increasingly more popular among system designers and timing oriented defects in this class of devices are becoming increasingly important, a need arises for more thorough studies into these problems.

REFERENCES

Abramovici M., et al. (1999). Using Roving STARs for On-Line Testing and Diagnosis of FPGAs in Fault Tolerant Applications. *Proc. IEEE Int. Test Conf.*, pp. 973-982.

Cheng K.-T. and H.-C. Chen (1993). Delay Testing for Non-Robust Untestable Circuits. *Proc. IEEE Int'l Test Conf.*, pp. 954-961.

Fuchs, K., F. Fink and M. H. Schulz (1991). DYNAMITE: An Efficient Automatic Test Pattern Generation System for Path Delay Faults. *IEEE Trans. on CAD*, **10**, pp. 1323-1335.

Harris, I.G., P.R. Menon and R. Tessier (2001). BIST-Based Delay Path Testing in FPGA Architectures. *Proc. IEEE Int. Test Conf.*, 2001.

Huang, W.K., F.J. Meyer, X.-T. Chen and F. Lombardi (1998). Testing Configurable LUT-Based FPGA's. *IEEE Trans. on VLSI Systems*, **6**, pp. 276-283.

Inoue T., S. Miyazaki, H. Fujiwara (1998). Universal Fault Diagnosis for Lookup Table FPGAs. *IEEE Design & Test of Computers*, **15**, pp. 39-44.

Itazaki N., Y. Matsumoto and K. Kinoshita (1997). BIST for PLBs of a Look-Up Table Type FPGA – A Comparator Based BIST Technique under Definite Fault Model. *Proc. 3rd On-Line Testing Workshop*, pp. 202-206.

Konuk H. (2000). On Invalidation Mechanisms for Non-Robust Delay Tests. *Proc. IEEE Int. Test Conf.*

Krasniewski A. (1999). Application-Dependent Testing of FPGA Delay Faults. *Proc. 25th EUROMICRO Conf.*, pp. 260-267.

Krasniewski A. (2000). Exploiting Reconfigurability for Effective Detection of Delay Faults in LUT-Based FPGAs. In *Lecture Notes in Computer Science*, **1896**, pp. 675-684, Springer Verlag.

Krasniewski A. (2001a). Elimination of Reconvergent Fanouts in a Network of LUTs for Effective Detection of FPGA Delay Faults. *Proc. IEEE DDECS*, pp. 47-51.

Krasniewski A. (2001b). Testing of FPGA Delay Faults in the System Environment Is Very Different from 'Ordinary' Delay Fault Testing. *Proc. 7th IEEE IOLTW*, pp. 37-40.

Krasniewski A. (2001c). Evaluation of Delay Fault Testability of LUT Functions for Improved Efficiency of FPGA Testing. *Proc. EUROMICRO Symp. on Digital Systems Design*, pp. 310-317.

Lam W.K., A. Saldanha, R.K. Brayton and A.L. Sangiovanni-Vincentelli (1995). Delay Fault Coverage, Test Set Size, and Performance Trade-Offs. *IEEE Trans. on CAD*, **14**, pp. 32-44.

Pomeranz I. and S. M. Reddy (1998). Design-for-Testability for Path Delay Faults in Large Combinational Circuits Using Test Points", *IEEE Trans. on CAD*, **17**, pp. 333-343.

Quddus, W., A. Jas and N.A. Touba (1999). Configuration Self-Test in FPGA-Based Reconfigurable Systems. *Proc. ISCAS'99*, pp. 97-100.

Renovell, M., J. Figueras and Y. Zorian (1997). Test of RAM-based FPGA: Methodology and Application to the Interconnect. *Proc. 15th VLSI Test Symp.*, pp. 230-237.

Renovell M. (2000). A Specific Test Methodology for Symmetric SRAM-Based FPGAs. In *Lecture Notes in Computer Science*, **1896**, pp. 300-311, Springer Verlag.

Sparmann, U., D. Luxenburger, K.-T. Chang and S.M. Reddy (1995). Fast Identification of Robust Dependent Path Delay Faults. *Proc. 32nd ACM/IEEE Design Automation Conf.*, pp. 119-125.

Stroud C., S. Konala, P. Chen and M. Abramovici (1996). Built-In Self-Test of Logic Blocks in FPGAs (Finally, a Free Lunch: BIST Without Overhead!). *Proc. 14th VLSI Test Symp.*, pp. 387-392.

Underwood B., W.-O. Law, S. Kang and H. Konuk (1994). Fastpath: A Path-Delay Test Generator for Standard Scan Designs. *Proc. IEEE Int'l Test Conf.*, pp. 154-163.

IFAC

Publications
www.elsevier.com/locate/ifac

DESIGNING FAIL–SAFE SYSTEMS WITH ERROR CORRECTION CAPABILITIES USING PROBABILISTIC ANALYSIS

Miguel Pereira[1], Enrique Soto[2]

*1 Intelsis Sistemas Inteligentes S.A. - R&D Digital Systems Department, Vía Edison 16
Polígono del Tambre 15890, Santiago de Compostela (La Coruña),
mpereira@intelsis.es; 2 Dept. Tecnología Electrónica, Universidad de Vigo, Apdo.
Oficial, 36200 Vigo, España, esoto@uvigo.es, http://www.dte.uvigo.es;*

Abstract: This paper proposes a general method for the design of fail-safe systems with error correction capabilities. A fail-safe system can detect an error in a transition between two states. With this method errors produced in a transition between different states can be corrected by a design based on the analysis of probabilities. Analyzing the transition probabilities, an error corrector system can be built from the original unsafe system. This corrector system takes the form of a combinational logic block added to the unsafe system. In this method the designer can adjust the complexity versus efficiency relationship of the corrector block. *Copyright © 2001 IFAC*

Keywords: fault-tolerant systems, fault tolerance, statistics, error correction, digital systems.

1. INTRODUCTION

Any state diagram can be specified with a state matrix on which transitions are represented with values that represent going from an i state (vertical axis) to a j state (horizontal axis). Unreachable transitions are positions in the matrix whose value are zero. If any transition of this kind occurs, in a fail-safe system it is possible to detect it and reinitialize the system indicating an error. Those errors that besides detecting them can be corrected allow the system to work on without any interruption, making the system, at least partly, fault tolerant [1]. There could be more than one transition from one state to other states and then there could be non-detectable errors. If a system can jump from an A state to either B or C states, jumping to B instead of C (or vice versa) would be a not detectable error. Following this example, if the system jumps from the A state to the D state because of an error, the system can be corrected with the method exposed in this paper. Basically, this method consists of taking the information of the current state and the information of the predicted following state, and checking if the transition is possible or not. The error correction will be implemented with a combinational logic block that checks both states and deciding whether if updating (correcting) the following state is necessary or not.

The table 1 shows the matrix of states of the state diagram of figure 1 with its respective transition probabilities for systems without errors.

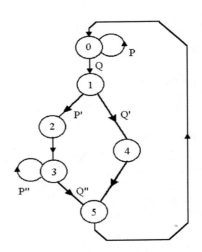

Figure 1 Example state diagram

Table 1 State transitions for example state diagram

	E_0	E_1	E_2	E_3	E_4	E_5
E_0	P	Q	0	0	0	0
E_1	0	0	P'	0	Q'	0
E_2	0	0	0	1	0	0
E_3	0	0	0	P"	0	Q"
E_4	0	0	0	0	0	1
E_5	1	0	0	0	0	0

Where P, Q, P', Q', P'' and Q'' are the transition probabilities in the states with multiple transitions. In a real system it is necessary to add the bit error probability P_b, that is, the probability of error in the value of a bit of the coded state. Assuming the following codification for the states:

E_0	000	E_4	100
E_1	001	E_5	101
E_2	010	E_6	110
E_3	011	E_7	111

The probability of being in a state and jumping to another is represented with the following equation:

$$P^e(E_j/ E_i) = \Sigma_{k=0}^{N-1} P^o(E_i / E_k) P_b^{dkj} (1 - P_b)^{nb-dkj}$$

$P^e(E_j / E_i)$ is the probability of jumping to the j state from the i state including the possible errors.
$P^o(E_j / E_i)$ is the probability of jumping to the j state from the i state without errors.
d_{kj} is the number of different bits between a k state and a j state.
nb is the number of bits in the state codification.
P_b is the probability of error in a coded bit of a state.
N is the number of states and it is always going to be less than or equal to $2^{nb} -1$.

2. ERROR ANALYSIS

2.1. Probability of being in a state

The probability of being in a i state is determined by the following system of equations:
$\Sigma_{k=0}^{N-1} P^e(E_k) = 1$, where $P^e(E_k)$ is the probability of being in the k state including errors.
$\Sigma_{k=0}^{N-1} P^e(E_k) P^e(E_i / E_k) = P^e(E_i)$,where $P^e(E_i)$ is the probability of being in the i state including errors.

2.2. Probability of being in a i state and jumping to the j state

$P^c(E_j , E_i) = P^e(E_i) * P^e(E_j / E_i)$, where $P^c(E_j , E_i)$ is the probability of being in the i state and jumping to the j state, without including the error corrector block.
$P^d(E_j , E_i) = P^o(E_i) * P^e(E_j / E_i)$, where $P^d(E_j , E_i)$ is the probability of being in the i state and jumping to the j state, including the error corrector block.
$P^o(E_j , E_i) = P^o(E_i) * P^o(E_j / E_i)$, where $P^o(E_j , E_i)$ is the probability of being in the i state and jumping to the j state without errors.
With these probabilities the matrix of probabilities M_c, M_d, and M_o, are defined , where $M_c(i,j) = P_c(E_j , E_i)$, $M_d(i,j) = P_d(E_j , E_i)$, and $M_o(i,j) = P_o (E_j, E_i)$.

2.3. Probability of error

The probability of an error in the system is determined by the following equation:

$$P_e = 1 - (1 - P_b)^{nb}$$

However this probability includes both the detectable errors and the non detectable errors. It is necessary to compare the matrix of probabilities M0 and Md to calculate the probability of error detection. The maximum probability of detecting an error Pde is determined by the following equations:

$$P_{de} = \Sigma_{i=0}^{N-1} \Sigma_{j=0}^{N-1} D(i,j)$$

$$D(i,j) = M_d(i,j) \delta(M_o(i,j))$$

D is the matrix that contains the probabilities of the wrong transitions. $\delta(x)$ is the delta function, and its value is 1 when x=0, and 0 in the rest of the cases. So $D(i,j)$ is not 0 when $Mo(i,j)$ is 0 (when the transition Ei, Ej is not allowed), and when $Md(i,j) > 0$ (when the not allowed transition occurs because of an error).

3. EXAMPLE

Returning to the diagram state of the figure 1, and the probability of jumping from a state in case of bifurcations is 0.5, and Pb = 0.1, the following matrix has been obtained:

Table 2. Matrix M_0. Probabilities of being in an i state and jumping to an j state without errors.

	E_0	E_1	E_2	E_3	E_4	E_5	E_6	E_7
E_0	0.167	0.167	0	0	0	0	0	0
E_1	0	0	0.083	0	0.083	0	0	0
E_2	0	0	0	0.083	0	0	0	0
E_3	0	0	0	0.083	0	0.083	0	0
E_4	0	0	0	0	0	0.083	0	0
E_5	0.167	0	0	0	0	0	0	0

Table 3. Matrix M_d. Probabilities of being in an i state and jumping to an j state including errors.

	E_0	E_1	E_2	E_3	E_4	E_5	E_6	E_7
E_0	0.135	0.135	0.015	0.015	0.015	0.015	0.002	0.002
E_1	0.014	0.001	0.062	0.007	0.062	0.007	0.014	0.002
E_2	0.001	0.007	0.007	0.061	0.0001	0.001	0.001	0.007
E_3	0.002	0.014	0.007	0.062	0.007	0.061	0.002	0.014
E_4	0.001	0.007	0.0001	0.001	0.007	0.061	0.001	0.007
E_5	0.122	0.014	0.014	0.002	0.014	0.002	0.002	0.0002

	E_0	E_1	E_2	E_3	E_4	E_5	E_6	E_7
E_0	0	0	0.015	0.015	0.015	0.015	0.002	0.002
E_1	0.0014	0.001	0	0.007	0	0.007	0.014	0.002
E_2	0.001	0.007	0.007	0	0.0001	0.001	0.001	0.007
E_3	0.002	0.014	0.007	0	0.007	0	0.002	0.014
E_4	0.001	0.007	0.0001	0.001	0.007	0	0.001	0.007
E_5	0	0.014	0.014	0.002	0.014	0.002	0.002	0.0002

Pd (Maximum probability of detection of errors)= 0.241

Pe (Probability of error without the detection block)= 0.271

Efd (Efficiency of the detection of errors)= 0.241/0.271 = 88.93%

In the table 4 the detectable error probabilities are shown, but not every detectable error can be corrected. An example could be the following one: if the system is in the E1 state, it can jump to the states E2 or E4. If an error has occurred and the system jumps to E3, is more likely the correct jump was to E2 and not to E4, therefore it can be corrected. If an error produces a jump to E7, this error cannot be corrected because the probability that the correct jump was to E2 is equal to the probability of the right jump (to E4). The error corrector block cannot estimate the right jump, but it can detect the error and reset the system jumping to an initial (safe) state E0. In the following table the detectable errors are shown with the respective corrections in brackets.

Table 5. Probabilities of correctable wrong transitions.

	E_0	E_1	E_2	E_3	E_4	E_5	E_6	E_7
E_0	0	0	0.015 (0)	0.015 (1)	0.015 (0)	0.015 (0)	0.002 (0)	0.002 (1)
E_1	0.014 (0)	0.001 (0)	0	0.007 (2)	0	0.007 (4)	0.014 (0)	0.002 (0)
E_2	0.001 (3)	0.007 (3)	0.007 (3)	0	0.0001 (3)	0.001 (3)	0.001 (3)	0.007 (3)
E_3	0.002 (0)	0.014 (0)	0.007 (3)	0	0.007 (5)	0	0.002 (0)	0.014 (0)
E_4	0.001 (5)	0.007 (5)	0.0001 (5)	0.001 (5)	0.007 (5)	0	0.001 (5)	0.007 (5)
E_5	0	0.014 (0)	0.014 (0)	0.002 (0)	0.014 (0)	0.002 (0)	0.002 (0)	0.0002 (0)

Pc (Probability of correcting an error) = 0.1559

Efc (Efficiency in the correction of errors)= 0.1559/0.271 = 57.53%

In the example, there is a 88.93% of detected errors, but there is only a 57.53% of corrected errors without interruptions. In the rest of the percentage up to 88.93% the system is reinitialized, or driven to a safe state.

4. CORRECTION OF ERRORS

With the probabilistic analysis resolved, the second step consists of designing the error corrector block with the complexity and efficiency chosen by the designer. Obviously, if the chosen efficiency is maximum, the complexity also is maximum. The design of the error corrector block is based on the elements of the D matrix because they are the probabilities of the forbidden transitions. The sequential system to implement including the corrector block is shown in figure 2.

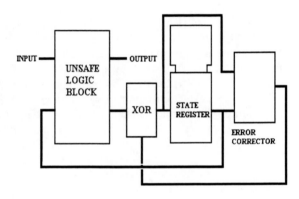

Figure 2 Sequential system with error corrector block

If the efficiency has to be the maximum efficiency with the scheme of the figure 2, the logic functions of the corrector are the following ones:

$$CQ_0 = (-Q_2 * Q_1 * -Q_0 * Q'_0) + (Q_2 * -Q_1 * -Q_0 * -Q'_2 * -Q'_0) + (Q_2 * -Q_1 * -Q_0 * Q'_1 * -Q'_0) + (-Q_2 * Q_0 * -Q'_2 * -Q'_1 * Q'_0) + (-Q_1 * Q_0 * Q'_0) + (-Q_2 * Q_0 * Q'_2 * Q'_1 * Q'_0) + (-Q_2 * Q_1 * -Q'_2 * Q'_1 * -Q'_0) + (-Q_2 * Q_1 * Q'_2 * -Q'_1 * -Q'_0)$$

$$CQ_1 = (-Q_2 * Q_1 * -Q_0 * -Q'_1) + (-Q_1 * -Q_0 * Q'_1) + (Q_2 * -Q_1 * Q'_1) + (-Q_2 * Q_0 * Q'_2 * Q'_1)$$

$$CQ_2 = (Q_2 * -Q_1 * -Q_0 * -Q'_2) + (-Q_2 * Q'_2 * Q'_1) + (-Q_2 * -Q_0 * Q'_2) + (Q_2 * -Q_1 * Q_0 * Q'_2)$$

Where CQi are the corrector bits of the state, Qi are the bits of the current state, y $Q'i$ are the bits of the predicted following state.

The error corrector block might be relatively complex if the efficiency must be maximum (Efd = 88.93%, Efc = 57.53%), but that complexity can be reduced, decreasing the efficiency as follows:

$$CQ_0 = (-Q_2 * Q_1 * -Q_0 * Q'_0) + (Q_2 * -Q_1 * -Q_0 * Q'_1 * -Q'_0) + (-Q_2 * Q_0 * -Q'_2 * -Q'_1 * Q'_0) + (-Q_1 * Q_0 * Q'_0) + (-Q_2 * Q_0 * Q'_2 * Q'_1 * Q'_0)$$

$$CQ_1 = (-Q_2 * Q_1 * -Q_0 * -Q'_1) + (-Q_1 * -Q_0 * Q'_1) + (Q_2 * -Q_1 * Q'_1) + (-Q_2 * Q_0 * Q'_2 * Q'_1)$$

$$CQ_2 = (Q_2 * -Q_1 * -Q_0 * -Q'_2) + (-Q_2 * Q'_2 * Q'_1) + (-Q_2 * -Q_0 * Q'_2) + (Q_2 * -Q_1 * Q_0 * Q'_2)$$

Efd = 81.37% , Efc = 49.63%

Table 6 shows the size of the implementations of the exposed example in two different programmable devices. The CPLD from Altera EPM3032, and the FPGA EPF6010 (also from Altera) have been chosen to make the comparison. It is necessary to remark that a CPLD is composed by a matrix of LABs (Logic Array Block), while a FPGA is composed by a matrix of CLBs (Configurable Logic Block). Every CLB in a FPGA is smaller than a LAB in a CPLD, so the number of used CLBs is greater than the number of used LABs.

The relation between the number of blocks used in an unsafe system, and the number of blocks used in a fault tolerant system is increased in a FPGA because of the maximum number of inputs on every CLB. This number is smaller than the number of inputs on a LAB, so more CLBs are included to implement logic products with a big number of inputs.

Table 6 Comparison between different programmable technologies

	EPM3032 (number of LABs)	EPF6010 (number of CLBs)
Unsafe system	9	11
Fault Tolerant System (FTS) Efd=88.93% Efc=57.63%	15	35
Reduced FTS Efd=81.37% Efc=49.63%	15	31

5. CONCLUSIONS

With the help of a probabilistic analysis of errors in a state diagram, and choosing the appropriated codification of states, an error detection and correction system can be developed, based in a combinational logic added to the original unsafe system. The main advantages consist of the inalterability of the original system, and the systematical design of the corrector block.

REFERENCES

[1] Parag K. Lala, *Self-Checking and Fault-Tolerant Digital Design*, 2001 Morgan Kaufmann Publishers.

ACKNOWLEDGEMENTS

This work was financed by the European Commission and the Comisión Interministerial de Ciencia y Tecnología (Spain) through research grant TIC 1FD97-2248-C02-02 in collaboration with the company Versaware S.L. (Vigo, Spain).

IFAC

Publications
www.elsevier.com/locate/ifac

FSM IMPLEMENTATION IN EMBEDDED MEMORY BLOCKS USING CONCEPT OF DECOMPOSITION

Mariusz Rawski, Tadeusz Łuba

Warsaw University of Technology, Institute of Telecommunications
Nowowiejska 15/19, 00-665 Warsaw, Poland

Abstract: Since modern programmable devices contain an embedded memory blocks, there exists a possibility to implement FSM using such blocks. The size of memory available in programmable devices is limited, though. The paper presents a general method for the synthesis of sequential circuits using embedded memory blocks. The method is based on the serial decomposition concept and relies on the decomposition of memory block into two blocks: a combinational address modifier and a smaller memory block. Appropriately chosen strategy of the decomposition may allow reducing required memory size at the cost of additional logic cells for address modifier implementation. This makes possible to implement FSM that exceed available memory through using embedded memory blocks and additional programmable logic. *Copyright © 2001 IFAC*

Keywords: digital circuits, logic minimization, implementation, sequential machines, programmable read only memory, Boolean functions

1. INTRODUCTION

Decomposition has become an important activity in the analysis and design of digital systems. It is fundamental to many fields of modern engineering and science (Brzozowski and Luba 1997; Hartmanis and Stearns 1966; Luba 1994; Zupan and Bohenec 1966a, b; Ross, *et al.*, 1991). Functional decomposition relies on breaking down a complex system into a network of smaller and relatively independent co-operating sub-systems, in such a way that the original system's behaviour is preserved. A system is decomposed into a set of smaller subsystems, such that each of them is easier to analyse, understand and synthesise.

By taking advantage of the opportunities the modern microelectronic technology provides us with, we are able to build very complex digital circuits and systems at relatively low cost. We can utilise a huge diversity of logic building blocks. The element libraries contain many various gates, a lot of complex gates can be generated in (semi-)custom CMOS design, and the field programmable logic families include different types of (C)PLDs and FPGAs. However, the opportunities created by modern microelectronic technology cannot be fully exploited

because of weaknesses of the traditional logic design methods.

Recently, new methods of logic synthesis based on functional decomposition are being developed (Luba, *et al.*, 1995; Chang, *et al.*, 1996; Burns, *et al.*, 1998; Jozwiak and Chojnacki 1999; Qiao, *et al.*, 2000). Unfortunately decomposition-based methods are mainly recognised as methods for implementation of combinational functions.

Modern FPGA architectures contain embedded memory blocks. In many cases, designers do not need to use these resources. However, such memory blocks allow implementing sequential machines in a way that requires less logic cells than traditional, flip-flop implementation. This may be used to implement "non-vital" sequential parts of the design saving logic cell resources for more important parts. However such an implementation may require more memory than available in a device. To reduce memory usage in ROM-based sequential machine implementations decomposition-based methods can be successfully used (Luba, *et al.*, 1992).

In this paper, once some basic information has been introduced, the application of the decomposition to implementation of sequential machines is presented. Subsequently, some experimental results, reached with a prototype tool that implements the functional decomposition, are discussed.

The experimental results demonstrate that the decomposition is capable of constructing solutions (utilising embedded memory blocks) of comparable or even better quality than the methods implemented in commercial systems.

2. BASIC NOTIONS

2.1 Functional decomposition

Let A and B be two subsets of X such that $A \cup B = X$. Assume that the variables $x_1,...,x_n$ have been relabelled in such way that:

$A = \{x_1,...,x_r\}$ and
$B = \{x_{n-s+1},...,x_n\}$.

Consequently, for an n-tuple x, the first r components are denoted by x^A and the last s components by x^B.

Let F be a Boolean function, with $n > 0$ inputs and $m > 0$ outputs. Let (A, B) be as above. Assume that F is specified by a set $\mathbf{F}$ of the function's cubes. Let G be a function with s inputs and p outputs; let H be a function with $r + p$ inputs and m outputs. The pair (G, H) represents a serial decomposition of F with respect to (A, B), if for every minterm b relevant to F, $G(b^B)$ is defined, $G(b^B) \in \{0, 1\}^p$, and $F(b) = H(b^A, G(b^B))$. G and H are called blocks of the decomposition.

Partition-based representation of Boolean functions can be used to describe functional decomposition algorithms (Brzozowski and Luba 1997; Luba and Selvaraj 1995; Luba 1994; Luba, et al., 1996; Rawski, et al., 1997).

If there exists an r-partition Π_G on $\mathbf{F}$ such that $P(B) \leq \Pi_G$, and $P(A) \bullet \Pi_G \leq P_F$, then F has a serial decomposition with respect to (A, B).

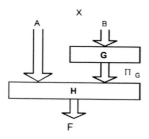

Fig. 1. Schematic representation of the serial decomposition

The serial decomposition process consists of the following steps: an input support selection (the most time-consuming part of the process), calculation of partitions $P(A)$, $P(B)$ and P_F, construction of partition Π_G, and creation of functions H and G (Luba and Selvaraj 1995).

2.2. Finite state machine

Let $A = \langle V, S, \delta \rangle$ be an FSM (completely or incompletely specified) with no outputs (outputs are omitted as they do not have essential impact on the method), where:

V – set of input symbols,
S – set of internal states,
δ – state transition function,

and $m = \lceil \log_2 |V| \rceil$, $n = \lceil \log_2 |S| \rceil$ denote the number of input and state variables respectively.

To describe logic dependencies in such an FSM special partition description (Brzozowski and Luba 1997) and special partition algebra (Hartmanis and Stearns 1966) are employed.

Let $\mathbf{K}$ be a one-to-one correspondence between the domain D_δ of transition function and $K = \{1, ... , p\}$, where $p = |D_\delta|$. The characteristic partition P_c of an FSM is defined in the following way:

$(k_1, k_2) \in B_{Pc}$ iff $\delta(K^{-1}(k_1)) = \delta(K^{-1}(k_2))$

Thus, each block B_{Pc} of the characteristic partition includes these elements from K which correspond to pairs (v, s) from the domain D_δ such that the transition function $\delta(v, s) = s'$ maps them onto the same next state s'.

A partition P on K is compatible with partition π on S iff for any inputs v_a, v_b the condition that s_i, s_j belong to one block of partition π implies that the elements from K corresponding to pairs (v_a, s_i) and (v_b, s_j) belong to one block of the partition P.

A partition P on K is compatible with partition θ on V iff for any states s_a, s_b the condition that v_i, v_j belong to one block of partition θ implies that the elements from K corresponding to pairs (v_i, s_a) and (v_j, s_b) belong to one block of the partition P.

In particular, a partition P on K is compatible with the set $\{\pi, \theta\}$ if it is compatible with both π and θ, while it is compatible with set $\{\pi_1, ..., \pi_\alpha\}$ of partitions on S (or set $\{\theta_1, ..., \theta_\alpha\}$ of partitions on V) iff it is compatible with $\pi = \pi_1 \bullet \pi_2 \bullet \pi_3 \bullet ... \bullet \pi_\alpha$ ($\theta = \theta_1 \bullet \theta_2 \bullet \theta_3 \bullet ... \bullet \theta_\alpha$).

3. ROM IMPLEMENTATION OF FINITE STATE MACHINES

The FSM can be implemented through the use of ROM (*Read Only Memory*) (Luba, *et al.*, 1992). Figure 2 shows the general architecture of such an implementation. State and input variables ($q_1, q_2, ..., q_n$ and $x_1, x_2, ..., x_m$) constitute ROM address variables ($a_1, a_2, ..., a_{m+n}$). The ROM would consist of words, each storing the encoded present state (control field) and output values (information field). The next state would be determined by input values and present-state information feedback from memory.

This kind of implementation requires much fewer logic cells than the traditional flip-flop implementation (or does not require them at all, if memory can be controlled by clock signal – no register required); therefore, it can be used to implement "non-vital" FSMs of the design, saving LCs resources for more important parts of the circuits. However, large FSM may require too many buried memory resources.

The size of memory needed for such an implementation depends on the lengths of the address and memory word.

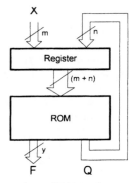

Fig. 2. Implementation of FSM using memory blocks

Let m be the number of inputs, n be the number of state encoding bits and y the number of output functions of FSM. The size of memory needed for implementation of such a FSM can be expressed by the following formula:

$$M = 2^{(m+n)} \times (n + y),$$

where $m + n$ is the size of the address, and $n + y$ is the size of the memory word.

Since modern programmable devices contain an embedded memory blocks, there exists a possibility to implement FSM using such blocks. The size of memory blocks available in programmable devices is limited, though. For example, in devices from Altera's FLEX family EAB (*Embedded Array Block*) is 2048 bits and device FLEX10K10 consists of 3

such EAB's. Functional decomposition can be used in order to implement FSM that exceed that size.

3.1. Address modifier

The FSM A defined by a given transition table can be implemented in a structure shown in Fig. 3 by means of an address modifier.

If $\pi_1, ..., \pi_n$ are partitions on S, $\theta_1, ..., \theta_m$ are partitions on V, and P_k is partition on K compatible with either π_i or θ_j then $P = \{P_1, ..., P_{m+n}\}$ is the set of all partitions compatible with $\{\pi_1, ..., \pi_m, \theta_1, ..., \theta_m \}$. Partitions $\pi_1, ..., \pi_n$ correspond to stare variables and $\theta_1, ..., \theta_m$ correspond to input variables. To achieve unambiguous encoding of address variables and at the same time consistency of relation K with the transition function δ, such partitions $P_1, .., P_w$ have to be found, that:

$$P_1 \bullet P_2 \bullet P_3 \bullet ... \bullet P_w \leq P_c.$$

This is the necessary and sufficient condition for $\{P_1, .., P_w\}$ to determine the address variables, because each memory cell is associated with single block of P_c, i.e. with those elements from K which map the corresponding (v, s) pairs onto the same next state.

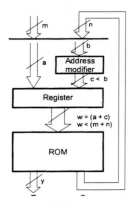

Fig. 3. Implementation of FSM using an address modifier

The choice of w ($w < n + m$) partitions from set $\{P_1, ..., P_{m+n}\}$ such that they will produce the simplest addressing unit is made thanks to the notion of r-admissibility (Luba 1994).

A set $\{P_1, ..., P_k\}$ is r-admissible in relation to partition P iff there is a set $\{P_{k+1}, ..., P_r\}$ of two block partitions such that the following condition holds:

$$P_1 \bullet P_2 \bullet P_3 \bullet ... \bullet P_k \bullet P_{k+1} \bullet ... \bullet P_r \leq P_c$$

and no set of $r - k - 1$ partitions exist which meets this requirement.

For partition $\rho \leq \sigma$ let $\sigma|\rho$ denote the quotient partition and $\varepsilon(\sigma|\rho)$ the number of elements in the largest block of $\sigma|\rho$. Let $e(\sigma|\rho)$ be the smallest

integer equal to or larger than $\log_2\varepsilon(\sigma|\rho)$ (i.e. $e(\sigma|\rho) = \lceil \log_2\varepsilon(\sigma|\rho) \rceil$). Then the r-admissibility of $\{P_1, ..., P_k\}$ is $r = k + e(\pi|\pi_f)$,
where π is the product of $P_1, ..., P_k$ and π_f is the product of π and P.

If $\boldsymbol{P} = \{P_1, ..., P_k\}$ is r-admissible in relation to P then each subset of $\boldsymbol{P}$ is r'-admissible, where $r' \leq r$.

The smallest partition i.e. one where each element is a separate block, will be denoted $P(0)$, $\pi(0)$, $\theta(0)$, etc.

3.2. Input/state encoding

The complexity of the address modifier is connected with address variables which depend on more that one input/state variable. Therefore it is important to choose such encoding of input and internal state symbols that we could obtain maximal set of partitions P (compatible with π or θ) whose r-admissibility in relation to P_c is w, where w is the number of address bits of given ROM block.

Appropriate encoding will be determined by generating partitions with knowledge that r-admissibility of a partition P compatible with partition $\pi = (B_1; ...; B_i; ...; B_\alpha)$ or compatible with partition $\theta = (B_1; ...; B_i; ...; B_\alpha)$ is:
$$r = \lceil \log_2\alpha \rceil + \lceil \log_2 \max |\delta(B_i)| \rceil,$$
where B_i is a block of partition π or θ, δ is the transition function, $\max |\delta(B_i)|$ denotes the number of elements in the most populous set $\delta(B_i)$, $i \in \{1... \alpha\}$. Because of the one-to-one correspondence between partitions P and π or θ, r-admissibility of π, θ or $\{\pi, \theta\}$ in relation to P_c can be considered.

For a given w necessary encoding that allows implementing FSM with use of address modifier can be found in following way:
1. find r_1 = r-admissibility of $\theta(0)$; find r_2 = r-admissibility of $\pi(0)$,
2. if $r_1 = w$ (or $r_2 = w$) then $a_1 = x_1, ..., a_x = x_x$ (or $a_1 = q_1, ..., a_x = q_q$) and further encoding partition are searched among $\pi(0)$ (or $\theta(0)$),
3. if both $r_1 > w$ and $r_2 > w$ then for subsequent steps θ if $|V| < |S|$ or π if $|V| > |S|$ is taken,
4. assume that θ was chosen in previous step; for $i = 1, 2, ...$ and $\alpha = 2^{m-i}$ such $\theta = (B_1; ...; B_\alpha)$ is looked for that $|B_1| + |B_2| + ... + |B_\alpha| = |V|$ and whose r-admissibility equals w.

In similar way $\pi = (D_1; ...; B_\beta)$, $\beta = 2^{n-j}$ $j = 1, 2, ...$ is looked for. The set $\{\pi, \theta\}$ must have r-admissibility of w. Partitions π and θ can be represented as follows:
$$\pi = \pi_1 \bullet \pi_2 \bullet ... \bullet \pi_k,$$
$$\theta = \theta_1 \bullet \theta_2 \bullet ... \bullet \theta_l,$$
where

$$k = \lceil \log_2\beta \rceil,$$
$$l = \lceil \log_2\alpha \rceil.$$

The encoding of remaining input and states variables can be obtained from following rules:
$$\pi_1 \bullet \pi_2 \bullet ... \bullet \pi_k \bullet \pi' = \pi(0),$$
$$\theta_1 \bullet \theta_2 \bullet ... \bullet \theta_l \bullet \theta' = \theta(0),$$
where π' and θ' represent partitions induced by those variables.

After variables are encoded the process may be considered as a decomposition of memory block into two blocks: a combinational address modifier and a smaller memory block. Decomposition is computed for partitions:
$$P(A) = \pi \bullet \theta = \pi_1 \bullet \pi_2 \bullet ... \bullet \pi_k \bullet \theta_1 \bullet \theta_2 \bullet ... \bullet \theta_l,$$
$$P(B) = \pi' \bullet \theta'.$$

Appropriately chosen strategy of decomposition may allow reducing required memory size at the cost of additional logic cells for address modifier implementation. This makes possible to implement FSM that exceed available memory through using embedded memory blocks and additional programmable logic.

4. EXPERIMENTAL RESULTS

The proposed method was applied to implement several examples from standard benchmark set in FLEX10K10 devices with use of ALTERA MAX+PlusII system. In Table 1 a comparison of different FSM implementation techniques are presented. In column falling under the *ROM Implementation* heading number of bits required to implement given FSM using ROM is presented. FLEX10K10 device is equipped only with 3 EAB memory blocks each consisting of 2048 bits. Most of presented FSM examples can not be implemented in this device, because their implementation requires much more memory resources than available. In the column falling under the *FF Implementation* heading number of logic cell is given, which is required to implement given FSM in "standard" way using flip-flops. To describe FSM for this kind of implementation special AHDL (*Altera Hardware Description Language*) construction was used. In the column falling under the *AM implementation* results of implementation of given FSM using concept of address modifier are presented. In this approach address modifier was implemented using logic cell resources and ROM was implemented in EAB blocks (Fig. 3). As implementation results number of logic cells an number of memory bits are given. It can be easily noticed that the application of decomposition can improve the quality of ROM as well as flip-flop implementation.

Table 1. Comparison of FSM implementation results of standard benchmarks in FPGA architecture (EPF10K10LC84-3 device). [1] Implementation not possible – not enough memory resources, [2] implementation not possible – not enough CLB resources

Benchmark	ROM Implementation #bits	FF implementation #LCs	AM implementation	
			#LCs	#bits
bbtas	160	10	7	80
beecount	448	32	14	112
d14	512	60	21	256
mc	224	14	2	56
lion9	320	24	1	80
train11	320	25	15	8
bbsse	22528 [1]	52	3	5632
cse	22528 [1]	92	2	5632
ex4	13312 [1]	28	2	3328
mark1	10240 [1]	40	2	5120
s1	24576 [1]	137	96	5632
sse	22528 [1]	52	3	5632
tbk	16384 [1]	759 [2]	333	4093
s389	22528 [1]	64	9	5632
Σ	156608	1389	510	41293
%	100 %	100 %	36.7%	26.4%

Table 2. Comparison of FSM implementation results in FPGA architecture (EPF10K10LC84-3 device). [1] FSM described with special AHDL construction; [2] decomposition not possible; [3] not enough memory bits to implement the project

Example	FF_MAX+PlusII		ROM		AM_ROM	
	LCs / Bits	Speed [MHz]	LCs / Bits	Speed [MHz]	LCs / Bits	Speed [MHz]
DESaut	46/0	41,1	8/1792	47,8	7/896	47,1
5B6B	93/0	48,7	6/448	48,0	– [2]	– [2]
count4	72/0 18/0 [1]	44,2 86,2 [1]	16/16384	– [3]	12/1024	39,5

The application of address modifier concept allows implementing FSM in such way that only about 37 % of logic cell resources required in flip-flop implementation and about 27% of memory resources required in ROM implementation is used. Application of address modifier concept allows implementing all of presented FSM using available memory and additional part (address modifier) implemented in CLBs.

In Table 2 results of implementation of several "real life" FSMs are presented. In experiment following examples were used:

- DESaut – the state machine used in DES algorithm implementation,
- 5B6B – the 5B-6B coder,
- count4 – 4 bit counter with COUNT UP, COUNT DOWN, HOLD, CLEAR and LOAD.

Each sequential machine was described by a transition table. As the results for each implementation method there are presented the number of logic cells and memory bits required (i.e. area of the circuit) and the maximal frequency of clock signal (i.e. speed of the circuit). The columns falling under the *FF_MAX+PlusII* heading present results obtained by the Altera MAX+PlusII system in a classical flip-flop implementation of FSM. The *ROM* columns provide the results of ROM implementation; the columns under *AM_ROM* present the results of ROM implementation with use of address modifier. Especially interesting is the implementation of the 4-bit counter. Its description with a transition table leads to strongly non-optimal implementation. On the other hand, its description when using a special AHDL construction produces very good results. The ROM implementation of this example requires to many memory bits (the size of required memory block exceeds the available memory), thus it can not be implemented in given structure. Application of address modifier concept allows reducing the necessary size of memory, what makes implementation possible. The performance of FSM implemented with use of address modifier concept is not significantly degraded.

5. CONCLUSIONS

Balanced decomposition produces very good results in combinational function implementation in FPGA-based architectures. However, results presented in this paper show that functional decomposition can be efficiently and effectively applied beyond the implementation of combinational circuits. Decomposition can be applied to implement large FSM in an alternative way – using ROM. This kind of implementation requires much fewer logic cells than the traditional flip-flop implementation; therefore, it can be used to implement "non-vital" FSMs of the design, saving LCs resources for more important parts of the circuits. However, large FSM may require too many buried memory resources. With the concept of address modifier, memory usage can be significantly reduced. The experimental results shown in this paper demonstrate that the synthesis method based on functional decomposition can help in implementing sequential machines using ROM memory.

REFERENCES

Burns M., Perkowski M., Jóźwiak L. (1998). An Efficient Approach to Decomposition of Multi-Output Boolean Functions with Large Set of Bound Variables. *Proc. of the Euromicro Conference*, Vasteras.

Brzozowski I., Kos A. (1999). Minimisation of Power Consumption in Digital Integrated Circuits by Reduction of Switching Activity. *Proc. of the Euromicro Conference*, pp.376-380. **Vol. 1**, Milan.

Brzozowski J. A., Luba T. (1997). Decomposition of Boolean Functions Specified by Cubes. *Research Report CS-97-01*, University of Waterloo, Waterloo; REVISED October 1998.

Chang S.C., Marek-Sadowska M., Hwang T.T. (1996). Technology Mapping for TLU FPGAs Based on Decomposition of Binary Decision Diagrams. *IEEE Trans. on CAD*, **Vol. 15**, No. 10, pp. 1226-1236.

De Micheli G. (1994): *Synthesis and Optimization of Digital Circuits*. McGraw-Hill, New York.

Hartmanis J., Stearns R.E. (1966). *Algebraic Structure Theory of Sequential Machines*. Prentice-Hall.

Jozwiak L., Chojnacki A. (1999). Functional Decomposition Based on Information Relationship Measures Extremely Effective and Efficient for Symmetric Functions. *Proc of the Euromicro Conference*, pp.150-159. **Vol. 1**, Milan.

Kravets V. N., Sakallah K. A. (2000). Constructive library-aware synthesis using symmetries. *Proc. Of Design, Automation and Test in Europe Conference*.

Luba T. (1994). Multi-level logic synthesis based on decomposition. *Microprocessors and Microsystems*, **18**, No. 8, pp. 429-437.

Luba T., Gorski K., Wronski L.B. (1992). ROM-based Finite State Machines with PLA address modifiers. *Proc. of EURO-DAC*, Hamburg.

Luba T., Selvaraj H., Nowicka M., Kraśniewski, A. (1995). Balanced multilevel decomposition and its applications in FPGA-based synthesis. In: *Logic and Architecture Synthesis* (G.Saucier, A.Mignotte ed.), Chapman&Hall.

Luba T., Selvaraj H. (1995). A General Approach to Boolean Function Decomposition and its Applications in FPGA-based Synthesis. *VLSI Design, Special Issue on Decompositions in VLSI Design*, vol. **3**, Nos. 3-4, pp. 289-300.

Luba T., Nowicka M., Rawski M. (1996). Performance-oriented Synthesis for LUT-based FPGAs, *Proc. Mixed Design of Integrated Circuits and Systems*, pp. 96-101, Lodz.

Luba T., Moraga C., Yanushkevich S., Opoka M., Shmerko V. (2000). Evolutionary Multilevel Network Synthesis in Given Design Style. *Proc. IEEE 30th Int. symposium on Multiple-Valued Logic*, pp.253-258.

Qiao J., Ikeda M., Asada K. (2000). Optimum Functional Decomposition for LUT-Based FPGA Synthesis. *Proc. of the FPL'2000 Conference*, pp. 555-564, Villach.

Rawski M., Jozwiak L., Nowicka M., Łuba T. (1997). Non-Disjoint Decomposition of Boolean Functions and Its Application in FPGA-oriented Technology Mapping. *Proc. of the EUROMICRO'97 Conference*, pp.24-30, IEEE Computer Society Press.

Ross T., Noviskey M., Taylor T., Gadd D. (1991). Pattern Theory: An Engineering Paradigm for Algorithm Design. *Final Technical Report*, Wright Laboratories, WL/AART/WPAFB.

Zupan B, Bohanec M. (1966). Experimental Evaluation of Three Partition Selection Criteria for Decision Table Decomposition. *Research Report*, Department of Inteligent Systems, Josef Stefan Institute, Ljubljana.

Zupan B, Bohanec M. (1966). Learning Concept Hierarchies from Examples by Functional Decomposition. *Research Report*, Department of Inteligent Systems, Josef Stefan Institute, Ljubljana.

IFAC
Publications
www.elsevier.com/locate/ifac

PARAMETERISED DECOMPOSITION OF USER-PROGRAMMED FPGAS FOR APPLICATION-DEPENDENT SELF-TESTING

Pawel Tomaszewicz

Warsaw University of Technology
Institute of Telecommunications

Abstract: A method for the development of a test plan for BIST-based exhaustive testing of a circuit implemented with an in-system reconfigurable FPGA is presented. A test plan for application-dependent testing of an FPGA is based on the concept of test groups and test sessions. Test groups must satisfy the testing time requirement. The decomposition of a user-programmed FGPA into test blocks is controlled by a parameter that describes the maximal length of an exhaustive test pattern generator. Experimental results on circuits taken from benchmark sets show a trade-off between the number of test sessions and the total testing time. *Copyright © 2001 IFAC*

Keywords: benchmark examples, digital circuits, fault detection, test generation, test length, testability.

1. INTRODUCTION

Testing of FPGAs and other user-programmable devices requires solutions that differ from those applicable to mask-programmable. Some of the proposed FPGA test techniques are based on the concept of built-in self-test (BIST). For example, with the techniques proposed in (Abramovici and Stroud, 2000; Das and Touba, 1999; Quddus, *et al.*, 1999; Stroud, *et al.*, 1996a; Tomaszewicz and Krasniewski, 1999; Tomaszewicz, 2000) selected programmable logic blocks (PLBs) are temporarily configured as generators or compactors to pseudoexhaustively test other PLBs (after the test is completed these test-supporting modules are configured back to their normal operation mode). This way self-test capability is achieved without any area overhead or performance penalty.

Testing techniques for FPGA are generally based on checking as many modes of FPGA operation as possible. These testing techniques give a reasonable chance that, after programming, the device will operate correctly in the field, regardless of a specific function implemented by the user. A chip manufacturer normally carries out such testing procedures and they are unrelated to the system design.

The user is interested whether or not an FGPA device, configured to implement a user-defined function, will operate correctly in the system. To satisfy user expectations, an FPGA should also be tested after programming. In contrast to the manufacturing test, such testing must exercise only one specific configuration of the circuit.

In this paper, we discuss several issues associated with application-dependent testing of FPGAs and show that BIST techniques are most suitable for such a testing. Based on the concept of C(combinationally)-exhaustive testing, we have developed a tool for parameterised decomposition of a user-programmed circuit into test sessions. We present results produced by this tool and discuss problems encountered during the circuit decomposition.

2. SELF-TESTING OF FPGA-BASED CIRCUITS

We present a method for the development of a test plan for BIST-based exhaustive testing of a circuit implemented with an in-system reconfigurable FPGA. We test only these resources which are used to implement the user-defined function. We focus on testing of internal part of an FPGA-based circuit (testing of Programmable Logic Blocks – PLBs and their interconnections). Testing of FPGA boundaries (pins, I/O cells, etc.) is not considered.

The test is performed in a number of test sessions, each of which consists of the following steps:
– programming of the device (configuring user-defined logic and BIST logic), which includes initialisation of the circuit registers (registers in the functional section and BIST registers),
– running the self-test,
– verification of the test result, i.e. comparison of the signature with its expected fault-free value.

In two or more test sessions, all the user-defined logic is examined. The execution of the test procedure (programming the device for each test session, providing clock signals, obtaining and verifying the signatures) is controlled by a simple external tester which can be integrated into in-system FPGA programming circuitry or system maintenance processor (Stroud, et al., 1996b).

In the proposed testing procedure, each part of user-programmable circuit is tested at normal operation speed. Therefore in each test session, we check not only functional correctness of a part of the user-programmed circuit, but we can also "by as a product" detect timing-related faults.

It should be emphasised that, for in-system programmable FPGAs, self-testability is not associated with an increased area or degraded performance of the circuit, as it is for non-reconfigurable circuits. This is because test pattern generators and test response compactors "disappear" after the test session is over.

For self-testing of an FPGA-based circuit, the critical requirement is to keep, during each test session, the operation conditions of the logic under test as close to the normal in-system operating conditions as possible. This means that after reconfiguring the FPGA into self-test mode, a portion of the FPGA being examined during that particular test session should remain physically intact, i.e. all PLBs must implement exactly the same functions as before the reconfiguration and the transformation of some portion of the FPGA into test resources should not have any impact on the logical structure and physical placement of that part of the FPGA which is examined during that particular test session.

For a conventional SRAM-based statically-reconfigurable FPGA, the requirement of not affecting the physical placement of the logic under test when configuring the FPGA into the self-test mode is difficult to meet if commercial CAD tools are used to automatically produce the full and minimal configuration bitmap from the high-level description of the circuit. Special options must be used to keep the selected portion of the circuit (section under test) at the same physical location for the two automatically synthesised circuits: the original design and its modified version in which some portions of the logic are converted into test recourses. Our experiments with the Altera MAX+plus II Baseline 10.0 system indicate that this requirement can easily be met with regard to the PLB blocks, but some problems may occur with preserving the physical structure of the interconnection network.

3. MODEL OF AN FPGA-BASED CIRCUIT

A circuit implemented with an FPGA is composed of a set of programmable logic blocks (PLBs). Each PLB has a combinational part, programmable flip-flop (one or more) that can be bypassed, and some extra logic. Every flip-flop can be clocked individually. We consider circuits which are fully synchronous and for which all memory elements are controlled by the same clock. A test unit represents the PLB.

The development of a test plan for application-dependent testing of an FPGA is based on the concept of a logic cone, defined for the circuit represented by graph (V, C). The circuit is decomposed into a set of test blocks (logic cones). Every test block consists of a set of test units (PLBs). A logic cone (LC) is a single-output subcircuit, whose inputs are fed by clocked outputs of PLBs or by the boundary (I/O logic) of the FPGA. The only memory element in a logic cone can be a flip-flop located at its output. If there is no such a flip-flop, then the output of the cone must feed the FPGA boundary. If a logic cone has an output flip-flop, then this flip-flop can feed some PLBs included in the cone. Thus, a logic cone is, in general, a sequential circuit. Each circuit can be uniquely decomposed into a number of logic cones.

In (Tomaszewicz and Krasniewski, 1999), a test procedure was proposed that assumes that during a single test session some number of logic cones are tested concurrently. The quality of test procedure can be described by the number of test sessions required to exhaustively test the circuit and the length of test pattern generators. The number of test sessions and the length of test generators are related to time required to perform an exhaustive test. Although algorithms using the concept of logic cones produce satisfying results for many circuits, in some cases the number of test sessions and/or the length of exhaustive test pattern generators ETPG do not satisfy test time limits. Thus, after logic cone computation, the segmentation procedure is performed. Those logic cones that do not satisfy requirement on the length of ETPG are decomposed into a set of subcircuits.

To decrease the test time and the number of test sessions, logic cones can be grouped. For this purpose, the procedure for linear segment computation is applied (Maximal Linear Segment-MLS procedure). We create a graph of logic cones (VLC, CLC), such that the set of vertices VLC consists of logic cones including IN and OUT vertices. An edge between two vertices from VLC means that there exists a connection between logic cones represented by these two vertices.

After the MLS procedure is completed, we have a set of maximal linear segments (MLSs). In a special case, when no linear segment greater than a single logic cone can be found, the set of MLSs is equal to the set of logic cones. The set of MLSs can be partitioned into two disjoint subsets: MLSs without feedback, and MLSs with feedback (Tomaszewicz and Krasniewski, 1999).

Some MLS can be too wide (have too many input variables) and thus not satisfy test time requirements. If a computed linear segment is too wide, such solution is turned down. So after MLS procedure, we have a set of segments, each satisfying the test time requirements.

4. COMPUTING TEST GROUPS

After MLSs are computed, self-test requirements have to be checked. There are two basic limitations of the self-test procedure: test time and memory necessary to store self-test session configurations. The test time consists of two main components: time required to load the configuration of the device and time required to perform test for loaded configuration. The first component depends on number of test sessions. The second one depends on a length of the longest test pattern generator. The amount of memory required to store configurations depends on the number of test sessions.

To minimise the number of test sessions, test blocks (maximal linear segments) can be grouped into sets of groups that can be tested concurrently. For this purpose the concept of single- and multiple-generator compatibility (s- and m-compatibility) are used (Tomaszewicz, and Krasniewski, 1999).

Two test blocks can be merged and supplied by a common generator, if they satisfy s-compatibility requirements. Two test blocks are s-compatible, if they do not feed each other. The concept of s-compatibility can be generalised to an arbitrary number of test blocks. If there is a group of test blocks and each two of these blocks are s-compatible, then this group is internally s-compatible.

In some cases, the length of a generator required to supply a group of s-compatible MLSs does not satisfy test time limits. In these cases, concurrent testing of several blocks using several test pattern generators can be performed.

Concurrent testing of several multi-output test blocks using several test pattern generators is related to the concept of multiple-generator compatibility (m-compatibility). Two multi-output test blocks are m-compatible, i.e. can be C-exhaustively tested concurrently with each block supplied by an individual exhaustive test pattern generator, if the blocks are internally s-compatible and if they are fully disjoint.

5. DEVELOPMENT OF A TEST PLAN

An outline of the proposed method for the development of a test plan for exhaustive testing of an FPGA-based circuit is shown in Fig. 1. The user-defined circuit is represented by graph (V, C), as described earlier. If a limitation on the test time is specified, then the width of an ETPG required for testing of each subcircuit must be examined to check if the time for exhaustive testing of this subcircuit does not exceed the limitation. The maximum width of the generator is specified by parameter w. The other limitation is a number of test sessions, which is related to the number of reconfigurations of FPGA required to perform the test. This limitation is not set and not checked, it is a product of a whole process.

The test schedule is based on the concept of test units and test blocks. Test units is a set of PLBs. A test block is a set of logic cones grouped into a linear segment. More than one test block can be tested at a time, thereby reducing the number of test sessions and test running time for a circuit. A test group is a set of test blocks that can be tested in parallel (concurrently). Each test session is responsible for testing a single test group.

Merging two test blocks into one test group is possible if they are in a compatibility relation and the maximal number of inputs to the test group, given by parameter w, is not exceeded. Otherwise, test blocks have to be tested in separate test sessions.

There are two strategies of test session development. The first approach is to find the minimal number of test sessions. For this purpose the s-compatibility relation concept is exploited. The result of s-compatibility procedure is a set of test groups, such that blocks belonging to each group can be tested concurrently. The most important parameter of the group is the length of the longest test pattern generator required to perform combinationally exhaustive test. In some cases, it is possible to reduce the length of the necessary test pattern generators by applying m-compatibility concept. This strategy is faster and generates less number of test sessions than the alternative next strategy. This procedure in not controlled by parameter w, so we can achieve results that do not meet testing requirements.

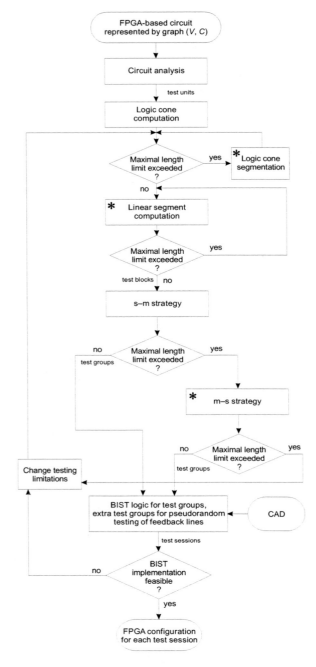

Fig. 1. Development of a test plan. Procedures controlled by parameter w are denoted with an asterisk (*)

The other approach is to apply the m-compatibility relation as the first step of test block grouping procedure. As a result we obtain, in contrary to the previous method, a larger number of test groups. However, the length of the longest test generator is as small as possible and is equal to the size (number of inputs) of the widest block. The size of each block is controlled by parameter w. In order to minimise the number of test groups (test sessions), the concept of s-compatibility is applied. The procedure of merging of s-compatible groups can increase the length of required test generators. To avoid an excessive increase of the test generator length, parameter w

controls the maximal acceptable length of the generator.

After the test group computation a procedure that generates BIST logic for every test group is performed and the resulting configurations are stored in separate files. Additionally, we have to generate BIST logic for extra test sessions to perform pseudorandom testing of feedback lines. At the end, we have to check if all test session configurations are feasible, i.e. can be implemented in an FGPA device.

6. EXPERIMENTAL RESULTS

The objective of our experiment is to find the minimal number of test sessions for a given user-defined circuit, so that to meet the test time constraints given by parameter w. The benchmark sets ISCAS'85 and ISCAS'89 were examined. Every design was given originally by a set of gates in a text file and translated into the AHDL format (Altera Hardware Description Language). Altera MAX+plus II ver10.0 Baseline system was used. All the designs used in the experiment were compiled for Altera FLEX EPF10K20RC240-4 device. All information about the design after synthesis and compilation was given as a network of FPGA cells in the report file.

In Table 1 the results of computing test sessions for ISCAS'85 circuits without constraints on the maximal length of ETPG are presented. For ISCAS'89 circuits the results are shown in Table 2. Some functions from the benchmark set, not listed in Table 1 and Table 2, were unsuccessfully compiled and did not fit in the specified device, because of insufficient number of PLBs or bi-directional inputs.

The column *Max length* in Table 1 and Table 2 shows the maximal length of ETPG. The test time was calculated under assumptions that the reconfiguration time of devices is 200 ms and the frequency of the system clock is 100 MHz. The results in column *Test time* are given for the shortest test time obtained in s-m and m-s strategies.

Table 1. Test sessions for ISCAS'85 circuits with no test time limitation

	Test blocks	Max. length	Test sessions	Test time [s]
c432	7	36	1	688
c499	32	41	1	$2,2\cdot10^4$
c880	26	60	1	$1,8\cdot10^{11}$
c1355	32	41	1	$2,2\cdot10^4$
c1908	25	33	1	86,3
c3540	22	50	1	$1,1\cdot10^7$
c6288	32	32	1	43,3

Table 2. Test sessions for ISCAS'89 circuits with no test time limitation

	Test blocks	Max. length	Test sessions	Test time [s]
s208.1	9	16	1	0,4
s298	20	13	6	1,4
s344	21	18	6	1,4
s349	21	18	6	1,4
s382	27	16	12	2,6
s386	13	13	7	1,6
s420.1	17	30	1	11,1
s444	27	14	12	2,6
s510	1	25	1	7,3
s526	24	14	12	2,6
s526n	27	14	12	2,6
s641	41	35	11	351
s713	40	35	11	351
s820	13	23	2	0,7
s832	13	23	2	0,7
s838.1	33	58	1	$2,9 \cdot 10^9$
s953	18	21	2	0,64
s1196	30	24	4	1,33
s1238	30	24	4	1,33
s1423	77	66	38	$7,6 \cdot 10^{11}$
s1488	25	14	7	1,6
s1494	25	14	7	1,6
s5378	152	72	1	$4,7 \cdot 10^{13}$
s9234.1	135	88	18	$1,5 \cdot 10^{15}$

Table 3. Test sessions for ISCAS'85 circuits with a 60 seconds test time limitation

	Test blocks	Param. w	Test sessions	Test time [s]
c432	21	30	5	46,8
c499	57	28	18	30,9
c880	33	29	12	38,2
c1355	56	28	18	35,7
c1908	64	28	23	53
c3540	94	27	33	33,3

Table 4. Test sessions for ISCAS'89 circuits with a 60 seconds test time limitation

	Test blocks	Max. length	Test sessions	Test time [s]
s641	41	32	10	48,6
s713	40	36	10	37,3
s838.1	36	32	3	49,4
s1423	91	28	29	44,3
s5378	152	33	3	3,8
s9234.1	176	29	29	54,1

As seen in Table 1 and Table 2 for some circuits the test time exceeds an acceptable limit. Therefore, for next experiment an assumption was made that the user test time could not exceed 60 seconds. Those benchmarks, which do not meet that constraint were decomposed using a lower value of parameter w. The experiment was done with the m–s strategy. The results are shown in Table 3 and Table 4.

With a decrease in parameter w (maximal length of ETPG), the number of test blocks and test sessions increases. The experiment was run for the range of values of parameter w from that achieved with no test time limitation (column *Max. length* in Table 1 and Table 2) to $w = 10$. The main observation is that the test time decreases which makes it possible to perform test within given time limitations. The major component of the test time for large w is the self-test time, and not the reconfiguration time. The next observation is that for $w < 25$ the major component of the test time is time required to load the configuration of the device. A more detailed description of the experiment with tables and charts is presented in (Tomaszewicz, 2001).

7. CONCLUSIONS

In this paper, a method for self-testing of circuits implemented with in-system reconfigurable FPGAs is proposed. The model of an FPGA-based circuit is developed and the test procedure is proposed. The test plan for application-dependent testing of an FPGA is based on the concept of test groups and test sessions. Test groups must satisfy the test time requirements. The decomposition of a user-programmed FGPA into test blocks is controlled by a parameter that describes the maximal length of an exhaustive test pattern generator and allows performing exhaustive testing for a given time limitation. Experimental results on circuits taken from benchmark sets show a trade-off between the number of test sessions and the total testing time.

REFERENCES

Abramovici, M., and Stroud, C. (2000). BIST-Based Detection of Multiple Faults in FPGAs, *Intn'l Test Conf. 2000*, pp. 778- 784.

Das, D., and Touba, N.A. (1999). A Low Cost Approach for Detecting, Locating, and Avoiding Interconnect Faults in FPGA-Based Reconfigurable Systems, *Proc. IEEE Intn'l Conf. on VLSI Design*, pp. 266-269.

Quddus, W., Jas, A., and Touba, N.A. (1999). Configuration Self-Test in FPGA-Based Reconfigurable Systems, *Proc. ISCAS'99*, pp. 97-100.

Stroud, C., Konala, S., Chen, P., and Abramovici, M. (1996). Built-In Self-Test of Logic Blocks in FPGAs, *Proc. 14th VLSI Test Symp.*

Stroud, C., Chen, P., Konala, S., and Abramovici, M. (1996). Evaluation of FPGA Resources for Built-In Self-Test of Programmable Logic Blocks, *Proc. ACM/SIGDA Int. Symp. on FPGAs*, pp. 107–113.

Tomaszewicz, P., and Krasniewski, A. (1999). Self-Testing of S-Compatible Test Units in User-Programmed FPGAs, *Proc. 25th EUROMICRO Conference*, **vol. I**, pp. 254-259.

Tomaszewicz, P. (2000). Self-Testing of Linear Segments in User-Programmed FPGAs, *Proc. 10th Int. Conf. Field Programmable Logic and Applications – The Roadmap to Reconfigurable Computing* (R. W. Hartenstein, H. Grunbacher), Lecture Notes in Computer Science, **1896**, pp. 169-174, Springer Verlag.

Tomaszewicz, P. (2001). *Self-testing of user-programmed FPGAs*. PhD Thesis (in Polish). Warsaw University of Technology.

IFAC

Publications

www.elsevier.com/locate/ifac

REALISATION OF MULTIPHASE OSCILLATORS OF PROGRAMMABLE ANALOG CIRCUITS

Lesław Topór-Kamiński, Tomasz Kraszewski

Silesian University of Technology
Institute of Theoretical and Industrial Electrical Engineering
ul.Akademicka 10, 44-100 Gliwice, Poland

Abstract: The structure, the way of use and the method of programming Programmable Analog Circuits (PAC) on the example of ispPAC10 were described. A theoretical possibility of its use in the construction of oscillators producing multiphase voltage signals with program tuneable parameters and a practical realisation of 2-, 3-, 4-, 6-phase oscillators using ispPAC10 were presented. Achieved parameters of practical circuits were analysed in the comparison to theoretical foundations. *Copyright © 2001 IFAC*

Keywords: programmable analog circuits, oscillators, operational amplifiers, circuit simulation

1 INTRODUCTION

The technological progress in manufacturing high-density electronic devices has caused the appearance of integrated circuits with properties programmable by a user. First of all they dominated digital technologies of signal processing and various directions of their technical solutions and uses develop still very dynamically. On the basis of experiences achieved during a construction of digital circuits, attempts of a realisation of more or less complex analog integrated structures with a design entry and parameters programmable by a user have appeared. They work on the basis of OTA op amps with digital tuneable properties of transconductance g_m - Lattice ispPAC devices (Lattice Semiconductor, 2000; Wójcikowski, *et al.*, 2000) or switched-capacitor circuits - Anadigm AN10E40. Additionally, they include programmable interconnect architecture with settings stored in electrically erasable CMOS memory and universal I/O blocks. The next step in the development of electronic programmable device is a use of systems consisting of co-operating digital and analog circuits.

An additional tool of programmable analog circuits is a development system software allowing to build analog circuits and simulate their work on a PC and then download fixed circuit parameters to a real programmable device by a use of a simple interface.

These circuits enable a realisation of typical analog filters with easily changeable amplitude-phase characteristics and gain, sample-and-hold circuits and circuits fitting processed signal to requirements of analog-to-digital converters by suitable programme interconnect architecture and setting of parameters. All these properties, easiness of parameter configuration and tuning and also a practical use cause that programmable analog circuits will become very competitive in comparison to classical analog solutions of signal processing, consisting of various separate electronic parts.

In the present paper the authors have made an attempt to extend uses of ispPAC10 programmable analog circuits, examining a simple oscillatory circuit and its use.

2 STRUCTURE AND PROPERTIES OF ISPPAC10 PROGRAMMABLE ANALOG CIRCUIT

IspPAC10 integrated circuit consists of four programmable macrocells called PACblocks emulating a connection of op amps, resistors and capacitors. In fact a single PACblock includes two differential input transconductance amplifiers OTA1 and OTA2, differential output op amp OA with transconductance amplifier OTAF and switched capacitors C_F in feedback.

Using Kirchoff's current law at the OA amplifier inputs A and B for a signal U_I and at OTA1 output only, there are two equations:

$$-U_{I1}g_{m1} + U_0 g_{mF} + (U_{0+} - U_A)sC_F = 0 \qquad (1)$$

$$U_{I1}g_{m1} - U_0 g_{mF} + (U_{0-} - U_B)sC_F = 0 \qquad (2)$$

Assuming that $\alpha \Rightarrow \infty$, then $U_A = U_B$, and

$$-U_{I1}g_{m1} + U_0g_{mF} + U_{0+}sC_F = $$
$$= U_{I1}g_{m1} - U_0g_{mF} + U_{0-}sC_F$$
(3)

which for $U_{0+} - U_{0-} = U_0$ gives a voltage transfer function:

$$\frac{U_0}{U_{I1}} = \frac{2g_m}{2g_{mF} + sC_F}$$
(4)

Input amplifiers have a programmable gain of $k*2\mu S$ (g_{m1} i g_{m2}) where k is an integer from –10 to 10. The feedback amplifier OTAF transconductance g_{mF} is fixed at $2\mu S$, but may be disabled ($g_{mF} = 0$) to open-circuit the output amplifier's resistive feedback. The programmable feedback capacitance lies in the range 1pF to 62pF. Taking into consideration all that, output voltage of a single macrocell is shown by the following relation:

$$U_0 = \frac{k_1U_{I1} + k_2U_{I2}}{1 + \frac{sC_F}{2g_m}}$$
(5)

A configuration of the required interconnects architecture and its parameters are made by a use of PAC-Designer development software. It includes a schematic window visualising ispPAC10 architecture shown in Fig. 2. and additional windows with parameters of circuit parts.

In this software reduced schematic symbolism is used which has replaced differential circuits by single-ended ones. Then a structure of one macrocell is like in Fig. 3. and will be represented in this way in further schemes.

3. A MODEL OF A FUNDAMENTAL 2- OR 4-PHASE OSCILLATOR WITH TWO INTEGRATORS

A realisation of simple 4-phase oscillator with a use of programmable analog circuit is based on two integrating circuits, which are fundamental components of its single macrocells. The oscillator structure is shown in Fig. 4.

Differential output voltage U_{01} and U_{02} of op amps on the basis of the relation (5) are joined by system of equations:

$$U_{01} = 2k_1 \frac{g_m}{sC_{F1}} U_{02}$$
(6)

$$U_{02} = 2k_2 \frac{g_m}{sC_{F2}} U_{01}$$
(7)

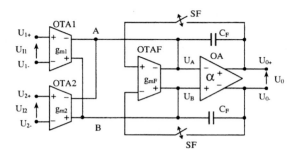

Fig. 1. Real structure of a single programmable macrocell of ispPAC10

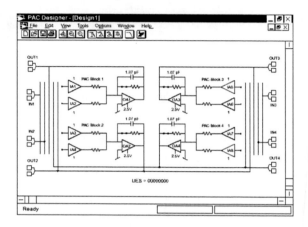

Fig. 2. PAC-Designer Schematic Window of ispPAC10

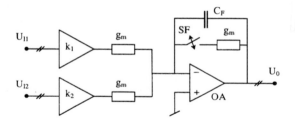

Fig. 3. Reduced equivalent circuit diagram of a single macrocell of ispPAC10

In relation to this a characteristic equation of the whole circuit is as follows:

$$s^2 - \frac{4k_1k_2g_m^2}{C_{F1}C_{F2}} = 0$$
(8)

That is why output angular frequency equals:

$$\omega_0 = 2g_m\sqrt{\frac{-k_1k_2}{C_{F1}C_{F2}}}$$
(9)

As it comes from equations (6) and (7) those oscillations are additionally shifted to each other for ¼ of an oscillation period, so an oscillator is a biquad one. While components of these voltages U_{01+}, U_{02+}, U_{01-}, U_{02-}, related to ground are successively shifted

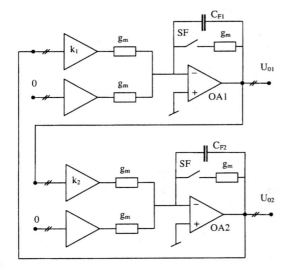

Fig. 4. A fundamental structure of 2- or 4-phase oscillator with two integrators using two ispPAC10 macrocells

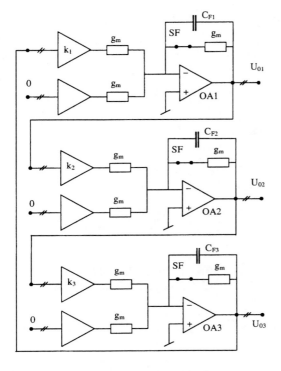

Fig. 5. A fundamental structure of 3- and 6-phase oscillator with three real integrating circuits built on three macrocells of ispPAC10

to each other for ¼ of a period, so they create four-phase signal circuit including a d-c component $U_R=2.5V$. For an identical value of coefficients $k_1=-k_2=k$ and capacitance $C_{F1}=C_{F2}=C_F$, an oscillator produces a harmonic oscillation with a angular frequency:

$$\omega'_0 = \frac{2kg_m}{C_F} \qquad (10)$$

with the smallest harmonic content. If a value of capacitance C_{F1} and C_{F2} differs significantly, eg $C_{F2} \gg C_{F1}$, the amplifier OA2 behaves as a comparator coming into a range of saturation with not large changes of input voltage. Therefore the time waveform of its output voltage U_{02} will become a trapezoidal close to a rectangular.

4. A FUNDAMENTAL MODEL OF 3- AND 6-PHASE OSCILLATOR

A realisation of 3- and 6-phase oscillator with a use of programmable analog circuit is based on three real integrating circuits, which are components of ispPAC10 macrocells. Its structure is shown in Fig. 5.

A single block of this circuit is described by transmittance on the basis of the relation (5):

$$\frac{U_{0i}}{U_{Ii}} = \frac{k_1}{1+\dfrac{s}{\omega_p}} \qquad (11)$$

Where:

$$\omega_P = \frac{2g_m}{C_F} \qquad (12)$$

Transmittance of the whole circuit with open-loop feedback is as follows:

$$L(s) = \left(\frac{k_1}{1+\dfrac{s}{\omega_P}} \right)^3 \qquad (13)$$

In order to make oscillations appear in a close-loop feedback circuit, transmittance should be equal 1. Then angular frequency ω_0 can be enumerated from the equation:

$$\left(1 + j\frac{\omega_0}{\omega_P} \right)^3 - k^3 = 0 \qquad (14)$$

When a real part of this equation is negative or equal zero then a condition of the appearance of oscillation is fulfilled. Therefore k gain of each block must be as follows:

$$k_2 \leq \frac{-1}{\cos\dfrac{\pi}{3}} \leq -2 \qquad (15)$$

Then frequency of achieved output oscillations of each macrocell equals:

305

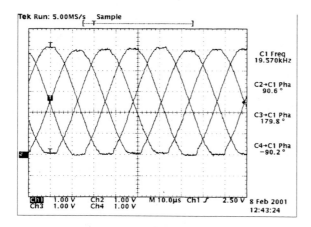

Fig. 6. Output voltage waveform of 4-phase oscillator for $C_{F1}=C_{F2}=29.95$pF and $k_1=-k_2=1$

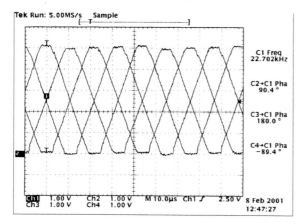

Fig. 7. Output voltage waveform of 4-phase oscillator for $C_{F1}=C_{F2}=29.95$pF and $k_1=-k_2=2$

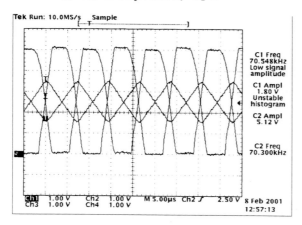

Fig. 8. Output voltage waveform of 2-phase oscillator for $C_{F1}\gg C_{F2}$ and $k_1=-k_2=2$

$$\omega_0 = \omega_p tg\frac{\pi}{3} = 2\sqrt{3}\frac{g_m}{C_F} \qquad (16)$$

Differential output voltage U_{01}, U_{02} and U_{03} are shifted to each other for $T/3$, so they make up 3-phase voltage circuit. However, their components U_{01+}, U_{03-}, U_{02+}, U_{01-}, U_{03+}, U_{02-} related to ground are successively

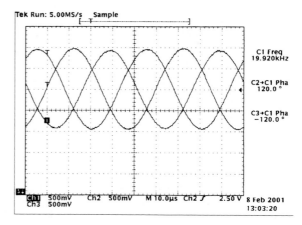

Fig. 9. Output voltage waveform of 3-phase oscillator for $C_F>30$pF and $k=-2$

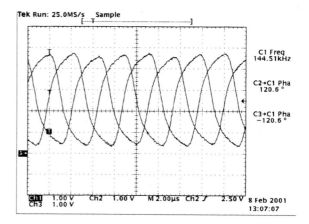

Fig. 10. Output voltage waveform of 3-phase oscillator for $C_F<30$pF and $k=-2$

shifted to each other for $T/6$, so they create six-phase signal circuit including a d-c component $U_R = 2.5$V.

5. EXAMINATION OF REAL MULTIPHASE OSCILLATORS BUILT ON ISPPAC10

An examination of two-integrator oscillator modelled in ispPAC10 programmable circuit has been started from equal values of $C_{F1}=C_{F2}=29.95$pF and coefficients $k_1=-k_2=1$. Voltage waveform U_{01+} and U_{02+} close to a sinusoidal wave has been achieved and is shown in Fig. 6. Their measured frequency is $f_{Z1}=19.57$ kHz and close to $f_{01}=21.17$ kHz defined by the relation (10)

An increase of coefficients k_i to a value $k_1=-k_2=2$ speed up saturation of amplifiers that is why wave of output voltage become more deflected (triangular, Fig. 7.), and their frequency increases to $f_{Z2}=22.7$ kHz. It differs significantly enough from theoretical $f_{02}=42.66$ kHz enumerated from the equation (10) what can be only explained by a strong impact of nonlinearity of an amplifier in the area of saturation, not taken into theoretical consideration.

If the capacitance C_{F2} decreases for $k_1=-k_2=2$, sharpness of a triangular wave U_{01} is achieved together with a decrease of its amplitude and an increase of its frequency, shown in Fig. 8. At the same time voltage waveform U_{02} becomes a trapezoidal close to a theoretical rectangular.

An examination of 3-phase oscillator in a scheme as in Fig. 5. modelled in ispPAC10 has been carried out for $k=-2$ and decreasing values C_F from 62pF to 1pF. For values $C_F>30$pF output voltage waveform have been achieved with a low distortion, presented in Fig. 9. However, for $C_F<30$pF a distortion of output signals appears, as shown in Fig. 10. They are shifted to each other for $T/3$.

6. CONCLUSION

The described programmable analog circuit ispPAC10 allows for a realisation of simple oscillating circuits consisting of 2 or 3 integrators. They produce multiphase voltage wave. Changing values of programmable parameters k_i and C_{Fi} shape and frequency of generated waveform can be changed. Achieved oscillators are 2- or 3-phase ones for symmetrical output voltages U_{0i} or 4- and 6-phase ones for output signals U_{0i+} and U_{0i-}.

REFERENCES

Lattice Semiconductor: *Specifications ispPAC10.*

Lee E. K. F. and P. G. Gulak (1991). A CMOS Field Programmable Analog Array. *IEEE Jour.Solid-State Circuits.* **Vol. No.12**.

Wójcikowski M., B. Pankiewicz, J. Jakusz and S. Szczepanski (2000). Programmable CMOS Operational Transconductance Amplifier OTA. *Conference proceedings ICSES 2000*. Ustroń, Poland.

Wu D. S., S. L. Liu, Y. S. Hwand and Y. P. Wu (1995). Multiphase Sinusoidal Oscillator Using the CFOA Pole. *IEE Proc. CDS.***Vol. No.1**.

AUTHOR INDEX